WEYERHAEUSER ENVIRONMENTAL CLASSICS

William Cronon, Editor

WEYERHAEUSER ENVIRONMENTAL CLASSICS
are reprinted editions of key works that explore human relationships
with natural environments in all their variety and complexity.
Drawn from many disciplines, they examine how natural systems
affect human communities, how people affect the environments
of which they are a part, and how different cultural conceptions
of nature powerfully shape our sense of the world around us.
These are books about the environment that continue to offer
profound insights about the human place in nature.

The Great Columbia Plain:
A Historical Geography, 1805-1910
by D. W. Meinig

Mountain Gloom and Mountain Glory:
The Development of the Aesthetics of the Infinite
by Marjorie Hope Nicolson

Tutira: The Story of a New Zealand Sheep Station
by H. Guthrie-Smith

A Symbol of Wilderness:
Echo Park and the American Conservation Movement
by Mark W. T. Harvey

WEYERHAEUSER ENVIRONMENTAL CLASSICS
is a subseries within Weyerhaeuser Environmental Books,
under the general editorship of William Cronon.
A complete listing of the series appears at the end of this book.

A Symbol of Wilderness

Echo Park and the
American Conservation Movement

Mark W. T. Harvey

Foreword by William Cronon

UNIVERSITY OF WASHINGTON PRESS

Seattle and London

A Symbol of Wilderness: Echo Park and the American Conservation Movement has been published with the assistance of a grant from the Weyerhaeuser Environmental Books Endowment, established by the Weyerhaeuser Company Foundation, members of the Weyerhaeuser family, and Janet and Jack Creighton.

Library of Congress Cataloging-in-Publication Data
Harvey, Mark W. T. (Mark William Thornton)
A symbol of wilderness : Echo Park and the American conservation
movement / Mark W. T. Harvey; foreword by William Cronon
p. cm. — (Weyerhaeuser environmental classics)
Originally published: Albuquerque : University of New Mexico Press, c1994.
Includes bibliographical references (p.) and index.
ISBN 0-295-97932-1 (alk. paper)
1. Dinosaur National Monument (Colo. and Utah)—History.
2. Nature conservation—United States—20th century.
3. Echo Park Dam (Colo.)—History. I. Title II. Series
F832.D5H37   2000   99-059057   978.8'12—DC21

Cover photo: Green River and Steamboat Rock, Echo Park (photo © 2000 by Jeff Gnass)

To my parents,
William and Dorothy Harvey

Contents

Illustrations

MAPS

Foreword: The Dam That Wasn't

WILLIAM CRONON

Among the more familiar but suggestive ways to narrate the history of conservation in the United States is to tell a tale of dams and the changing ways Americans have perceived them. The canonical beginning of that story, of course, is Hetch Hetchy and the great struggle during the first two decades of the twentieth century over whether the city of San Francisco should be permitted to convert a remote valley in Yosemite National Park into a reservoir to supply water for drinking and fire-fighting. Hetch Hetchy has become an almost mythic symbol of the seemingly opposing impulses that have typified American environmental politics ever since: toward responsible use of natural resources on the one hand, and toward preservation of wild nature on the other, epitomized in the antagonistic figures of Gifford Pinchot and John Muir. Muir and his Sierra Club lost the battle to save Hetch Hetchy, but the dam that now floods that beautiful valley (a dam that surprisingly few Americans have ever seen) has forever after served as a battle cry in subsequent controversies about reclamation and wilderness preservation in the United States.

But Hetch Hetchy was only the first of a long series of dams that have provided benchmarks for American environmental politics since the start of the twentieth century. During the New Deal, Boulder, Bonneville, and Grand Coulee, and the many dams of the Tennessee Valley Authority, served as triumphal declarations that an aggressively activist federal government could deploy its immense financial, human, and scientific resources to tame wild rivers in the service of human need. The widespread public celebration of these feats of engineering stands in stark contrast to the very different political response to dam construction that characterized the second half of the century. The Glen Canyon Dam of the early 1960s would be among the last mega-dams constructed on the heroic model of the 1930s, and by then the old triumphalist rhetoric had begun to wear thin. Glen Canyon would be mourned by environmentalists as "The Place No One Knew," and the tone of elegiac loss conveyed by those words would stiffen resistance to subsequent projects. When the Bureau of Reclamation proposed to dam even the Grand Canyon itself, David Brower's Sierra Club leapt into the fray with its now famous *New York Times* advertisement asking the rhetorical question, "Should we also flood the Sistine Chapel so tourists can get nearer the ceiling?" The campaign was successful, and the Canyon remained undammed. In the 1970s, the Tellico Dam in Tennessee would furnish a crucial test case for the new Endangered Species Act. And by the 1980s and 1990s, more than one Secretary of Interior would provoke heated controversy by suggesting in the name of environmental protection that the time had come to consider dismantling dams whose value had once seemed compellingly self-evident.

Over the course of the century, in other words, dams came to stand as potent icons both of progress and of desecration, immense physical structures that were both deified and demonized as symbols of what the United States should or should not become. They were not merely objects. They stood for aspirations and fears, for dreams and nightmares, and they cannot be understood apart from the mythic meanings that came to be attached to them. Although they by no means constitute the whole story of environmental conflict in the twentieth century, it is no accident that they loom so large in that narrative.

Among these many dams, one of the most important, perhaps second only to Hetch Hetchy in its narrative significance, is the dam that was never built at an obscure place called Echo Park located within the boundaries of Dinosaur National Monument, on the Utah-Colorado border. Echo Park was to be the first defeat of a major dam project that the Bureau of Reclamation ever experienced in what had previously been a steadily onward and upward march of heroic regional and national progress. The Bureau had enjoyed for decades an enviable reputation as a federal agency capable of delivering great good to the people whose lives it touched: water for agricultural irrigation, water for growing cities, water for cheap clean kilowatts of electricity, water to make the desert bloom. The Bureau was thus unprepared to find itself cast in the unfamiliar role of villain in a high drama that focused on a small national monument that few people had ever visited and that most Americans had never even heard of. But the struggle to stop the dam at Echo Park would become a defining moment in the emergence of a new post-war environmental politics in which the protection of sacred nature and of recreational land would move ever higher on the national agenda. From this moment on, dam-builders would gradually find themselves in ever more defensive positions, so much so that many of their most hoped-for projects eventually became untenable.

To understand the history of modern environmentalism, then, one has to understand the debate over Echo Park. There is no better guide for doing so than Mark Harvey's *A Symbol of Wilderness*, now offered for the first time in a paperback edition. Harvey has combed the archives and produced what will surely remain the standard scholarly account of this vital controversy. Because the debate turned on highly technical issues like the permeability of aquifers and the evaporation rates of impounded water exposed to the hot sun and arid winds of the southwestern landscape, Harvey has to work in this book to help the reader understand the language not just of activists and politicians but of engineers and bureaucrats. He does this with aplomb and great teacherly skill.

Although Mark Harvey's personal sympathies are clearly with those who fought the dam at Echo Park, he is scrupulously even-handed in his effort to make sure that all sides of the debate are fairly represented. This is important, because these arguments of the 1950s have continued to echo through the subsequent decades. Best of all, Harvey is a gifted teacher and storyteller, so that material which in less skillful hands might seem dull becomes in this book not just an engaging narrative but an intricate and appropriately ambiguous moral fable. In a very real sense, Echo Park was

merely the opening dialogue of a national conversation that is unlikely to reach a definitive conclusion during the lifetime of anyone who reads this book. We will be asking these questions—about
dams and wilderness, about development and preservation, about
human need and natural integrity—again and again. Knowing how
an earlier generation of Americans confronted similar dilemmas
cannot help but give us bearings as we try to find the best way
forward ourselves.

Preface

In the middle of the 1950s the best known conflict over public lands and scenic preserves in the United States centered on a proposal to build a large dam near Echo Park inside Dinosaur National Monument. First suggested by the Bureau of Reclamation in the 1940s, the dam threatened this remote and nearly unheard of part of the national park system which spanned the border of Utah and Colorado. The Bureau wanted to build this dam on the Green River in a high walled canyon near the center of the monument. Echo Park, a picturesque valley just upstream from the dam site, along with miles of nearby Lodore and Yampa Canyons, would have been inundated beneath several hundred feet of water. Beginning in 1949 wilderness enthusiasts and conservation organizations rushed to defend the scenic preserve by mounting a national campaign against the dam. Determined to prevent the magnificent scenery from being flooded they sought to reaffirm one of the founding principles of the national park system, that the scenery should be left unimpaired for future generations. Led by Howard Zahniser of the Wilderness Society, Ira Gabrielson of the Emergency Committee on Natural Resources, Fred Packard and Sigurd Olson of the National Parks Association, and David Brower of the Sierra Club, conservationists badgered Congress to find an alternate site for the dam. Finally in 1956 Congress agreed. Echo Park Dam would not be built.

More than four decades earlier, a dam in the Hetch Hetchy Valley of Yosemite National Park had not been stopped, and to preservationists like John Muir it seemed a tragedy that such grand beauty must be inundated beneath a reservoir. Defenders of Echo Park very much had the ghost of Hetch Hetchy Valley in mind, and this time the outcome proved different. Preservation had seemingly come of age, for the dam builders met their match in a more invigorated, popular based movement to guard parks and wilderness lands. That movement was fueled in part by a travel boom to the national parks and monuments which helped strengthen organizations devoted to their protection, and, in a larger sense, by a growing interest in natural areas and a desire to escape from cities which blossomed after the Second World War.

As the Echo Park controversy unfolded, thousands of Americans who belonged to sporting and wildlife clubs and to national organizations like the Audubon Society, Izaak Walton League, and National Parks Association expressed their opposition to the project. Late in 1955, after two hard fought sessions of Congress, their efforts bore fruit when they forced removal of the dam from the legislation and won a guarantee against the building of dams in any area of the national park system along the Colorado River. By forcing the Bureau of Reclamation to back down, conservationists secured a major victory on behalf of the national park system and wilderness preservation in general. As Roderick Nash and other historians have recognized, the defeat of Echo Park Dam proved to be a milestone to the fledgling wilderness movement and a critically important episode prior to the Wilderness Act of 1964.

Historians have been aware of the significance of the Echo Park battle for some time, although none have fully analyzed its origins or adequately placed it in historical context, either in terms of wilderness or national park history or in the context of the American West after the Second World War, which provides the key to understanding the dynamics of the long battle. Like many other states in the West, Utah, Colorado, Wyoming, and New Mexico experienced considerable economic growth during the war, a fact which increased their appetite for water and power and spurred demands for dams along the upper Colorado River. States on the lower Colorado River, namely Nevada and California, had already benefitted economically from the water and power provided by Hoover Dam and various aqueducts and canals built in the 1930s and 1940s. As urban and industrial growth sprouted in the upper basin states during the war, residents there fervently hoped that their own turn for large scale water development had come. The Bureau of Reclamation soon responded to these demands with an ambitious plan called the Colorado River Storage Project (known by its acronym CRSP), a team of dams and power plants to be constructed along the upper Colorado and its tributaries in Wyoming, Colorado, New Mexico, and Utah. Echo Park Dam was to be a major facility in the overall plan and a key component of the Central Utah Project, one of the major "participating projects." Hydropower from Echo Park Dam stood to benefit rural residents and urban communities across the four states, while its storage reservoir would aid upstream states in regulating the river. Soon the dam became a much touted instrument of economic growth and diversification, a symbol of progress and prosperity to one of the West's emerging regions.

While the boom sparked by World War II gave rise to this upper Colorado River project, it also created stresses and strains over water within the West and between the West and other regions of the country. For the defenders of Dinosaur who lacked recourse to environmental impact statements or other legal weapons available to a later generation of environmentalists, political allies proved vitally necessary to their campaign. Such allies came from lawmakers in eastern, midwestern and southern states who raised questions about the merits of adding to an already burdened agricultural economy strained by surpluses, and who asked whether the funds from hydropower sales would be sufficient to pay the cost of this large multi-purpose Colorado River Storage Project with its huge dams slated for Echo Park and Glen Canyon and its many "participating projects." In addition, the CRSP disturbed California which considered its own claims to the Colorado River endangered by upper basin demands. At times the controversy over Echo Park Dam appeared to be subsumed to an even fiercer struggle over the Colorado River between upstream and downstream states. The story, then, has various layers and the controversy over Dinosaur must be seen as part of a larger struggle over water development in the West.

Yet the heart of the story centers on the dispute over the Bureau's proposed dam at Echo Park. American conservationists considered the dam a dire threat to the national park system and to the larger aim of preserving wilderness. Their concerns grew out of the growing pressures on public lands in the West during the postwar economic spurt. From the middle to late 1940s conservationists witnessed a number of threats to national parks and monuments, including Olympic and Kings Canyon National Parks and Jackson Hole National Monument, the forerunner of Grand Teton National Park. Logging firms, resort developers, and other interests also sought to exploit resources in primitive and roadless areas in national forests in Minnesota and California as well, prompting the Wilderness Society and its allies to press for added protection of such areas. Meanwhile, a "landgrab" movement arose in Wyoming, Nevada and other ranching states, adding to conservationists' fears about the public lands in the West and the new pressures being exerted upon them. All of this caused growing anxieties among a gamut of organizations who became determined to take a strong stand against the Bureau of Reclamation's proposal to dam the Green River in Dinosaur National Monument. The conflict at Dinosaur became a great test to the sanctity of the park system.

In the broadest sense, the controversy over Echo Park revealed a generational clash in American conservation and environmental history, a clash that became a prominent theme in the history of the American West in the postwar era. Beginning with the Echo Park battle, the West since World War II has witnessed a pattern of conflict between resource oriented conservation agencies and their constituents, and "preservationists" or environmentalists as they are commonly called today. Earlier in the twentieth century, federal conservation agencies like the Bureau of Reclamation, Forest Service, and Bureau of Land Management had become deeply enmeshed in the economy of western states, setting the stage for a series of sharp conflicts with preservationists who became substantially stronger in the postwar years. The Echo Park controversy is to be understood as the first of these conflicts in the postwar era, the initial sign that the West would be a constant battleground for two distinct generations in the nation's environmental history.

The Bureau of Reclamation was deeply rooted in the Department of the Interior and part of the bedrock of the nation's conservation establishment. Created in 1902, the agency had been a keystone in the Progressive Era conservation movement. By erecting dams and canals and various irrigation and power facilities, it sought to make efficient use of the West's water resources. Here was a perfect example of the kind of utilitarian resource management that marked the first wave of American conservation history. In the aftermath of World War II such a managed view of rivers began to come under fire from a new generation of preservationists, represented by such groups as the Wilderness Society and the Sierra Club. These and other organizations advocated greater protection of national parks and wilderness areas and sought to take advantage of changing economic and social trends including a rise in educational levels, greater geographic mobility, and steady growth in the numbers of people moving out of cities and into suburbs and expressing a fondness for natural beauty. As Samuel Hays has argued, these societal changes after the war created new demands for "amenities" like wildlife protection and access to parks and wilderness areas, and established the foundation for what later came to be termed the environmental movement.

Although those who battled against the Echo Park Dam did not refer to themselves as "environmentalists" (the term was not in use then), they represented a growing segment of the populace which embraced a broader definition of "conservation" than that held by the Bureau of Reclamation. The battle over Echo Park was the first major clash between preservationists and the dam builders in the postwar American West, and it established a major theme in the region's recent history.

There can be little doubt that the Bureau of Reclamation had the upper hand when the Echo Park controversy surfaced in the early Cold War period. By the 1940s the agency had attained a substantial presence in the West with regional offices in Billings, Denver, Salt Lake City, and other cities, and had earned a strong reputation for "harnessing" western rivers and stimulating economic growth. Its formidable power also derived from its staff of professionally trained geologists, hydrologists, and engineers. Steeped in a tradition of management of natural resources by experts, the Bureau could not easily be challenged by those who lacked technical knowledge of irrigation, hydropower, and cost-benefit ratios. Opponents of Echo Park Dam had to teach themselves how the Bureau's mind worked and how the many facets of a multiple purpose plan of river basin management were related. They had to come to grips with the interrelationship of the many dams and power plants in the proposed CRSP and attempt to persuade the upper basin states that Echo Park Dam could be eliminated from the overall plan without adversely affecting it. Their efforts to do so eventually weakened the Bureau's tight grip on information about the Colorado River and water "development," and a core of wilderness activists emerged with an understanding of evaporation rates, the legal complexities of the Colorado River Compact, and the economics of large water and power projects. The Echo Park controversy gave birth to a generation of preservationists who had substantial expertise of their own, who knew how to scrutinize the Bureau's plans, and this body of knowledge carried over into subsequent wilderness and water controversies.

Still, while gaining an understanding of the Bureau proved helpful in future clashes, the Bureau in many ways set the terms of the Echo Park debate and for subsequent development of the Colorado River. Despite their eventual triumph in eliminating the Echo Park Dam from the larger project, conservationists proved unable and unwilling to block approval of Glen Canyon Dam, and, much to their regret, some priceless scenery was inundated beneath its reservoir, adding to their bitterness toward the Bureau. Further controversy ensued when conservationists accused the Bureau of reneging on its agreement to protect Rainbow Bridge from Lake Powell. And in the middle of the 1960s another major challenge came to the fore as the Bureau sought to construct two dams in the Grand Canyon. All of this testified to the Bureau's substantial power in the American West in the decades following World War II. The Echo Park controversy gave birth to an icy relationship between the Bureau and champions of wilderness, and by the 1960s and 1970s it had grown into a virtual cold war.

Despite disappointments over Glen Canyon and Rainbow Bridge, the battle for Echo Park opened an exciting new region of wild and scenic wonders to many of those involved in the campaign. This controversy brought the Colorado River and Plateau into the consciousness of the Sierra Club, National Parks Association, Wilderness Society, and other groups. Here was a region largely unknown to most wilderness enthusiasts who were more familiar with forested and alpine areas like the Adirondacks, Lake of the Woods, the Grand Tetons, and the High Sierra. Dinosaur Monument and Glen Canyon offered a new kind of wilderness terrain, one dominated by canyons, mesas, and a vast array of geologic features carved in the earth's surface by millions of years of river flows. The Dinosaur controversy—and the battles it ignited at Glen Canyon and Grand Canyon—brought into public view the canyons and rivers of the American Southwest, and the region thereafter became a focal point of environmental activism in the American West.

I have sought to tell this story in the larger context of Western and environmental history, and have organized the book accordingly. Given the importance of the controversy to the evolution of the national park system, I begin by analyzing the origins and early history of Dinosaur National Monument, and suggest that the preserve lacked a strong identity as well as public support. The monument evolved in two stages with an initial preserve created in 1915 to protect dinosaur bones. In 1938 President Roosevelt enlarged the monument by more than 200,000 acres to protect the Green and Yampa River Canyons, making it one of the largest areas in the entire national park system. Yet despite its size, the public had little involvement in its evolution and Dinosaur Monument was largely unknown. Further, the National Park Service lacked funds to build roads or trails into the preserve, leaving it underdeveloped and vulnerable. Even its name proved problematical since most of the preserve did not contain dinosaur bones. In short, the seeds of the controversy were planted long before the newly visible preservationists came onto the scene in the 1950s.

I then widen the lens and examine the economic growth and aspirations of the region surrounding the monument, focusing on the massive changes that descended on four states which shared upper Colorado River water during the Second World War. Their sudden growth fueled their appetite for water and hydropower and gave birth to the Bureau of Reclamation's proposed dams near Echo Park, Glen Canyon, and Flaming Gorge. Throughout the book I try to make clear how Echo Park Dam and the larger project symbolized the economic aspirations of this part of the West in the postwar era. Water and power were considered key instruments of growth and stability. Against this backdrop, I examine how conservationists became alarmed with the sudden pressures on parks, forests, and other public lands after the war and how this shaped their determination to block the proposed dam in Dinosaur. To them the dam was part of a disturbing pattern of threats to areas in the national park system, and seemed ominous for the future of wilderness preservation. By the end of chapter three we will have seen the primary conditions and forces that gave birth to the controversy: a weak and underfunded national monument, a growing region in the intermountain West eager to develop water and power, and a newly strengthened but rather worried conservation and wilderness movement.

Chapter four begins to examine the politics of the controversy in more detail. The conflict over Echo Park erupted most noticeably first in the Department of the Interior in Washington, where tensions between the National Park Service and Bureau of Reclamation boiled over in 1950 and 1951, culminating with the resignation of NPS Director Newton Drury. Drury's departure brought the growing pressures on national parks and monuments into public view and, at the same time, effectively removed the Park Service from any public role in opposing the dam. Now the defense of the monument fell entirely to the Wilderness Society, the Sierra Club, the Izaak Walton League, and other organizations, and the gamut of conservationists had to find a strategy for winning. They did so by questioning the need for the dam and proposing alternate sites, mounting an advertising campaign for the remote and isolated Dinosaur Monument, and establishing ties with political forces in Congress that sought to quell such large water and power projects.

All of this took time and substantial energy, and the Bureau proved most reluctant to surrender the dam. At a crucial point in the controversy (the subject of chapter seven), conservationists revealed errors in the Bureau's computations of the evaporation rates of the reservoirs, and this provided momentum at a critical stage of their campaign. When the battle at last reached Congress in 1954 and 1955 (described in chapters eight through ten), conservationists took advantage of their momentum to forge a political coalition with other opponents of the CRSP. As the controversy peaked the political tensions between the Colorado basin states and between the West and other regions of the country came to the fore and helped bring an end to the Echo Park Dam proposal.

It should be emphasized that this controversy took place long before the "environmental revolution" of the 1960s, and that the opponents of the dam had no legal weapons to block or delay the project. While a later generation of conservationists delayed Tellico Dam with the Endangered Species Act and the Central Utah Project and Garrison Diversion Unit with the National Environmental Policy Act, no such legislation existed in the early 1950s. Opponents of the Echo Park project had little choice but to operate solely in the political arena. This battle could not have been won without attaining allies and a large part of the story centers on the making of that coalition.

My interest in the Echo Park story began when Robert Righter first suggested the topic to me at the University of Wyoming. He knew of my interest in national park history, and indicated that although the battle had been briefly covered in other works it deserved a full length scholarly treatment. While I have Bob Righter to thank for introducing me to the subject, my interest in the national parks—and in Western and environmental history generally—originated with my parents who took my three sisters and me to the Rocky Mountain West in 1962. Our family's trip to the Grand Tetons that year proved to be the beginning of a long love affair with the western parks. For the next two decades we returned each summer to camp, backpack, climb, and canoe in the Tetons, Glacier, and Yellowstone National Parks and Glacier Bay, Alaska, as well as in various national forests. Later I landed a summer job with a private boating concession at Jenny Lake in the Tetons, a job which lasted for eight wonderful years. At that time I knew little about the broader history of the national parks, but every day at Jenny Lake taught me that the parks provided enjoyment and pleasure for millions of people around the world. When I arrived in Laramie for graduate school in 1980 I took naturally to scholarly work on the parks and the broader environmental history of the West, and in so doing quickly realized the merits of Bob Righter's suggestion to explore the Echo Park controversy in depth.

In subsequent summers I worked for the Park Service at Mesa Verde National Park in Colorado and at Dinosaur National Monument, where at last I had the great pleasure of seeing Echo Park up close. The road into Echo Park is long and bumpy but the reward is great. Here, beneath the massive and sheer east wall of Steamboat Rock, the Green and Yampa Rivers converge in the heart of Dinosaur National Monument. As I stood there one spring day in 1984, waiting to board a raft for a float down the Green River, it was difficult to think objectively about the great battle that had raged over this place for more than half a decade in the 1950s. For my part I was thankful that the Sierra Club, the Wilderness Society, and their friends had fought the battle and won. Echo Park is an unforgettable place and I count myself among those who are grateful that it was not lost beneath a reservoir.

Still, while my love of Echo Park as well as my environmental sympathies may well be evident to those who read these pages, I must emphasize that my primary interests are with the history of the American West, and as is true with most historians, I have an abiding interest in appreciating the past on its own terms. I have striven hard in the pages that follow to tell the story of this contentious battle with a full accounting of the historical forces at work, to see the controversy as a significant episode not only for the national park movement but for the American West in terms of the themes I have emphasized above. How well I have done so will be left for readers to judge.

Mark W. T. Harvey
Fargo, North Dakota
May 1993

Acknowledgments

During the years that this study has evolved from a master's thesis to a published book, I have benefitted from the counsel and aid of many people and institutions.

I must first thank Robert Righter who introduced me to the Echo Park controversy and encouraged me to undertake a full fledged history of the battle. Bob understood how the topic nicely matched my interests and, while neither one of us realized at the time just how long it would take for the book to emerge, I will always be grateful to Bob for helping put me on the path of the environmental history of the American West. Bob's insistence on literary excellence was not something that I always welcomed during the dissertation stage, but in the years since leaving his tutelage, I have come to appreciate his devotion to literary quality in ways that I did not then. I hope that he will find the final product to be worthy of his counsel.

Besides Bob Righter, I wish to thank some of my teachers from the history department at Wyoming in the middle 1980s, especially Roger Williams, David Robson, and William Gienapp. While none of them are historians of the West, they each taught me a great deal about the art and craft of history, and their influence in my work has been far reaching. David Walker and Dan Flores both passed through Laramie for teaching sojourns, and they read various drafts of my thesis and dissertation and offered valuable suggestions. Peter Iverson, both at Wyoming and from his position at Arizona State University, has long been a source of encouragement and counsel, and among other things I wish to thank him for guiding me to the University of New Mexico Press.

Aside from my earliest mentors, I have benefitted from several western and environmental historians, who have read my work in various stages. Tom Dunlap helped me to see the need for illuminating the story I was trying to tell. William Cronon did the same, and his suggestions and insights in correspondence and through his own work have been an inspiration. Donald Worster sharpened my ideas by asking searching questions that helped me to broaden my focus and uncover the deeper themes of the study. His NEH seminar in Logan, Utah in 1988 was invaluable as I was then beginning to undertake the revisions to turn the dissertation into a book. Finally, I am more than grateful to Donald Pisani, whose generosity, good cheer, humor, words of encouragement, and scholarly example have sustained me throughout.

Some of those who helped were active in the controversy itself. David Brower offered his memories and insight in two interviews, while his friend and colleague Dan Luten did the same. John B. Oakes took valuable time from his editorial work at the *New York Times* to speak with a young historian over the phone in 1986, and provided additional information by correspondence. The late Wallace Stegner also offered his thoughts. His responses to my inquiries were a pleasure when they arrived, and are now something I greatly treasure.

Richard Bradley provided a tremendous amount of information and insight through several interviews at his home. His recounting of the evaporation debate greatly aided in the writing of chapter seven. He also showed me his father's 1952 home movie of the Green and Yampa Rivers, which was instrumental in drawing the Sierra Club into the fight. Ric and Dorry Bradley shared many memories of Echo Park, Glen Canyon, and other wilderness battles, and they read parts of the manuscript and saved me from several errors. They also supplied some of the illustrations for this book.

Roy Webb, an historian of the Green River, carefully checked my account of the river trips through Dinosaur. He shared his knowledge of the river and of riverman Bus Hatch, and he helped locate relevant material in the Untermann Papers at the University of Utah. I thank him for his careful reading of chapter six and, on a trivial but not unimportant matter, for reminding me that the trees that grow in Echo Park are box elders, not willows.

Peter Blodgett has been a friend and confidant as well as a fellow historian of the national parks whose advice and counsel I highly value. Together with Bill Frank, he helped me find my way around the Huntington Library. Other archivists and librarians who deserve my thanks include Emmett Chisum and Rick Ewig at the University of Wyoming, and the staffs at the Harry S. Truman Library, Dwight D. Eisenhower Library, Bancroft Library, Department of Special Collections at Stanford University, the National Archives, and the Library of Congress. The Bureau of Reclamation in Washington D.C. supplied microfiche records from the late 1940s and early 1950s, which proved invaluable for writing the early chapters. Eleanor Gehres, Barbara Walton, and the superb staff at the Western History Department of the Denver Public Library were always courteous and helpful. My thanks also to archivists at the Department of Special Collections at Utah State University, University of Utah, Chicago Historical Society, University of Arizona, and Arizona State University. Tom Durant, director of the National Park Service Photographic Collection, tracked down many of the illustrations that appear in the book. I wish also to thank the editors of *Utah Historical Quarterly* and *Pacific Historical Review* for permitting me to republish portions of my articles, which appear in chapters one, seven, and eight. Thanks also to the Bancroft Library for permission to quote from oral histories conducted through the Regional Oral History Office.

A number of friends and relatives outside of academia have also been of great help. They include Gary Lessing, whose generous support of my work at an early stage was unmatched; Charles and Wendy Webster, who furnished wonderful accommodations during my research at the Bancroft Library; and Linda West, Lynn Loetterle, and Dan Chure from Dinosaur National Monument, who taught me about the world of dinosaurs and made numerous suggestions for improving chapter one. Larry Anderson has taught me a great deal about the Wilderness Society, and I appreciate his generosity in sharing his own research on Benton MacKaye. Barry Mackintosh, chief historian of the National Park Service, helped me find my way around the research enclaves of Washington D.C.

My colleagues in the history department at North Dakota State

University have been supportive and generous with their time and collegiality. I am particularly grateful to Larry Peterson, chair of the department, for furnishing release time from teaching to work on the manuscript. The College of Humanities and Social Sciences at NDSU has provided financial support for several research trips. David Holtby of New Mexico Press has helped me survive the challenges of publishing with his grace, wit, and enthusiastic dedication to the book.

My sisters Kathleen, Pam, and Patricia have supported me in many ways over the years, and a special thanks goes to Pam for helping me choose illustrations for the book from the Park Service Photo Collection. Gretchen has helped in more ways than I can say. I thank her for her keen eye in writing and editing, for her insights into history and the human situation in the past and present, and for her love. My parents, William and Dorothy Harvey, have given freely of their resources, strength and wisdom from the beginning. They introduced me to the West as a young child, and their keen interest in the region's history and its environment provided a foundation for my work. Both of them read various parts of this book and both greatly improved its prose and substance. My father's mathematical background helped me understand the evaporation controversy, while his eye for language caught a number of awkward phrasings. My mother's own environmental work on the Central Utah Project proved invaluable. She read many drafts of the whole study and gave it the benefit of her own vast knowledge of water and environmental politics in Utah and the West. This book is dedicated to them, with thanks for all the memories of our travels to the West.

A Symbol of Wilderness

Echo Park and the
American Conservation Movement

ONE

The Peculiar Past
of a National Monument

Olaus Murie, executive director of the Wilderness Society, had become frustrated one January day in 1954, and he decided that Wyoming's Senator Frank Barrett should be the first to know. Barrett was scheduled to appear later that month before a House subcommittee to testify on behalf of the Colorado River Storage Project, and to urge approval of its most controversial feature, the dam slated for construction near Echo Park in Dinosaur National Monument. Murie had been reading the Utah newspapers, whose editorial pages reflected that state's overwhelming support for the project. To Murie's way of thinking, the *Salt Lake Tribune* and *Deseret News* deliberately confused the issue with headlines and editorials proclaiming that Echo Park Dam would not flood any dinosaur bones. Murie wanted Barrett to know that for conservationists the fossils were not—and had never been—the point of contention.

Ever since the beginning of the controversy, defenders of the monument had tried to make clear that the dam would not harm the precious fossils—that this was impossible since the quarry of dinosaur remains rested many miles from the dam site. But the prospect of the dam inundating miles of the Green and Yampa rivers and scenic canyons did very much alarm many of the nation's conservationists, who considered this an unwarranted intrusion into Dinosaur Monument and a threat to the national park system. Any mention of flooded bones merely detracted from this more fundamental issue, and was, as Murie told Senator Barrett, "merely a straw man put up for effect."[1]

About a year later, as the Echo Park controversy crested in the spring of 1955, Murie found himself scolding the other Wyoming senator out of a similar frustration. Democrat Joseph O'Mahoney had argued the case for Echo Park Dam in the Senate, in the middle of April, taking pains to point out that no one should fear the flooding of fossils. "We know that dinosaur remains are not endangered and that is *not* one of our arguments," Murie wrote in exasperation. "To continually dig up this dinosaur bone argument is not a fair procedure, on the floor of the Senate of all places."[2]

Wyoming's two senators and most residents of Wyoming, Utah, Colorado, and New Mexico had an obvious interest in raising the issue of the bones. Doing so helped to calm public fears about the dam ruining this part of the national park system, which spanned the border of Utah and Colorado. After all, the very name of the preserve suggested that precious beds of fossils might be at stake, and if they were not, then those who denounced the proposed dam and warned of a desecrated preserve were simply misinformed. And if that were true, it merely confirmed what most residents of Wyoming and surrounding states who stood to benefit from the dam had thought all along: that opponents of the project were driven more by sentiment than by logic. Indeed, the notion that conservationists had bigger hearts than minds was one shared by many Americans in the 1950s. Early in 1954, *Time* magazine referred to opponents of Echo Park Dam as "professional nature lovers."[3]

That comment alone suggests the difficulty of challenging a water project in the era of economic expansion in the United States following the Second World War. Indeed, as this book endeavors to show, the battle to keep the Bureau of Reclamation from having its dam in Dinosaur was enormously difficult for conservation and wilderness groups to win, and most leaders of those organizations did not know if they had the power to do so. Certainly, they lacked the legal means of challenging the project, means that environmentalists of a later generation came to rely on as a matter of course. The 1950s was a vastly different time in the nation's environmental affairs. Even the word *environment* had not entered the lexicon. More to the point, the Echo Park controversy came along a decade before the Wilderness Act, and nearly two decades before the Wild and Scenic Rivers Act and the National Environmental Policy Act.

As a result, opponents of the dam did not have the legal means at their disposal to delay or to block its approval. No environmental-impact statement could be requested from the Bureau of Reclamation, no lawsuit could be brought. The controversy—it should be made clear at the outset—played out in an almost purely political arena, and the political odds appeared in many ways to favor the Bureau of Reclamation rather than the National Parks Association, the Sierra Club, or the Wilderness Society.

So the challenge facing Olaus Murie and the gamut of conservationists who sought to defend Dinosaur Monument seemed great. Faced with an expanding national economy, steady population growth in the American West, and a powerful development agency in the Bureau of Reclamation, conservationists could only hope to persuade the public and their political representatives in Congress to reject this plan for a dam in a part of the national park system. Still, as Murie's frustration with Senators Barrett and O'Mahoney suggests, this was no easy prospect. Dinosaur was a remote and virtually unknown national monument, with a somewhat misleading name. What was there other than a desolate and barren landscape with a few remnants of ancient beasts? Would the dam flood the dinosaur quarry? Would the reservoir cover up age-old skeletons that modern science required for study? If not, then what was the whole fight about? What would be flooded anyway, and why was this important? Why should Americans care about it?[4]

For five and a half years, members of the Wilderness Society, the Sierra Club, the National Parks Association, the Izaak Walton League, and dozens of other organizations struggled to make clear the answer to these questions. They contended that safeguarding the monument from the dam was necessary in order to reaffirm the principles on which the national park system had been founded, as well as to foster protection of a broad range of wilderness lands in the United States. As the controversy unfolded, Echo Park became a prime symbol of the American wilderness, at once magnificently beautiful as well as in danger of disappearing under development pressures.

But Dinosaur National Monument did not become a symbol of wilderness without substantial effort. The monument was hardly well known, even among Americans who traveled regularly to the national parks—as increasing numbers began to do after the war. Here was a part of the national park system known to few Americans, located in an isolated corner of Colorado and Utah. The

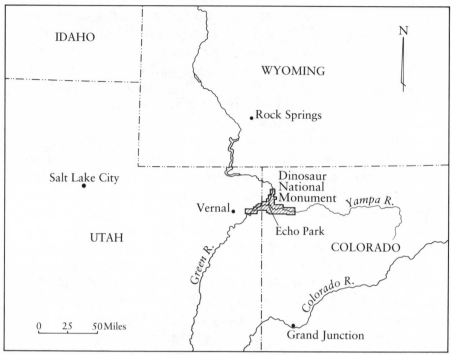

The Location of Dinosaur National Monument

Green and Yampa rivers and surrounding canyons possessed scenery that many came to consider undeniably grand, but Dinosaur did not have the name recognition of more popular destinations such as Yellowstone, Yosemite, or Grand Canyon national parks. More importantly, as the confusion over the fossil bones reveals, the monument did not possess a clear identity like these other well-known areas in the national park system, and the lack of a distinct identity can be traced to the very beginnings of the preserve. In fact, it was the lack of identity that first gave rise to the suggestion of a dam at Echo Park (and another at Split Mountain). If Dinosaur had been one of the better-known parks, such as Yellowstone or the Grand Canyon, this controversy might not ever have come about. But Dinosaur? What did it matter if a dam were built there? Who cared about a few old fossils?

The problem of having an uncertain purpose and a hazy public identity began in 1915. In that year, President Woodrow Wilson, under authority of the Antiquities Act, proclaimed the new monument around eighty acres of public domain in northeast Utah in order to protect a bed of well-preserved dinosaur skeletons. Wilson's proclamation followed by six years a remarkable scientific discovery—one of the most remarkable in the American West and one of the most memorable in the history of North American paleontology.[5]

Supplied with funds from Andrew Carnegie, who dreamed of obtaining a dinosaur skeleton for his museum in Pittsburgh, paleontologist Earl Douglass located, one day in the summer of 1909, a fantastic graveyard of dinosaur fossils in the purple, gray, and red hills that made up the Morrison formation north of tiny Jensen, Utah. Over the next fifteen years, Douglass dedicated himself to the Carnegie Quarry, as it came to be known. The quarry contained the remains of some 400 Jurassic dinosaurs, of which nearly twenty skeletons were complete enough to be mounted. Douglass's discovery became renowned as one of the best windows ever opened into the middle period of dinosaur history. Writing to W. J. Holland, director of the Carnegie Museum, Douglass exclaimed that "so far as I know no dinosaur quarry like this has ever been found before. . . . One couldn't have prayed for anything better and have been at all reasonable in his requests."[6]

The fossil quarry sparked great interest in the sparsely populated Uinta Basin of northeast Utah. Residents of nearby Vernal regularly visited Douglass during the excavations, sold him food and supplies, and occasionally pitched in and assisted him. In 1916, one year after Wilson's proclamation, the newly created National Park Service took jurisdiction over Dinosaur National Monument, raising hopes with local residents, businesses, and the Vernal Chamber of Commerce. They hoped that the Park Service would allow some of the fossil bones to remain in place for public viewing, and they urged that a display be erected that would help to attract tourists. Earl Douglass himself expressed interest in such a project; indeed, he became a spokesman for local business interests.

However, when President Wilson established this monument, he was less interested in preserving the fossil bones in their place than in encouraging the Department of the Interior to regulate excavations—a step that Wilson had been persuaded was needed in order to prevent the Carnegie Museum from monopolizing the quarry.[7] Then, in subsequent years, despite his intention, the Department of the Interior and the National Park Service continued to permit the Carnegie Museum to remove skeletons, and by 1923 they were nearly gone. By the middle of the 1920s the quarry had been mined of its precious fossils, leaving a national monument with little to show for itself.[8] Having been in the spotlight for some years during the excavations, northeast Utah's little communities felt all but forgotten by museums and by the Park Service. Once the skeletons had been carted off to Pittsburgh, Washington, D.C., and Salt Lake City, Dinosaur National Monument seemed to have become a public preserve whose existence could not be justified.[9] Local residents were understandably disillusioned, and they had to wait until the early 1950s before the Park Service began to excavate additional fossil bones and place them on display in a new visitor center.

Yet despite deep frustration with the National Park Service, hopes of attracting tourist dollars by no means disappeared, and in the 1930s there began a campaign to protect a large portion of public land north and east of the dinosaur quarry. Here lay a vast, rugged landscape dominated by the Lodore and Yampa canyons and two great tributaries of the Colorado River, the Green and Yampa. A showcase of geologic history, the river canyons provided a haven for wildlife and held innumerable geologic and archeological wonders, including petroglyphs and pictographs decorating the canyon walls. In prehistoric times, these canyons had been home to the Fremont people who inhabited portions of the Colorado Plateau from about 500 A.D. to 1250 A.D. Remnants of their culture were evident in pithouses, petroglyphs, and pieces of pottery and corn.[10] In later years, the Ute, who occupied much of present-day Colorado in the eighteenth and early nineteenth centuries, may have hunted and gathered in the canyons.[11] During historic times, the canyons did not have permanent inhabitants and seemed to have been primarily a passageway for fur traders and explorers who floated the rivers. Fur trader William Ashley carved his name on a rock wall in 1825, in Red Canyon, north of Lodore Canyon and beyond Browns Park. Ashley descended the Green River and made his way through Echo Park and Split Mountain—as yet unnamed—but left no further sign of his presence.[12]

John Wesley Powell left a more lasting mark on the region. A geologist by trade, Powell had become enamored with the geology and topography of the West, particularly the Colorado River and Plateau, a region largely unexplored by white Americans in the mid-nineteenth century. With financial assistance from the Illinois Natural History Society and the Chicago Academy of Sciences, Powell launched himself and nine other men in four wooden boats on the Green River in May of 1869.[13] Powell successfully descended the Green and Colorado rivers in 1869, and again in 1871. On the first expedition, one of the four boats called *No Name* wrecked in a furious rapid in upper Lodore Canyon. The three-man crew survived uninjured, but two thousand pounds of provisions and weapons were lost. Appropriately, Powell and his men dubbed the rapid Disaster Falls, and it remains to this day one of the most exciting among many rapids on the Green River in Dinosaur National Monument. Powell named a number of other places inside the present monument, including Lodore, Split Mountain, and Whirlpool canyons, and Island and Rainbow parks.[14]

On June 18, 1869, Powell rested his crew on a small beach located at the confluence of the Green and Yampa rivers. Here was a perfect site to recuperate. Away from Lodore's rapids, the men relaxed in the grass beneath the cool shade of box elder trees. Surrounding cliffs hovered overhead, and across the Green River from their resting place loomed an eight-hundred-foot sandstone wall, massive and perfectly vertical and so high they had to crane their necks to see the top. The wall dominated the whole scene and, to their astonished delight, sent their voices back across the river in loud echoes. Powell recorded his impression in his daily journal: "Standing opposite the rock, our words are repeated with startling clearness, but in a soft, mellow tone, that transforms them into magical music. Scarcely can you believe it is the echo of your own voice." Powell decided to call the little glade Echo Park, and the name found its way from his journal to government maps. "Echo Rock" later took the name Steamboat Rock because it resembled a giant ship's prow. Powell and his men treasured their stay in Echo Park, and in 1871 Powell's party camped there for almost a week.[15]

Following Powell's journey down the rivers, the region surrounding Lodore and Yampa canyons became cattle country, giving rise to small towns such as Maybell, Lay, Hayden, Colorado, and Baggs, Wyoming. A broad, grassy valley known as Browns Park was the center of the industry and a thoroughfare for Butch Cassidy, Tom Horn, Isom Dart and other famous and infamous persons connected to the cattle trade. Two of the so-called outlaw trails between Montana and Texas crossed through Browns Park, skirting the Lodore and Yampa canyons in what later became Dinosaur National Monument.[16]

Despite the success of the cattle industry, this border region of Colorado and Utah remained isolated from the corridors of transportation and the centers of population in the nineteenth-century West. Vernal, Utah, the small town in the Uinta Basin closest to the canyons, lay nearly two hundred miles east of Salt Lake City and four hundred miles west of Denver, Colorado. In such a marginal region, hopes ran high for new settlements and for still undiscovered sources of wealth. Such hopes appeared in the attempt to establish a colony of homesteaders near Craig, Colorado, in 1915 and in the frequent outbursts of boosterism from local newspaper editors.[17]

But no amount of hope could overcome the primary obstacle to prosperity—the lack of reliable transportation to the outside world. After the turn of the century, excitement grew with the prospect of a railroad across northwest Colorado, to be built by Denver businessman David Moffat. Moffat promoted his railroad with great enthusiasm, and he had the capital to allow construction to begin. Unfortunately, Moffat's death in 1911 dashed hopes and soon afterward the Denver and Rio Grande Railroad purchased the Moffat line and built the so-called Dotsero Cutoff, a move that left Craig, Colorado, as the western extent of the railroad. Northwest Colorado remained isolated from a good railroad. Residents of Rangely, Colorado, Vernal, Utah, and other towns near the border remained isolated and set apart from the more prosperous communities of Colorado's western slope and Utah's Wasatch Front.[18]

Given their economic woes, it is not surprising that local residents considered tourism as a prime source of income, and in the 1920s interest grew in attracting visitors to the Lodore and Yampa canyons. Archeologists, explorers, and a handful of river runners came into the area and helped to spark excitement with their writings. Then, in 1928, one A. G. Birch, an employee of the *Denver Post,* took a much-publicized river trip down the Yampa, which gave rise to a movement to set aside the river and surrounding canyons as part of the national park system. Birch had an urge to explore and a strong interest in promoting desirable tourist attractions in Colorado. Knowing how much "wealthy sportsmen" enjoyed remote spots, Birch believed that Yampa Canyon offered a vacationer's paradise. For three weeks *Denver Post* readers followed his journey, tantalized by dramatic headlines: "Post's Expedition Is Nearing Perilous Trip Down Cañon" and "Expedition to Risk Death in Wild Region."[19] Given the Birch party's inexperience, the *Post* headlines probably did not exaggerate the risks. Birch and his companions were lucky to complete the trip safely.

But Birch could not contain his enthusiasm for the scenery. "There is nothing like the Yampa river cañon that I have ever seen," he wrote, sounding like a seasoned world traveler. Back in Denver, rested and fed, he waxed eloquent about the landscape: "Imagine seven or eight Zion cañons strung together, end-to-end; with Yosemite valley dropped down in the middle of them; with half-a-dozen 'pockets' as weird and awe-inspiring as Crater Lake; and a score of Devil's towers plumbed down here and there for good measure—then you will just begin to get some conception of Yampa cañon." As for its tourist potential, Birch suggested that if the inaccessible gorge could ever be opened, "it will be impossible to stem the tide of visitors . . . Colorado has nothing else like it."[20] In Birch's mind, this strikingly scenic corner of the American West had been overlooked.

The publicity surrounding the Birch expedition awoke many in Colorado and Utah to the scenic beauties within Lodore and Yampa canyons, and spurred the interest of tourist promoters in both states as well as that of the National Park Service. Gradually, the two came together. In the fall of 1933, Park Service Director Arno Cammerer requested an inspection of the two canyons from Roger Toll, the superintendent of Yellowstone National Park. In a lengthy report prepared for the NPS, Toll explained that the canyons had much to offer with their archeological sites, human and natural history, wildlife, and stark beauty. As for Yampa Canyon, Toll suggested that it "has a scenic individuality that is different from any other area and it is outstandingly beautiful, rugged, and picturesque."[21]

Toll saw Powell's beloved Echo Park as well, referring to it by the less romantic name "Pat's Hole," which came from one Pat Lynch, a hermit who had once lived near the confluence of the Green and Yampa:

> Pats Hole is one of the most beautiful and impressive places in the area. Here on the banks of the Green River, where the Yampa joins it, one can easily visualize the parties of early explorers on their way

down the river. The same scene that they saw is there today, unchanged. On the outer side of the loop that the Green River makes
is a park-like meadow, with large cottonwoods and box elders,
golden in their autumn foliage, at the time of our visit. Across the
river stands Steamboat Rock, a great sentinel, with a smooth cliff
that suggests a small El Capitan, in color, or perhaps the mystery of
the Enchanted Mesa. It is a beautiful place to camp, and the Indians
doubtless found it so for there are pictographs nearby. Steamboat
Rock throws back an exceptionally perfect echo, and repeats some
six or seven words, spoken rapidly. No wonder that Major Powell
called it Echo Park. Its equal is hard to find.[22]

Toll found historic and natural features sprinkled throughout
the canyons that convinced him of the merits of protecting this
portion of the Green and Yampa rivers by the National Park Service. Along with plentiful evidence of the Fremont culture, he saw
an array of wildlife including Rocky Mountain mule deer, bighorn
sheep, bobcats, coyotes, cougars, and gray foxes, along with por-
cupines, beavers, and smaller animals. There were Canadian geese,
golden eagles, pinyon jays, woodpeckers, and mountain chicka-
dees. The vegetation resembled that of other southwestern parks,
with pinyon-juniper forests and mountain mahogany, serviceberry,
shadscale, greasewood, and the ubiquitous yucca. Toll concluded
that "the area lends itself to popular education along the lines of
geology, erosion, archaeology, botany, and other natural history
subjects, as well as having fascinating historical background."[23]

Toll believed the Park Service ought to act quickly to protect this
area as a national monument, and he thought this could be done
rather effortlessly due to the almost complete absence of private
land within the 250–square-mile area. Most ranchers lived in
Browns Park, well outside the canyons. A few ranchers, including
Charlie Mantle, Jim Monaghan, and Ralph Chew, ran cattle on
the land that Toll recommended for preservation, and not wishing
to irritate them, Toll thought they should be permitted to continue
grazing their cattle after the Park Service moved in.[24] Hearings with
ranchers and Park Service officials subsequently took place, and
several months passed before boundary lines and other arrangements could be finalized. As matters developed, the Park Service
agreed to let the federal Grazing Service continue to issue permits
on the new monument land, while ranchers won a promise that
their grazing rights would be maintained until the 1980s.[25]

However, Roger Toll's conviction that the Park Service should acquire this land and protect it as a national monument generally accorded with local sentiment in Colorado and Utah. W. D. Rishel, president of the Utah Automobile Association, anticipated the tourist dollars that the Green and Yampa canyons would help to supply. Dr. J. E. Broaddus of Salt Lake City held six hundred people spellbound at a local high school, in 1937, with a lecture and slide program of a trip he took down the Yampa River. Broaddus predicted that tourists would eagerly flock to the canyons, and a local newspaper echoed his remarks in claiming that "with the development of roads and trails and properly supervised river trips, visitors would be attracted to this region by the tens of thousands."[26] (Broaddus, interestingly enough, later joined in advocating approval of Echo Park Dam.) Utah Congressman Abe Murdock claimed to be "overwhelmed with the scenic grandeur" of Split Mountain Gorge, and spoke publicly on behalf of the proposed monument.[27] In Colorado, a group of Denver businessmen enthusiastically backed the plan, aware that the Lodore and Yampa canyons lay at the center of a circle of national parks in Wyoming, Colorado, Utah, and Arizona. Their spokesman, Frederick Reid, mentioned Yampa Canyon in a letter to Undersecretary of the Interior Oscar Chapman, noting that "its strategic location gives it unusual importance in any plan to develop a park-to-park highway." Colorado Congressman Edward Taylor, whose district encompassed Yampa and Lodore canyons, likewise supported the idea.[28]

The prospect of a new national monument also fit in nicely with plans of the Park Service during the 1930s. During that decade the agency underwent considerable expansion, acquiring a variety of national military parks, cemeteries, memorials, and sites in the nation's capital. According to one historian, these additions "changed assumptions of what the National Park System should contain by adding new kinds of areas not previously considered as either parks or monuments."[29] With a diversity of sites now under its care, the Park Service became fearful of drifting away from its original mission of protecting scenic landscapes, and therefore sought to reaffirm its interest in such places. The new Wilderness Society, formed in 1935, signified a small but devoted group of Americans who regarded such lands with reverence. Therefore, a scenic national monument in the West seemed an attractive addition.[30] Secretary of the Interior Harold Ickes also supported the enlargement of the monument. Ickes, as Barry Mackintosh has pointed out, became increasingly devoted over the course of the New Deal to preserving the aesthetic qualities of the great scenic parks, and expressed interest in safeguarding their wilderness character by resisting pressures for more roads and concessions.[31] Writing to President Roosevelt in April, 1938, Ickes recommended expansion of Dinosaur National Monument around Lodore and Yampa canyons, emphasizing their rugged beauty and "frontier" past.[32]

Everything seemed ready. With local support for the new monument strong and cattle ranchers satisfied with grazing rights, the only question that remained was what to call the new preserve. For a time, the National Park Service considered Green River National Monument and Yampa Canyon National Monument. But Wyoming and Utah each had towns called Green River, and the Park Service considered it wise to avoid that name. Finally, the NPS decided for the sake of convenience to graft onto Dinosaur National Monument the section of the Green and Yampa rivers that it sought to protect. On July 14, 1938, President Roosevelt signed the necessary proclamation.[33] Now a huge landscape—well over 200,000 acres—was attached to the original monument surrounding the dinosaur quarry. Literally overnight, a tiny national monument in northeast Utah became one of the biggest areas in the national park system, with more than 360 square miles. The National Park Service rejoiced. A new and magnificently scenic national monument had been created in the American West, one that could complement the array of battlefields and historic birthplaces in the East.

Certainly, the local populace was pleased. The new preserve in their backyard promised to beckon tourists, and its establishment just happened to coincide with the completion of a new highway. In 1938, the old "Victory" Highway through Colorado and Utah was superseded by U.S. 40, a far superior road that provided a direct connection from Denver to Salt Lake City. Upon completion of U.S. 40, the staffs of two local newspapers, the *Vernal Express* and *Roosevelt Standard,* collaborated on a "scenic edition" insert section. Lauding the possibilities of tourism in the region, the papers pointed to scenic wonders and historic attractions in this remote part of the West, including Park City, an important mining town in Utah; the Uinta Mountains; and Rocky Mountain National Park. As for Dinosaur National Monument, its "possibilities are almost unlimited, and whether it be fishing, bathing, scenery, history, seclusion, or study, the tourist may find it there."[34]

Yet the eagerness of residents could not overcome the fact that the new monument had been born with severe handicaps, all of them related to its lack of a clear public identity. In the first place, the decision to retain the original name, while convenient to the Park Service, produced confusion that persisted for many years—confusion that appeared repeatedly in the 1950s with charges that Echo Park Dam would not flood dinosaur bones. The preserve covered more than 360 square miles of northeast Utah and northwest Colorado, yet only a small corner contained deposits of dinosaur fossil bones. "Dinosaur" was simply a misnomer. In terms of geologic age, the canyons were much younger than the Morrison Formation only a few miles to the west and they held no dinosaur remains. Since the name of the preserve provided no clues to what might be found there, most visitors believed that the quarry was the only attraction and many overlooked the canyons entirely.

Secondly, Dinosaur did not have the same high status that the more exalted "national park" label brought. National monuments captured less public attention than parks and usually received substantially less funding.[35] More to the point, Congress enjoyed no role in creating Dinosaur because there had been no national movement to set aside the Green and Yampa canyons. Although local residents overwhelmingly supported the new monument, its presence could not be attributed directly to their support, compared to nearby Rocky Mountain National Park. National monuments, of course, were created by proclamation rather than by Congress, and in this instance the difference proved to be critical. Because Dinosaur arose from the Department of the Interior during the Great Depression—to protect a large portion of public land that remained largely unknown—it received little fanfare outside Colorado and Utah and went all but unnoticed by the traveling public. And while the Park Service was glad to have the new monument, the agency could not afford to build roads or properly administer the new preserve for a number of years after the proclamation.

Still another reason for its lack of identity was that few devotees of national parks outside the government had anything to do with its establishment. Conservation activists, except for a handful in Colorado and Utah, did not notice this new preserve. Groups who played leading roles in the Echo Park controversy, such as the Sierra

Club and the Wilderness Society, had no interest in Dinosaur in the 1930s and had not been involved in its creation. The Sierra Club was then uninterested in regions outside California, while the Wilderness Society had just been established in 1935, had barely a few hundred members, and had more pressing concerns elsewhere. In time, these organizations and others cast Echo Park as a national symbol of endangered wilderness, but they had little or no awareness of the new preserve at the time of its creation in 1938.

But the most important explanation for the new monument's lack of identity was President Roosevelt's proclamation itself, which contained a couple of simple phrases that made it so. The proclamation had been carefully crafted by various public agencies following months of negotiations and bargaining. Those negotiations took place in Washington among personnel in the Park Service, the Bureau of Reclamation, and the Federal Power Commission, who debated how the land and water in the new preserve should ultimately be used. The three agencies each wanted something from the presidential proclamation, and when it appeared on President Roosevelt's desk for signature in the summer of 1938 each could point to clauses recognizing their interests. Since no agency wanted to preclude the interests of any other— and since the Park Service lacked political power to stand up to the other two—the proclamation reflected various agendas for water and power. Both the Federal Power Commission and the Bureau of Reclamation obtained crucial wording that established the possibility of dams and power plants in later years. As a result, the monument was given an ambiguous legal status from the start. Roosevelt's proclamation, in effect, perpetuated this already peculiar national monument and planted the seeds of the Echo Park Dam controversy.

By the early 1930s it was clear that Echo Park and other locations in these rugged and steep-walled canyons of Utah and Colorado had great potential for generating power, and while no large market for that power yet existed, the mere expression of interest proved to be significant. When Roger Toll filed his report on the Green and Yampa region in 1933, he indicated having been informed by an official of the Bureau of Reclamation in Denver that several dam sites along the Green and Yampa rivers were regarded as "very important sources of future power."[36] Toll stated in his report that Utah Power and Light Company had a strong interest in a power plant near Echo Park, and had applied to the Federal Power Commission in Washington for a license to develop power at the site and at other locations in the canyons of northwest Colorado, including Blue Mountain and Lilly Park on the Yampa River.[37] He concluded that it might "be a generation before these dams are built, and it seems desirable to go ahead with the national monument program, without reference to this possibility, in the expectation that the dams if built will not seriously interfere with the successful recreational use of the area by the public."[38]

With a private firm taking interest in the potential of hydropower, the Federal Power Commission became an important player in Dinosaur's early history. Created by Congress under the Federal Water Power Act of 1920, the FPC maintained jurisdiction over power development on public land and navigable streams. In order to prevent private firms from monopolizing power resources, the FPC possessed the authority to determine who could build power plants.[39] In the early 1920s, the FPC established "power withdrawals" within the Lodore and Yampa canyons, thereby asserting its authority to license power firms that undertook to build dams and power facilities in the region.[40] FPC Chairman Frank McNinch, in correspondence with the Park Service in the early 1930s, quoted estimates of Utah Power and Light and wrote that "the Green and Yampa Rivers present one of the most attractive fields remaining open for comprehensive and economical power development on a large scale." He referred specifically to the Echo Park site as especially excellent for generating power, and made it clear that the FPC was well aware of its potential.[41]

As matters developed, Utah Power and Light withdrew its application with the FPC in 1935. Apparently, the firm lacked the financial resources to build power plants at Echo Park or at Blue Mountain, because its own power market had sagged during the Depression. Park Service officials now wondered if the FPC would agree to cancel its "power withdrawals" so that the new national monument could be established without any mention of future power development. Secretary of the Interior Harold Ickes wrote to the FPC with this request.[42]

Yet the Utah company's action did nothing to alter the FPC's interest in the Green and Yampa rivers. McNinch informed Ickes that although Utah Power and Light's application had been rescinded, the FPC wished to maintain its power withdrawals. Al-

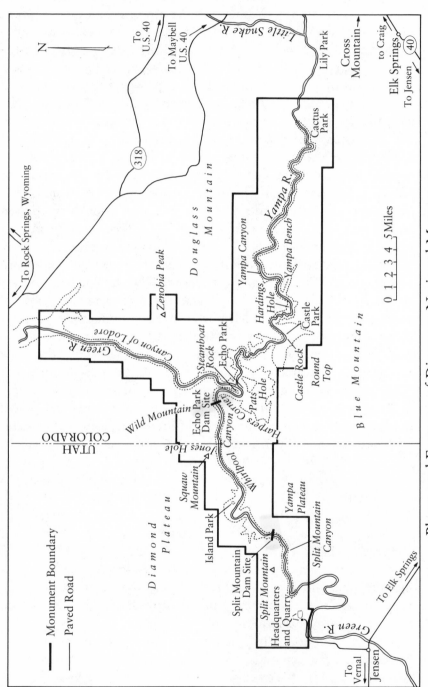

Places and Features of Dinosaur National Monument

though he did not say so directly, it was clear that the FPC did not want to surrender its jurisdiction over rivers that were potential sources of power, especially in light of the Depression and the Roosevelt administration's inclination to regard rivers as tools for economic growth. As for the Park Service's interest in a national monument to preserve the region, McNinch indicated that the FPC had no objection, so long as the executive proclamation creating the preserve specifically recognized "that power development under the provisions of the Federal Water Power Act will be permitted."[43] Roger Toll had apparently been right to recognize the importance of power resources along the Green and Yampa, and it now seemed that the interest in power would have to be recognized in the executive proclamation. Aware now of the FPC's concerns, Park Service officials continued to work their way through various political and bureaucratic avenues to prepare the required executive proclamation.[44]

As they did so, Congress amended the Federal Water Power Act of 1920. Enacted in 1935, this amendment had great bearing on the Echo Park controversy two decades later, although it drew little attention when it first appeared. Under the amendment, lawmakers curtailed the FPC's authority to issue licenses for power development on public land designated as national parks or monuments. The evidence is sketchy at best, but it appears that McNinch and personnel at the FPC did not realize that this amendment had been enacted, and the agency therefore continued to demand that its authority to license power dams be recognized in the executive proclamation being drawn up for the new Dinosaur Monument.[45] This amendment did not escape the notice of the National Park Service, however, and NPS officials now felt confident that the FPC would lack authority to issue power permits in the Green and Yampa canyons once they became part of the national park system.[46] From a legal point of view, it seemed clear that the FPC's "power withdrawals" would become ineffective once the Green and Yampa canyons were designated as a national monument.

Nevertheless, Park Service personnel knew that there had been no strong public outcry for the amendment and that it had gone unnoticed even by most members of Congress. In the future, if a private power company sought to build a dam in a national park or monument a sympathetic Congress might still approve of such a plan and conveniently ignore such "a technicality in the law."[47]

As a result of these considerations, NPS officials agreed to allow the FPC to insert a clause recognizing its authority into the presidential proclamation for the new monument; above all, they realized that satisfying the FPC must be done if the proclamation were to move forward. Roosevelt's proclamation thus indicated that the new monument "shall not affect the operation of the Federal Water Power Act of June 10, 1920, as amended and the administration of the monument shall be subject to the Reclamation withdrawal of October 17, 1904, for the Brown's Park Reservoir Site in connection with the Green River Project."[48] Contained in these few words were the seeds of the great battle over Dinosaur to come.

As indicated above, the Bureau of Reclamation also had a hand in formulating the proclamation. In the 1930s, the Bureau was beginning to take steps toward planning for dams along the upper Colorado River, and though a great deal of survey work needed to be done and specific plans formulated, the Bureau's negotiators also insisted on "protective language" in the proclamation. One Bureau official suggested that the proclamation read that "all necessary rights of way are reserved to government agencies, including the Bureau of Reclamation, requisite for the full development of potential power possibilities within the monument area . . . and for the conveyance of water for irrigation and other use beyond the monument boundaries."[49]

The Park Service resisted this kind of blanket reservation, as it would have compromised the purposes of the monument from the start, but NPS negotiators looked for some way to acknowledge the Bureau's ongoing studies. After further discussion, the Park Service agreed to recognize the so-called Browns Park dam site located in the northern end of Lodore Canyon. This site, first noticed by the Bureau many years earlier, lay far to the north of Echo Park but within the proposed boundary of the new monument. By recognizing the Browns Park site in the proclamation, the Park Service avoided the more sweeping language initially sought by the Bureau. At the same time, the "Browns Park" clause, according to records in the National Archives, represented "a commitment by the National Park Service that if and when a definite project has been determined upon by the Bureau of Reclamation the necessary adjustments in the status of the monument area would be made to permit water power development."[50] It was not the last time that the agency took such a position. Only a few years later, the director of the National Park Service signed an agreement with the Bureau to the same effect.

If all of this seems strangely out of place or a rather inappropriate kind of bargaining, it is because the general atmosphere surrounding the national park system and the legal framework of regulations protecting parks from particular kinds of uses have changed greatly since the 1930s. Like many national parks and monuments established in the late nineteenth and early twentieth centuries, Dinosaur was created with provisions allowing for economic uses—in this case, grazing and water development. With few legal barriers to prohibit this, other agencies with jurisdiction over land and water in the West—namely, the FPC and the Bureau of Reclamation—could thereby demand that the Park Service allow "multiple use" of parks or monuments.

Moreover, NPS officials felt that they had little choice but to consent to their demands. The Park Service did not have the political strength during the Depression to stand up to powerful agencies like the Bureau of Reclamation or the FPC. Certainly, conservation and wilderness groups did not have the same powerful constituencies as water and power interests. In the dark days of the Depression, the construction of dams and hydroelectric-power plants appealed to the American public far more than preserving national parks. Without support from the FPC and the Bureau of Reclamation, expansion of the Dinosaur Monument could be delayed, and NPS officials considered that unwise because it could derail the entire process well under way for creating the new monument. Lawmakers from Colorado and Utah had consented to the presidential proclamation, and ranchers had been satisfied that grazing rights could be protected. The Park Service did not want to lose that momentum and felt that the time appeared to be right for moving ahead with the proclamation. It was therefore willing to concede to the language permitting dams and power projects.

The result of these competing interests in the use of the Green and Yampa rivers and the land surrounding them was a kind of modus vivendi in which the Park Service obtained a spectacular new national monument in a remote corner of the American West, one made subject to the possible construction of dams and power plants inside its boundaries and at an indefinite point in the future.

It should be said here that the proclamation of 1938 became one of the focal points of dispute in later years, during the public phase of the controversy. Conservationists and supporters of Echo Park Dam then debated fiercely the meaning of Roosevelt's decree, especially the legal status of the power withdrawals following expansion of the monument. But when President Roosevelt affixed his signature to the proclamation in July 1938, no one could have imagined the conflict that would develop in later years over Echo Park Dam. The nation was struggling with the Great Depression, and the establishment of a national monument in a remote corner of Colorado and Utah went unnoticed to all but a few. Nor could anyone then foresee the Second World War, which dramatically changed the economy of Utah, Colorado, and their neighbors who shared the water of the Colorado River, and escalated the Bureau's plans for building dams to capture it.

As the Depression waned and the war began, the states that shared the upper Colorado River began to experience tremendous changes, and in the decade following Roosevelt's proclamation, a new era of prosperity and population growth arrived in the region. These changes generated increasing demands for water and power across Wyoming, Utah, Colorado, and New Mexico, states that had not yet utilized their share of the Colorado River. The Bureau of Reclamation happily responded to their demands with an ambitious blueprint of dams known as the Colorado River Storage Project. A giant plan of water impoundments and power plants, the project included a proposal for two dams in Dinosaur National Monument, one near Echo Park and another in Split Mountain Canyon. Upon its creation in 1938, the new Dinosaur National Monument lacked a clear identity and faced an ambiguous future, but in the following decade these remote canyons of the Green and Yampa rivers became highly important for their water and power resources. As they did, a new identity for Dinosaur Monument emerged, not from national park or wilderness lovers, to be sure, but from the Bureau of Reclamation's office in Salt Lake City.

T W O

The Seeds of Controversy

If the prophet Moses and the legendary King Midas were to appear in the state of Utah, there is little doubt whom we would follow. There might be a handful who would prefer the touch of gold. But the multitudes would follow Moses, preferring that our rocks bring forth life-giving water rather than that they be turned to gold.
—Utah Congressman Wallace Bennett,
U.S. Senate, June 1954[1]

Some fundamental questions might well have occurred to the casual observer who noticed the proclamation enlarging Dinosaur National Monument. What would be the future of this national monument? Would it take its place alongside Yellowstone, Yosemite, Glacier, or other famous parts of the national park system? Would it become a great attraction for travelers and contribute to the economy of Utah, Colorado, or Wyoming? And given the proclamation itself, with its language about reclamation and power projects, would the national monument be held in perpetuity for the enjoyment of future generations? Or would it become dominated by dams and reservoirs, and ultimately become known as a prime recreation area for boaters and fishing enthusiasts?

The answers to these questions did not appear very clear to anyone in 1938, nor, for that matter, were they made clear at all until another decade and a half passed and the Echo Park controversy came to an end. Dinosaur National Monument had been created ambiguously, with the executive proclamation allowing the possibility of one or more dams and power plants inside the preserve in the years ahead. Despite being a part of the national park system, the enlarged monument lacked a clear identity and therefore faced an uncertain future. Moreover, the Park Service had few funds for building roads or public facilities, and it could do little to "develop" the place in the same manner as Yellowstone or its cousins.

This was partly because national monuments customarily received less funding and were generally less well respected than national parks. For many years, monuments languished as "second class sites," as Hal Rothman has pointed out, and did not garner the same degree of respect as the grand national parks.[2] Dinosaur was no exception.

The outbreak of World War II compounded the problems of the newly enlarged monument in Utah and Colorado. The war had a powerful effect on the National Park Service, which found itself shunted aside as wartime agencies sprouted in Washington. Agency headquarters moved to Chicago, and Park Service Director Newton Drury learned to make do with skeleton staffs and a slashed budget. The war left the national park system in a holding pattern, with little to do but wait until the conflict ended. With a strictly limited annual budget, the Park Service had precious little funding for sites like Dinosaur National Monument. Lacking access roads into the canyons or any kind of interpretive center, the preserve languished, with its vast landscape remaining practically untouched, unexplored, and unseen by most tourists who preferred a trip to one of the "big name" parks like Yellowstone or Yosemite. With the old dinosaur quarry undeveloped as well, only a few hundred visitors per year saw any part of the preserve. The potential of Dinosaur for becoming a shining star in the national park system did not come to fruition because of its location and the timing of its entry into the park system. In short, its "identity crisis" persisted for many years after President Roosevelt's proclamation in 1938.

With the Park Service playing a passive role at its newest holding, the Bureau of Reclamation gained the offensive in setting an agenda for how the preserve might ultimately be used. While the National Park Service suffered from the Great Depression and the Second World War, its sister agency at the Department of the Interior blossomed and quickly became a major force throughout the American West. Together with the Army Corps of Engineers, the Bureau of Reclamation became a visible presence in western states, active in "harnessing" major rivers. While the Army Corps undertook construction of Bonneville Dam on the Columbia River in Washington and Fort Peck Dam on the Missouri River in Montana (symbols of New Deal spending in those states), the Bureau began to build the Colorado–Big Thompson Project, the Central Valley

Project in California, as well as Grand Coulee Dam on the Columbia and Boulder Dam on the lower Colorado.[3] The great dam-
building era in the West—what Marc Reisner calls "the go-go
years" of the Bureau of Reclamation—had begun, and few were
inclined to block its expansion or question its merits. With giant
dams like Grand Coulee, Boulder, and Fort Peck, the nation found
proof that science, technology, and a dash of ingenuity could lift
the country out of its doldrums. In his well-told history of water
and the West, *Cadillac Desert,* Reisner claims that "symbolic
achievements mattered terribly in the thirties, and the federal dams
going up on the western rivers were the reigning symbols of the
era."[4]

During the late 1930s and 1940s, economic forces in the West
gave birth to a large blueprint for harnessing the Colorado River,
and planted the seeds of the controversy over Echo Park. The Great
Depression first stirred interest in the upper Colorado River by
New Mexico, Utah, Colorado, and Wyoming. Like much of the
American West, these states had been devastated by years of
drought and depression, primarily because their agricultural base
had fallen into desperate straits. The ranching industry suffered as
prices plummeted on cattle and sheep, and stockowners had no
choice but to allow the government to buy their animals at rock-
bottom prices. Farmers saw wheat and other commodity prices
plunge. As historian Robert Athearn has aptly stated, "western
agriculture was carried out under marginal conditions, and when
all of the above negatives were present the additional blows of
drought and dust spelled the difference between a momentary defeat and absolute disaster."[5] The Rocky Mountain states also
slumped due to their reliance on extractive industries such as mining and oil. This remote area in the American West could survive
only through federal aid and subsidy. New Deal programs like
AAA, FERA, and WPA provided the critical sustenance needed for
these states to weather the Great Depression.[6]

As their economic woes deepened, these four small states increasingly regarded water within their borders as a valuable resource offering hope and economic stability. Ranchers, farmers,
and communities from southwest Wyoming to eastern Utah to
Colorado's western slope clamored for the development of streams
and rivers that were tributaries to the Colorado River. Throughout
this sparsely populated and economically marginal region, interest

began to grow for dams and power plants along the Colorado and tributaries like the Green, Gunnison, San Juan, and Dolores rivers. Building dams to regulate the water and generate power promised benefits to ranchers who sought to irrigate dry lands, farmers who belonged to rural electrical cooperatives, and business people in Denver, Albuquerque, and Salt Lake City. Storage dams and irrigation projects offered hope for the future and reassurance in the aftermath of the dry and depressed conditions of the thirties. Although the Colorado River did not match the Missouri or Columbia, either in size or in strength, it became a critically important resource to states in the central Rockies and Southwest in the waning years of the Great Depression. They now considered it their lifeblood.

Enhancing the value of Colorado River water to Utah, Wyoming, Colorado, and New Mexico was the Bureau of Reclamation's efforts to tame the lower stretches of the river to benefit downstream states. Congress approved the Boulder Canyon Project Act in 1928, authorizing construction of giant Boulder Dam in Black Canyon, southeast of Las Vegas. Later renamed for President Hoover, the dam, completed in 1935, blocked the lower Colorado River behind one of the world's great man-made structures. Hoover Dam protected southern California from ravaging floods that had devastated farms and communities earlier in the twentieth century. It created Lake Mead, an exciting new playground of water in the desert Southwest, and it supplied power to southern California and southern Nevada.[7] Los Angeles soon began to attract a host of manufacturing concerns, while tiny Las Vegas found itself transformed overnight by construction of the dam and power plant, receiving a massive economic stimulus that helped transform it into a major resort city in the Southwest.[8]

The Bureau of Reclamation lost no time in following up on the mammoth Hoover Dam. To provide additional flood control and power as well as a diversion dam for the Colorado River Aqueduct, which was needed to send water to Los Angeles, the Bureau constructed Parker Dam, downstream from Hoover. In 1941, water began to flow along the 242–mile aqueduct toward Los Angeles, which, in the words of Norris Hundley, Jr., "obliterated any sense of restraint about Los Angeles's capacity to absorb ever more people and industries." Meanwhile, completion of the All-American Canal in 1942 furnished water to the Imperial Valley, fostering agribusiness on a large scale.[9] By the onset of World War II, Colorado River water had begun to fuel industrial and agricultural expansion throughout southern California and to transform the economy of the American Southwest.

All this flurry of construction of dams, power plants, and aqueducts on the lower Colorado caused trepidation in the sparsely populated states upstream. With California and her neighbors beginning to utilize the lower Colorado for irrigation and hydropower, residents of Utah and neighboring states hundreds of miles upstream began to realize that downstream states had an ever-increasing appetite for the great river, and they began to fear that their future could be adversely affected unless they found ways to capture and use their own part of the Colorado River. As they watched the Bureau tame the lower Colorado, in response to rapid growth of southern California and the Southwest, upper-basin states grew restless.

Of course, from one standpoint, upstream states did not need to worry about losing Colorado River water to California and its neighbors, at least not in a legal sense, for all seven states that shared the great river had agreed some years earlier to the Colorado River Compact, the Magna Carta governing its use. (Arizona did not actually ratify the Compact until 1944.) The Compact created an upper and lower basin and permitted each to use half the water available annually in the river system.[10] But while the Compact guaranteed water for all concerned states, it also required that the upper-basin states see to it that 75 million acre-feet of water reach the lower basin every decade, or in more practical terms, 7.5 million acre-feet every year. In order to accomplish such a feat, as well as to be able to utilize their own allotment, the upper basin had to be able to regulate the river with dams above Lee's Ferry, the dividing point of the upper and lower basins. Only by capturing surplus water in wet years could the upper basin make use of its own share while also meeting its delivery requirements under the Compact.[11]

By the late 1930s the upper-basin states were urging the Bureau to pay attention to their own water demands. Water must be stored in the upper basin, too, so that the whole Colorado River system could be properly regulated to the benefit of all. Here was the most compelling reason behind the upper basin's determination to put its water to use, and the fundamental motive giving birth to the

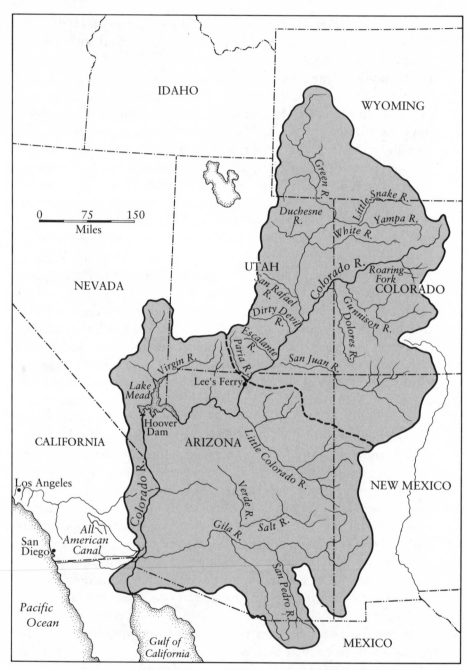

The Colorado River Basin,
with the dividing point at Lee's Ferry, Arizona

plan to build Echo Park Dam in Dinosaur National Monument. From the beginning, interest in dams in Dinosaur sprang from an ardent desire to capture part of the Colorado River.[12] In March 1939, an editorial in the *Salt Lake Tribune* urged citizens of Utah to "become active in the matter of utilizing their water resources, [or else] they are likely to awaken to find their water in California."[13]

In 1939 the Bureau began to examine locations for dams along the upper Colorado River and its tributaries. Authorized to do so under a provision of the Boulder Canyon Project Act of 1928, engineers and geologists scurried throughout the upper basin, searching for the best sites to build storage dams and power plants.[14] Fanning out into this region of swift rivers and high narrow canyons, the Bureau's surveyors examined dozens of dam sites, including Flaming Gorge on the Green River in northern Utah, and Glen Canyon on the main stem of the Colorado, just south of the border between Arizona and Utah, both of which had appealing qualities. Other potential dam sites lay within the boundaries of recently enlarged Dinosaur National Monument, including the Browns Park site, in the northern end of Lodore Canyon—noted in Roosevelt's proclamation—and another in Split Mountain Canyon, near the monument's east end.[15]

Bureau personnel quickly recognized the merits of a dam near Echo Park as well. Surveyors exploring the inner gorge deep inside the Green and Yampa canyons saw that the narrow canyon below Echo Park—aptly named Whirlpool by John Wesley Powell—offered a superb place for a dam. The Green River narrows sharply in the northern end of Whirlpool Canyon, where sheer cliffs only a few hundred feet apart flank the swiftly moving water. The rock formation is Uinta Mountain Group Quartzite, a solid, dependable geologic layer ideal for withstanding the stress of a huge dam and reservoir. A dam at this location would effectively trap the Green and Yampa rivers only a short distance below their confluence at Echo Park, and that proved crucial to the state of Utah, which later obtained water rights to the Yampa River from the state of Colorado. By accumulating water from two major tributaries of the Colorado, a dam below Echo Park would create a sizable reservoir ideal for water storage as well as for power generation.

Evaporation of water always worried the Bureau's engineers, and from that point of view the Echo Park location had major advantages as well. The rugged inner gorge of the monument made for an excellent storage reservoir. Instead of water spreading out on a broad plain, the narrow shape of the pool would help to minimize evaporation, perhaps the greatest enemy the Bureau faced in the arid West. In this regard the Echo Park site was superior to the Browns Park site mentioned in the 1938 proclamation. A dam there—in upper Lodore Canyon—would back water north into Browns Park, where the reservoir would expand over a wide area and greatly increase evaporation. During World War II, when population and economic growth brought greater demands for power, the Bureau had further reason to favor Echo Park over the Browns Park location. The physical advantages of the Whirlpool Canyon site seemed indisputable. As a Salt Lake City newspaper aptly put it, "in the process of gouging half-mile deep canyons, the [Green and Yampa] rivers did two things, they created colorful scenery with wild abandon, and they made the traps by which man can snare them." [16]

Surveyors and geologists took their enthusiasm back to the Bureau's offices in Vernal and Salt Lake City. There, engineers began to write detailed blueprints of Echo Park Dam, one of a series of dams and power plants that the agency planned along the upper Colorado. In the following years, the Bureau undertook further studies of other locations for dams, analyzing potential storage capacities, power capabilities, and "cost-benefit" ratios. The Bureau had in mind a large-scale plan to regulate the upper Colorado River and its tributaries, a plan that eventually came to be called the Colorado River Storage Project. Such an ambitious project promised to take years to bring to fruition, and the Bureau had extensive research to do before it could make public a final blueprint. Much of this took place during the Second World War in the Bureau's Regional Office in Salt Lake City.

In 1939, along with its surveys, the Bureau of Reclamation constructed a primitive road thirteen miles long which ran from the Iron Springs bench—the high plateau far above the rivers—down into Echo Park, and managed to do so without obtaining clearance from the National Park Service. The Park Service did not have personnel stationed at Dinosaur National Monument at that time, and the preserve fell under the administration of Rocky Mountain National Park two hundred miles to the east. David Canfield, superintendent at Rocky Mountain, wrote to the Bureau of Reclamation in Salt Lake City, in 1940, to remind the agency of Park Service jurisdiction at Dinosaur. He asked that permission be obtained prior to building roads or for entering the monument to undertake surveys. This was the first of many exchanges between the two Interior Department agencies over access to the monument, and it signaled the great clash to come.[17]

Such conflicts might suggest that the controversy over Echo Park resulted from a "turf fight" over the public land and waters of the Colorado River Basin. With the Bureau interested in a dam and with the Park Service lacking a strong presence at Dinosaur Monument, friction between the two agencies was bound to occur. Furthermore, the Bureau and the Park Service had decidedly different missions, so disagreement over the merits of a dam in a national monument might have been inevitable. Yet it must be remembered that the Bureau and the Park Service *cooperated* in planning the use of the upper Colorado River throughout the war years. While conflict over jurisdiction occurred at the "field level," the two agencies worked closely together higher up in the bureaucracy, a fact which helps to explain the roots of the controversy.[18]

In the early 1940s, the Park Service found itself engaged in recreational planning across the country. The agency had undergone a major reorganization in the 1930s and it had taken administrative control over a variety of "recreational sites." Under the Park, Parkway, and Recreation Area Study Act of 1936, Congress had asked the Park Service to help coordinate the country's "recreational resources," including national parks, monuments, historic sites, and a variety of federal and state landholdings.[19] Under this new law, the NPS took responsibility for managing reservoirs like Lake Mead; the Boulder Dam National Recreation Area became one of the first "man-made" areas under Park Service jurisdiction. The NPS subsequently began to plan recreational facilities for reservoirs anticipated along the Missouri River.[20] Not surprisingly, the Bureau and the Park Service soon found themselves coordinating plans along the Colorado River.

At a meeting in Denver in 1941, representatives from each agency conferred over the future of Dinosaur Monument and other public lands along the river. The meeting produced an interagency "memorandum of understanding," an important milestone on the road to the Echo Park controversy. Signed by S. O. Harper, an engineer with the Bureau, and Frederick Law Olmsted, Jr., planner and consultant to the Park Service, the agreement called for a joint survey by the two agencies of the "recreational resources" along the Colorado River. Consideration would be given to recreational opportunities afforded by Lake Mead as well as to potential recreation areas in Dinosaur and Grand Canyon national monuments. The Park Service, with its mandate to manage national recreation areas, agreed to consider the possibility of administering such an area in Dinosaur Monument, should a dam be constructed along the Green or Yampa rivers. According to the agreement, Dinosaur should be examined carefully so that it could "effectively serve the purposes and objectives of both bureaus."[21] By agreeing to this memorandum, the Park Service signified that if the Bureau ultimately decided on a dam at Echo Park, a change could be made in the status of the monument, and if Congress approved it could be designated "Dinosaur National Recreation Area."[22]

The willingness of the Park Service to consider a change in status for Dinosaur came at a time when conservationists did not seriously question the merits of reservoirs under the agency's care. Prior to World War II the Park Service did not fear a hostile reaction from conservationists for entering into such agreements with the Bureau; and investigating "recreational resources" at Dinosaur National Monument and elsewhere in the Colorado River Basin seemed appropriate. The NPS, at the time, operated much as it had during the Depression, when parks, by beckoning tourists, were considered assets to local economies. Such a role continued to be important in the early 1940s. In addition, as Susan Rhoades Neel argues in her dissertation on the subject, NPS Director Newton Drury agreed to the interagency memorandum and played a pivotal role in establishing a cooperative relationship with the Bureau of Reclamation.[23] Drury's role in the controversy has been a source of great confusion; it is commonly said that he knew nothing about Echo Park Dam until he saw a story in a Salt Lake City newspaper in 1943.[24] In truth, Drury gave his blessing to the 1941 memorandum and agreed to additional joint studies of the upper Colorado River in subsequent years.

Drury knew how much the Park Service stood to benefit in its jurisdiction over Dinosaur Monument, in part because the agency possessed little knowledge of its newest holding, particularly the scenic and scientific values of Lodore and Yampa canyons. Most of the land remained unsurveyed and largely unknown, as little had been done to explore the Green and Yampa since Roger Toll's inspection in 1933.[25] When the survey work began in 1942, archeologists descended into the deep inner gorge and began to record the location of petroglyphs and Fremont sites.[26] Drury also supported the survey because he wished to gather sufficient information in order to make future decisions. Neel argues that Drury shrank from confrontation and sought to develop sound public policy based upon solid factual knowledge. Most of all, Drury wished to accommodate a variety of competing interests in the Colorado River Basin, including water users, recreational concerns, and national park devotees.[27]

This should not be taken to mean that Drury surrendered to the Bureau early in its process of planning for the upper Colorado River. But faced with severe budgetary pressures during the war as well as with possible demands from the government for greater hydroelectric-power production, Drury did not find himself in a position to ignore the Bureau or other competing interests in the region. So he took a cooperative approach, and even signed additional agreements with the Bureau in the following years. In 1944, Drury agreed to accept dams in Dinosaur when and if Congress determined that the national interest warranted them. In that event, this agreement stated, "the status of the [monument] should be changed to that of a multiple-purpose area in which water-control for the generation of power would be the principal use, and recreation the secondary but also important use."[28]

Drury and the Park Service adopted a "wait and see" position, a practical stance during the war years. Let the Bureau complete all of its surveys, let the engineers investigate Echo Park and other likely dam sites inside and outside the monument, let the dam builders draw their blueprints, then let Congress determine whether a dam should be built. The American people could then have their say about whether to allow a dam in a part of the national park system. To make a strong stand against the dam during the war seemed inappropriate. By adopting such a cooperative attitude, the Park Service made itself a friendly partner of the Bureau. In retrospect, it is clear that Drury unwittingly provided the Bureau with encouragement to proceed with its plans for Echo Park Dam and with time to inform residents throughout the upper basin. Drury did not assume that a dam at Echo Park was inevitable, but neither did he foresee the strong pressure that the Bureau later brought to bear on behalf of the project at Interior. In general, Drury and other Park Service officials did not worry about making such agreements during the war because they knew the dam could not be undertaken without the funds and permission of Congress.[29]

From the viewpoint of the Bureau of Reclamation, however, these agreements in 1941 and 1944 merely continued a process begun in the 1930s during negotiations leading to Roosevelt's proclamation of 1938. At that time, the NPS had agreed to permit dams in Dinosaur if a future Congress consented, and the recent agreements simply reaffirmed that position.[30] For their part, Bureau officials had little reason to think the Park Service objected to an Echo Park Dam, and in correspondence exchanged between the two agencies, the general assumption appeared to be that it might well exist sometime in the future.

On June 17, 1943, the Bureau made a move to bring the dam a step closer to reality. It filed a "reclamation withdrawal" with the Department of the Interior to protect the Echo Park site and other potential dam sites in the monument. Decision makers at the Bureau probably had no second thoughts because the move seemed consistent with the general understanding about the future of Dinosaur. Furthermore, once a dam site had been selected on public domain such "withdrawals" were routine to prevent other land claims which might later stand in the way of a dam.[31]

But the Bureau's move sparked fears among a number of Park Service officials who understood that Dinosaur National Monument was not merely "public domain." A flurry of correspondence went back and forth. Why had the Bureau made such a move? Was it legal? Why had it not consulted with the Park Service? Did it mean to challenge its jurisdiction over Dinosaur National Monument? Some believed the withdrawal had gone too far, that the Bureau had taken advantage of the cordial relations between the agencies and simply assumed that the Park Service would not object to necessary legal steps for eventual construction of a dam. Such an assumption disturbed a number of officials in the Park Service, who began to lose trust in the dam builders.[32] NPS Regional Director Lawrence Merriam questioned the withdrawal, as did Chief of Lands Conrad Wirth, and Acting Director Hillory Tolson. None of them could fathom how the Bureau had managed to file for the withdrawal without approaching the Park Service for its permission.[33] Daniel Beard, the lone park ranger at Dinosaur during the war, thought the Bureau's move was quite ominous. "This Service has been treated as if it was not here at all," he wrote to the regional director, "as though the area as a national monument instead of part of the public domain makes no difference." Beard thought the two agencies had skirted around the issue of dams too long, and that the time for vague speculation about the future had ended. He advocated a long-term "master plan" for the monument. "In our nebulous planning to date we have tried to wrap our schemes around what we think the Reclamation Service is likely to do . . . [I think] we must have a show down and I rather suspect this as opportune a moment as any."[34] Nevertheless, the showdown that Dan Beard wanted did not come to pass. The decision not to challenge the 1943 withdrawal gave the Bureau of Reclamation additional room to lay the groundwork for the dam in Dinosaur. In 1948, Park Service Associate Director Arthur Demaray wrote that "we see no advantage to be gained now in questioning the legality of the withdrawal. To do so undoubtedly would be extremely embarrassing to the Department [of the Interior.]"[35]

For the historian looking back on these bureaucratic agreements and legal maneuvers, it appears that the Bureau and the Park Service played a cat and mouse game over Echo Park and the future of Dinosaur National Monument throughout World War II. Yet despite all the jockeying for position during the early 1940s, Echo Park Dam remained a point of contention only within federal agencies. Residents of the upper Colorado basin had heard little about the controversial dam because the Bureau had not yet made its proposals public; nor, for that matter, had the economic and political forces which gave rise to the proposed dam in Dinosaur fully coalesced.

The Second World War proved to be the catalyst for the Bureau of Reclamation's planning, as it had a tremendous effect on the economy of Utah, Wyoming, Colorado, and New Mexico. Each of these states underwent sweeping economic changes during the war and became eager to build dams and generate hydroelectric power. Mobilization for the war brought a massive influx of federal money into the West, and, in the words of Gerald Nash, "the colonial economy of the region, heavily dependent on raw materials production before 1941 now became increasingly diversified and self-sufficient."[36] The war stimulated an economic boom, which caused great changes to small states in the upper Colorado basin. Camp

Carson and Ent Air Force Base in Colorado Springs and Warren Air Force Base near Cheyenne became important training sites, while the Rocky Mountain Arsenal in Denver quickly became a primary employer in the Colorado capital. In New Mexico, the federal government's support for science helped to establish the premier atomic-research center at Los Alamos, while firms like the Zia Corporation in Albuquerque engaged in rocket research.[37] These economic developments focused ever-increasing attention on water and power supplies, and elevated the value of the Colorado River.

While interest in capturing the Colorado River grew throughout the upper basin, it reached a fever pitch in Utah. The Beehive State experienced unparalleled change during the war, benefiting from the federal government's interest in dispersing military facilities and industries across the nation. Utah acquired ten military bases, an army hospital, and various defense industries, such as the Utah General Depot, the nation's largest quartermaster depot. Industry and manufacturing sprouted along the Wasatch Front as well, especially in such endeavors as oil refining and the mining of manganese, copper, and zinc. Kennecott Copper Corporation at Bingham operated the world's largest open-pit copper mine, while smelters and mills processed raw ores and turned them into finished products. By far the most prominent symbol of Utah's wartime growth was the huge Geneva Steel plant in Utah County. Built by the federal government and sold to U.S. Steel following the war, the Geneva plant became the centerpiece of Utah's industrial boom.[38] Its ovens and blast furnaces each produced more than one million tons of pig iron and steel annually. During and after the war, Geneva Steel supplied raw materials for plants in and out of Utah and acted as a magnet for other industrial concerns.[39] Salt Lake City fast became a major regional economic hub in the intermountain West.

Utah's industrial development and population growth during and after World War II was nothing sort of astonishing. The war created 50,000 jobs in the Beehive State, more than 100,000 people moved there, and cities along the Wasatch Front boomed, including Ogden, Logan, Salt Lake City, and Provo. All of this created new demands for inexpensive electrical power and agricultural produce. Governor Herbert Maw estimated that 450,000 acres of land could be irrigated and its produce immediately put to use.[40] Eastern Utah, an important breadbasket of the state, had a special need for irrigation, while industrial growth increased the demand for inexpensive electrical power.

Utah was not alone in seeking new sources of low-cost power. A shortage of power had been a burden to Wyoming for years, curtailing industrial development and perpetuating a boom and bust economy. Industry had avoided the Cowboy state, and an economy based on agriculture and mineral extraction had kept Wyoming underdeveloped. Colorado had greater wealth and diversity, but the growth of Denver and Colorado Springs during the war stimulated demand for power. Industry in New Mexico likewise sought low-cost power to fuel its population growth based on its atomic industry. Rural electrification, having made some progress since the 1930s, remained an unfinished task, and REA cooperatives in the mountain states supported dams and power plants to generate additional supplies. According to the *Salt Lake Tribune,* "without power the upper basin's future probably always will be just around the corner."[41]

In Utah, where the search for power was especially pronounced, the legislature established the Water and Power Board, and it quickly became a voice for business and industry as well as a powerful entity in the state. Its director, George Clyde, heard the cries of newly sprouting industry for more power and he believed that with a strong industrial base and a dynamic and diversified economy Utah could do its part to strengthen the nation during the growing Cold War. Before long the Water and Power Board became a spearhead of economic growth and diversification, as well as the major force in promoting state and federal efforts to develop Utah's water and generate power. It also became deeply committed to the Bureau of Reclamation's plans for the upper Colorado, and took the position that the dam site at Echo Park was vitally important to the entire plan's success. Up until now the major dam site in Dinosaur Monument had been just one of many that the Bureau of Reclamation had surveyed, but Utah's rapid growth and escalating demands for power elevated its value greatly.

Echo Park Dam soon became an integral component in the Central Utah Project, a plan designed and vigorously promoted by the Water and Power Board. Studies undertaken by the Board revealed that the primary sources of water for Salt Lake City and the Bonneville Basin—the Bear and Weber rivers—were no longer adequate and that new sources must be tapped.[42] The Board recom-

mended a plan to capture the Green River, the foremost tributary
to the Colorado River in the upper basin, which joined with the
Yampa River at Echo Park, just across the Colorado line and about
two hundred miles from the Wasatch Front and Salt Lake City. The
Green offered the water and power that a burgeoning Salt Lake
Valley yearned to have. Although the precise amount of water that
Utah could claim from the Colorado River system remained unde-
termined, plans for the CUP moved forward quickly, and eventually the project emerged as an elaborate network of aqueducts,
storage dams, and power plants with Echo Park Dam in Dinosaur
Monument being one of the key components.[43] By means of the
CUP, water from the Green River would be transported across the
mountains to the Bonneville Basin and Wasatch Front, from
sparsely populated eastern Utah into the Salt Lake Valley. According to the *Deseret News,* "the project will accomplish a veritable transformation along the old pioneer route to Utah's Dixie. But
what is much more important, it will help form a safe foundation
for the industrial empire which lovers of the state are planning to
erect."[44]

Utah's interest in capturing the Green River by means of the CUP
coincided perfectly with the Bureau of Reclamation's larger Colorado River Storage Project (CRSP). In fact, the CUP became a key
"participating project" of the CRSP, and the Water and Power
Board became a close partner of the Bureau's regional office in Salt
Lake City. While the Water and Power Board provided Utah residents with a voice of their own, George Clyde recognized that a
project as large as the CUP could not be funded by the state of
Utah alone. Therefore, he looked to the Bureau of Reclamation to
incorporate the CUP into its upper-basin plans. Utah's interest in
Echo Park Dam and the CUP naturally converged with the entire
upper basin's interest in putting its share of the Colorado River to
consumptive use.[45]

To repeat—Echo Park Dam was considered a key part of the
plan. The CUP was designed to take water originating high in the
Uinta Mountain range which normally flowed off the mountains
southward into the Uinta Basin and divert it to the west along a
one-hundred-mile aqueduct into the Strawberry Reservoir. From
there, water would be channeled south and west through tunnels
and canals into power plants in Diamond Fork Canyon, and finally
into Utah Lake on the Wasatch Front. The stream water off the

Uinta Range diverted away from the Uinta Basin would have to be replaced to satisfy irrigators in that part of Utah, and the reservoir behind Echo Park Dam was designed to be a source of this replacement water. Actually, the initial plans for the CUP and the CRSP called for construction of two storage dams on the Green River: one at Flaming Gorge north of Browns Park (well to the north of Dinosaur National Monument) and a second just below Echo Park, and under this blueprint the Bureau gave itself a choice of utilizing either one of these reservoirs to replace water in the Uinta Basin. In many ways Flaming Gorge appeared to be the obvious choice because the reservoir, considerably higher in elevation than the Uinta Basin, could be tapped at low cost simply by allowing the water to flow downhill into the basin. Since water from Echo Park Reservoir would have to be pumped, a much costlier proposition, the Bureau favored the Flaming Gorge Reservoir as a replacement source.[46]

Yet it soon became clear that irrigators in the Uinta Basin did not want Green River water sent to them from Flaming Gorge, because they suspected that its quality had been tainted. Several phosphate plants had been constructed along the upper Green River in Wyoming, and they discharged salts and other pollutants into the river. Uinta Basin farmers expressed their fears to their political representatives and insisted on a source of clean replacement water from the Colorado River system. To them, the Yampa River, originating in Colorado near Steamboat Springs, was far superior to the Green.

As a result of their misgivings about the Green River, the state of Utah took steps to obtain water rights to the Yampa, negotiating for such rights between 1946 and 1948 when the upper-basin states bargained over how to divide their half of the Colorado River. It took months of hard bargaining, but Colorado at last agreed. Under Article XIII of the Upper Colorado Basin Compact of 1948, the state of Colorado agreed to allow 500,000 acre-feet of Yampa River water to flow into Utah each year. The provision of the Compact sprang directly from concerns of Uinta Basin irrigators, who now realized that capturing the Yampa was of cardinal importance, and Echo Park Dam would make it possible to do so. In short, the specific location of the dam site below the Green and Yampa confluence now became paramount for Uinta Basin irrigators.[47] It had the great advantage over Flaming Gorge of capturing two rivers instead of one, and its reservoir would mix clear Yampa water with possibly tainted Green River water, making it especially attractive to the designers of the CUP.[48] The location of the dam site below the Green and Yampa confluence goes far to explain why residents of Utah and the Bureau stridently resisted utilizing any alternate sites, which conservationists advocated in the years to come as the controversy unfolded.

Echo Park Dam also took on great importance because the state of Utah and its neighbors had suddenly acquired an insatiable appetite for low-cost power. In the years to come, proponents of the big dams on the upper Colorado made much of their role in controlling water and minimizing evaporation, but power generation was a major purpose of Echo Park and other dams from the beginning. The Bureau of Reclamation hoped to win approval for Echo Park, Glen Canyon, and other main-stem dams so as to generate large blocks of power, the sales of which would fund the associated reclamation projects in individual states. As Bureau Regional Director E. O. Larson said in a letter to the Park Service, "the primary function in the initial stage will be power generation to obtain the maximum revenues."[49] The Echo Park power plant was slated to have an installed capacity of 200,000 kilowatts; sale of this power would provide funds to help complete the Central Utah Project and other so-called participating projects in the CRSP. Bureau engineer P. R. Neeley told a group of businessmen in Vernal, in 1947, that Echo Park's power sales were absolutely critical for swift completion of the CUP.[50] Since the early 1900s, many of the Bureau's big projects had relied on hydroelectric-power sales for funding, and in this case, the Bureau could appeal to nascent upper-basin industries, REAs, and manufacturing interests.

Given the power demands after the war the Bureau did not worry about finding an adequate market, although this did not mean that all of the power could be sold locally. But neither the Bureau nor the Water and Power Board felt troubled by the absence of an immediately available market, in part because the power generated from Echo Park and other dams was expected to help create its own market. In that sense, the construction of dams for hydropower production can be regarded as the twentieth-century equivalent of the transcontinental railroads, whose effect on the West in the nineteenth century was enormous. Railroads were built in advance of population and markets and were designed to create them. In the twentieth-century West, construction of hydroelectric-power plants often sprang from a similar rationale. Thus, in post-war Utah, power from dams along the Colorado River would not only feed the existing market along the Wasatch Front but would provide the needed lure for more industries and guarantee further population growth. George Clyde made this quite clear in the Water and Power Board's annual report of 1948:

> It would indeed be poor and short-sighted policy to restrict potential development of hydroelectric power sources in the State to such industries which can now be visioned as definitely potential of establishment. It must be remembered that new industries cannot wait for power supplies to be developed when all other factors are favorable. Thus new power capacity and transmission systems must be built well in advance of actual market definition; this principle is well recognized in other parts of the United States, particularly in the West." [51]

Such an emphasis on generating large blocks of power helps to explain the Bureau's rejection of the so-called Browns Park dam site in upper Lodore Canyon, mentioned in President Roosevelt's proclamation of 1938 enlarging Dinosaur National Monument. It will be recalled that the Park Service had agreed to include the "Browns Park" clause in the proclamation in order to signal its willingness to permit the Bureau to proceed with its surveys in Lodore and Yampa canyons, and the Bureau had, by that time, surveyed the Browns Park site and considered it a likely possibility. [52] But the boom on the Wasatch Front during the war caused the Bureau to alter course and to favor a dam at Echo Park over Lodore Canyon. Deeper water could be stored in the narrow canyons near the Green and Yampa confluence, giving more capacity to Echo Park's generators. [53] A dam at Lodore could not store as much water as the one below Echo Park—three and a half million acre-feet compared to six and a half million acre-feet—primarily because the Echo Park site was below the confluence of both the Green and Yampa rivers. The lower storage capacity at Lodore, in turn, would "considerably reduce the power potential." Power output at Lodore would be so low that it could only be sold at a very high price, somewhere around ten mills per kilowatt, in order "to effect repayment." [54] That price could not compete with power sold from a conventional steam plant. In short, the Bureau's engineers and planners now rejected the Lodore site because of its high-cost power and lower generating capacity. Although they could not say for sure that all of Echo Park's power could immediately be absorbed, the lower cost of its power gave it the advantage of being able to compete with conventional power sources, or so the Bureau said.[55] None of these considerations had been foreseen in 1938, when President Roosevelt included the Browns Park site in his proclamation; Bureau engineers could not have known that the site in upper Lodore would become obsolete in future years. In any event, the Echo Park site had now been accorded top priority in the Bureau's plans due to its greater power capacity.[56]

Most of the public along the Wasatch Front and throughout the upper basin knew little about such engineering details, but felt confident in the wisdom of the Water and Power Board and the Bureau of Reclamation. With plans to harness the upper Colorado River taking shape, residents came to understand that their region's economic growth depended heavily on federal "development" of that water. Utah's interest in Echo Park Dam may well have been the strongest since it stood to benefit most directly from its water and power, but the dam soon came to be seen as an instrument of economic growth across the region. Throughout the upper basin, growing numbers heard about the grand plans for the upper Colorado River and its tributaries. They read about Echo Park Dam regularly in the newspapers and saw it as a great bonanza. As spokesmen of the Bureau vigorously promoted the Central Utah Project and Echo Park Dam in Salt Lake City, Denver, and other upper-basin cities and towns, residents became eager for the plan to be implemented, and anticipated quick approval of Echo Park Dam by the Interior Department and the Congress. Businesses prepared bids on construction contracts, while newspapers like the *Salt Lake Tribune* and *Denver Post* heralded Echo Park and other dams with great regularity.[57]

Shortly after the war, the Bureau's regional office in Salt Lake City finalized its blueprint of dams, irrigation works, and power plants, and in 1946, published a detailed description of the CRSP in an impressive report entitled "A Natural Menace Becomes a Natural Resource." Nicely capturing the agency's philosophy, the report pointed out that the Colorado River, left in its natural state, was destructive and wasteful. "Yesterday the Colorado River was a natural menace," it declared. "Unharnessed it tore through desert, flooded fields, and ravaged villages. It drained the water from the mountains and plains, rushed it through sun-baked thirsty lands, and dumped it into the Pacific Ocean—a treasure lost forever." [58]

This report also described a number of reclamation projects that upper-basin states had been demanding since the drought years of the 1930s, including the Central Utah Project and others like the LaBarge, Lyman, and Seedskadee projects in southwest Wyoming, which would supply water to one hundred thousand new acres, and the Paonia, Blue Mesa, and San Juan Chama Projects in Colorado and New Mexico. The CRSP, then, amounted to a giant package of reclamation projects, storage dams, and power facilities that promised to transform this part of the American West, much as TVA had done for an important region of the South.

To justify the huge project the Bureau promised many economic benefits, and did so with a type of glowing rhetoric perfectly in tune with the nation's outlook at the dawn of the Cold War. Private enterprise, the report intoned, had done all it could to provide irrigation and water supplies in this part of the West, and the time for a federally sponsored plan had arrived. Low-cost hydroelectric power was needed to develop "the vast mineral resources." Range land in Wyoming, Colorado, and northern Utah had suffered from overgrazing due to ranchers' inability to grow sufficient hay; irrigated land would thereby help ranchers to become more productive and maintain their ranges, and improved farm incomes would enhance the tax base of rural communities and bring improved services to residents. The Colorado River report abounded with the innumerable benefits of the upper-basin project. [59]

The Bureau could not ignore the possible adverse effects of this blueprint on other realms of federal concern along the upper Colorado. Thus the report commented on Indian tribes and their water demands, the potential effect of the dams on mining, and on possible damage to areas in the national park system, including Dinosaur and Grand Canyon national monuments. As for Dinosaur the report noted that, following construction of Echo Park Dam, "a good deal that is of geological and scenic interest would remain. . . . Some parts would be wholly unaffected, such as Jones Hole, the precipitous canyons of the Yampa River near Thanksgiving Gorge and Cactus Park. . . . In some sections, the reduction in the visible height of canyon walls would be a relatively small proportion of the present total height, and some of the canyons would still be impressive."[60] Toward the end of the 1940s (as we shall see in greater detail in the next chapter) the National Park Service, led by director Drury, strenuously objected to such claims partly because the agency found itself on the defensive after the war and also because a bitter controversy had only just concluded over the merits of including Jackson Lake Reservoir inside Jackson Hole National Monument. That conflict caused Drury and the Park Service to rethink their earlier commitments to dams in Dinosaur Monument and to confront the Bureau's proposed Echo Park Dam.

However, neither the Park Service nor conservation interests posed the main obstacle to the Bureau' plans just yet. For the time being, the major impediment to the plan was that upper-basin states had not agreed on dividing their rights to the upper-basin water. In 1947, the Bureau had sent its report to the upper-basin states and other federal agencies for comment. Upon receiving these comments in 1947, Interior Secretary Julius Krug sent a summary of the Bureau's plans to Congress. It became an "interim report" on the Colorado River Storage Project, and signaled the Bureau's readiness to move ahead with Echo Park and other dams once the upper-basin states agreed on how to divide up their own half of the Colorado River.[61] But Krug's report did not officially recommend any of the dams. Although upper-basin states had been guaranteed the right to use 7.5 million acre-feet of Colorado River water under the 1922 Compact, no steps had been taken to allocate the total amount among the four states. The Bureau of Reclamation well knew that legal problems invariably developed when dams were built prematurely. Arizona and California had recently become embroiled in a dispute over rights to the lower Colorado— a fact that did not escape the Bureau's attention as it proceeded with its plans for dams on the upper Colorado. Therefore, the Bureau had no choice but to deliver a warning in its report: The upper basin had too many projects on its wish list and the upper Colorado River system did not have sufficient water to allow all of them to be built. Nothing could be done, in other words, until the states determined their own water rights.[62]

Meetings and negotiations began in 1946. State water engineers, a federal representative, and a newly established Upper Colorado River Commission spent two years hammering out an agreement for a second compact. Among other considerations, the Upper Basin Compact could not violate the 1922 Compact, and making sure of that kept lawyers and politicians busy for months. They also struggled over other matters. How much water did each state intend to use in the future and for what purpose? How much of the upper Colorado River originated in each state, and how much could each state fairly claim for itself? The state of Colorado produced the lion's share of the upper basin's total supply, somewhere around 70 percent, but delegates from the state realized that they had to compromise. Each state tried to estimate its future economic growth and thereby determine the amount of water needed to ensure it.[63] The negotiators worked hard to settle these questions. They had little choice so long as they wanted the Bureau to build their dams.[64]

Besides, without a compact upper-basin states feared the disagreeable prospect of federal management of the river. Few in the upper basin wanted any federal entity the likes of a valley authority, and President Truman and various members of Congress were indeed advocating such plans for several river basins. Private power companies in the upper Colorado basin stridently opposed valley authorities, for they realized that federally marketed electrical power threatened their business. Ranchers who held water rights under state law likewise abhorred the thought of surrendering them to a federal agency. In short, while the upper-basin states courted the Bureau of Reclamation to erect dams and power plants, they were united against a valley authority, wanting no truck with public-power schemes and federal corporations managing the entire river basin.[65] They knew they could prevent establishment of a TVA-style agency if they could succeed in writing an upper-basin compact.

Finally, the job was done. Under the final terms of the Upper Basin Compact, Colorado received rights to 51.75 percent of the upper basin's share, Utah 23 percent, Wyoming 14 percent, and New Mexico 11.25 percent. Utah negotiators rejoiced in obtaining water rights to 500,000 acre-feet of the Yampa River from Colorado, and this strengthened the Beehive State's determination to win approval of Echo Park Dam. The new Compact had to be approved by the four state legislatures and then by Congress. Senators and representatives from lower-basin states of California,

Arizona, and Nevada combed every paragraph to make sure that the upper basin could not take away any of their water under the new compact.

When the Upper Basin Compact at last became law in 1948, it stood as the second major legal document governing the Colorado River.[66] It also removed the central obstacle blocking the Bureau's plan for dams and thereby paved the way for the upper basin to use its half of the great river. After completing the Upper Basin Compact, water officials and politicians from Wyoming, Utah, Colorado, and New Mexico began to develop legislation to make all of these projects a reality, and the Bureau began to work with members of each state's congressional delegation.

As upper-basin lawmakers prepared to introduce legislation in Congress, partisan politics came to the surface. Lawmakers from the upper basin were eager to introduce legislation for the CRSP, and a number of them held powerful positions on the all-important Interior and Insular Affairs committees in the House and Senate where the project would first be considered. Republican Senators Arthur Watkins of Utah and Eugene Millikin of Colorado became leading proponents of the Colorado River Storage Project and enthusiastic backers of Echo Park Dam, doing their best to make it appear that the Republicans spearheaded the new project. But Democrats backed the plan just as eagerly. In the Senate, Joseph O'Mahoney of Wyoming and Clinton Anderson of New Mexico joined Watkins and Millikin in urging quick passage of the CRSP. These lawmakers had tremendous power in moving the project through the political process, especially on the Senate committee because the West had proportionally greater influence in the upper chamber. As we shall see later in the story, western states needed to muster all the unity behind the plan that they could, because objections to this massive water project arose from other sections of the nation and posed a strong challenge to the bill, a fact which proved particularly helpful to the conservation groups battling to defeat the Echo Park project.

But that was all to come. Echo Park and Glen Canyon Dams and the Colorado River Storage Project emerged from the postwar climate of growing prosperity and the Cold War, as the region's rapid population growth and strategic importance to the nation were escalating rapidly. Industrial growth and increased manufacturing in Utah and the intermountain West helped give rise to the upper-basin project, while the West's population growth gave it momentum as it was taken up by lawmakers in Washington. In the first five years following the war the Bureau of Reclamation had enough support from both parties in Congress to attain ever bigger budgets annually, and the record budgets elevated the agency's already dominant place in the Department of the Interior, giving it an advantage over lesser agencies like the National Park Service.

Due to its dominant position at Interior and the inclination of the Bureau's Salt Lake City Office to publicize the project, spokesmen from the agency became vigorous campaigners for the CRSP. Together with state water engineers and with members of the Upper Colorado River Commission, they traveled about the upper basin and met with local chambers of commerce and business groups.[67] In these gatherings, Bureau spokesmen indicated that much of the groundwork for dams and power plants had been laid. Surveys of dozens of dam sites throughout the upper basin had been completed; rock foundations had been tested; detailed blueprints had been drawn; and official reports had been written. The Bureau had begun to work closely with the Utah Water and Power Board and other water agencies in New Mexico, Wyoming, and Colorado. Meanwhile upper-basin newspapers, notably the *Salt Lake Tribune, Deseret News,* and *Denver Post,* publicized the project lavishly and generated substantial support across the region.[68]

Eagerness to capture the upper Colorado and its tributaries escalated due to rapid growth in southern California, which had been taking water from the lower river with the recently completed All American Canal and Colorado River Aqueduct.[69] With lower-basin cities growing fast and demanding increasing amounts of the Colorado River, it seemed as if the upper basin's contribution to the Colorado River system was being drained down an ever-larger opening year by year. Echo Park and Glen Canyon Dams were designed to be giant plugs on the drain—holding water so that it could be regulated for upper-basin needs and lower-basin demands.

Residents of the American West generally prided themselves on their "pioneer spirit," on their self-reliance and disdain of big federal expenditures, but few could resist the appeal of the Bureau of Reclamation's grand plans for the upper Colorado River, whose benefits to the entire regional economy seemed so potentially great.

Construction of Echo Park Dam alone promised to spark the biggest bonanza in northeast Utah and northwest Colorado since the Moffat tunnel, and usher in a new era of economic growth. The building contracts promised to deliver an economic boom and to create hundreds of well-paying jobs. Federal dollars would flow into Denver and Salt Lake City, as well as to smaller towns close to Dinosaur like Craig and Rangely, Colorado, Roosevelt and Vernal, Utah, and Rock Springs, Wyoming. Once built, the recreation area would attract campers, boaters, and fishing enthusiasts, who would spend their dollars in local communities along the way. Many anticipated a boom from the newly available power at the dam, including members of rural electrical cooperatives, residents of towns trying to encourage new industry, and booming industries along the Wasatch Front. Irrigators and local communities understood that the revenues from power sales would help complete the Central Utah Project. "What we are interested in is a business boom," a Vernal gas-station attendant told Devereux Butcher of the National Parks Association.[70]

Certainly, a great deal had changed in the upper Colorado basin since the days of Butch Cassidy and the range wars. In some respects, Echo Park Dam and the CRSP symbolized the changes. This area in the American West no longer wanted to be left outside the mainstream of economic growth. The Bureau of Reclamation promised residents a bright future. People knew that water had magical qualities in their arid region, and in Echo Park Dam and the Colorado River Storage Project they planned to put the magician to work.

Donald Worster has argued that the American West became a decidedly more powerful region after World War II as population, money, and pacesetting residents moved to the region in ever-increasing numbers.[71] In Worster's mind, "the command over water in the region" by the Bureau of Reclamation was central to forging a newfound sense of regional identity, or as he puts it, "the western empire came at last into its own."[72] There is surely truth in the notion that the massive water development that Congress eagerly supported for the West in the postwar years brought enormous changes to the regional economy and identity; whether or not such changes should be accorded "empire status" can certainly be debated. But there can be no doubt that farmers, ranchers, and urban residents across the upper Colorado River Basin regarded the Bureau's upper Colorado project as an instrument of growth, a well-deserved influx of federal money that would help to remake their regional economy and, at last, bring an end to old patterns of boom and bust. If Worster is accurate in claiming that water underpinned the West's economic awakening after the war, there was no other part of the West where that was as true as the sparsely populated upper-basin states.

Still, while the upper Colorado basin states looked hopefully to the future they soon realized that all the momentum that had been generated on behalf of Echo Park Dam and the CRSP was inadequate to gain quick approval of the plan by Congress. In part, this was due to the strains placed on the federal budget by American involvement in the Korean War, which began in early 1950. In fact, there was to be no Colorado River Storage Project until the end of that war. But a larger factor was the sudden appearance of various conservation organizations on the scene, all of whom raised loud protests against the dams slated for construction inside Dinosaur National Monument. Historians of the West have generally regarded the war as most important on the region's economy, but less has been said on how the economic changes arising from the wartime boom proved to be a catalyst for conservation interests. After the war, public lands across the West became increasingly subject to mining, logging, and grazing, and these growing pressures soon provoked a number of conservation battles. While upper-basin states considered the Echo Park Dam a symbol of their aspirations, it became a symbol of an entirely different kind to national conservation forces who quickly raised their battle flags. Fierce controversy was not long in coming.

THREE

Primeval Parks and the Wilderness Movement

Dinosaur National Monument, in the writer's opinion, is second to no other area of the national park and monument system in its magnificence of scenic grandeur; and its unique scenery is duplicated nowhere else in the system.
—Devereux Butcher[1]

It must be admitted that the Dinosaur National Monument is an area originally selected presumably for excellent reasons, to be saved as . . . [one of the] samples of original America.[2]
—Olaus Murie

At the same time that the Bureau of Reclamation looked to build dams and power plants along the upper Colorado River after the Second World War, conservation groups interested in national parks, forests, wildlife, and wilderness preservation also became a more visible presence in the West and in the nation at large. *Conservationists,* to employ the word in vogue at that time (as opposed to *environmentalists*), became more prominent in the postwar years, partly because of increased travel and growing public interest in the outdoors. Rapid economic expansion and a growing appetite for natural resources caused increased pressures on public lands in the West from loggers, mining firms, ranchers, and other interests and sparked a number of conflicts with conservationists over various federal and state preserves. Organizations such as the Izaak Walton League, the Sierra Club, and the National Parks Association soon found themselves fending off threats to national parks and monuments as well as wilderness and primitive areas in national forests. When the Bureau of Reclamation suggested a dam in Dinosaur Monument, these and other organizations considered the proposal a symbol of mounting pressures on the national park system, and a challenge to the very notion of wilderness preservation.

Roderick Nash, who first described the Dinosaur battle in his now-classic work, *Wilderness and the American Mind,* pointed to the Echo Park controversy as a milestone in the American wilderness movement.[3] While a spectrum of conservationists took part in defending the monument from the dam, the battle proved especially significant to tiny organizations like the Sierra Club and the Wilderness Society, with the latter focusing its attention on the last remaining and highly vulnerable "wild" lands within national parks and forests. With the Bureau of Reclamation determined to build Echo Park Dam, wilderness groups quickly denounced the proposal with a flourish of publicity, which brought this hardly known part of the national park system into public view. The battle to save Dinosaur soon became a prominent story in the *Sierra Club Bulletin, Living Wilderness,* and *National Parks Magazine,* and eventually the controversy found its way into the *New York Times,* the *Washington Post,* and the *Los Angeles Times,* along with the *Saturday Evening Post,* and *National Geographic.* Such publicity transformed the largely unknown Dinosaur National Monument into the most potent symbol of wilderness since Hetch Hetchy, the valley in Yosemite National Park inundated by a dam early in the twentieth century. "Not since Hetch Hetchy had so many Americans so thoroughly debated the wisdom of preserving wilderness," wrote Nash, and in summarizing the victory conservationists ultimately achieved he concluded that "the American wilderness movement had its finest hour to that date."[4]

There can be no doubt that the Echo Park controversy was a key episode in the American wilderness movement, the central event between the founding of the Wilderness Society in 1935 and the passage of the Wilderness Act of 1964.[5] The controversy erupted at a time when wilderness activists were emerging as a new dimension of the conservation movement, energetically promoting their ideas and urging stronger protection for wild lands. In 1947, the Wilderness Society Council adopted a resolution calling for a national wilderness system, and soon after, Executive Secretary Howard Zahniser asked the Congressional Reference Service to aid in assembling a report on wilderness. Appearing in 1949, the study served as a basis of much of the society's efforts for the next decade and a half. In a way, the Echo Park battle rudely interrupted such efforts, forcing the Wilderness Society and its allies into a defensive posture, where they remained for the following six years until the threat to Dinosaur Monument had been erased.

Yet, despite being placed in such a position, wilderness activists sought to use the conflict to their advantage in order to make the value of wilderness known to the nation at large. In its own contribution to the Legislative Reference Report, the Wilderness Society spelled out in some detail a definition of wilderness, which included an absence of mechanized travel routes, an area "impressing the visitor with a sense of remoteness," and boundaries "as far-flung as possible" to teach visitors "their relationship to other forms of life, and to afford those who linger an intimation of the interdependence of all life."[6] Echo Park and the Green and Yampa rivers qualified on all counts. Dinosaur had unique scenic qualities that this generation of wilderness lovers sought to highlight and promote for the emerging generation of outdoor enthusiasts. All the major conservation journals at the time repeatedly published pictures of Steamboat Rock and dramatically beautiful Echo Park, making Dinosaur Monument a symbol of wilderness.

While the scenic wonders of the monument helped them to press the case for guarding remaining wilderness lands, the sheer length of the controversy also proved invaluable. Spanning nearly six years in the 1950s, the battle became the best-known conservation issue of the decade, a prime topic of debate in four separate congressional hearings, and a featured story in countless newspaper articles and editorials. Conservationists had many opportunities to urge that the project be scrapped and to make their case that national parks, monuments, and other preserves must be guarded in order to satisfy the public's growing appetite for outings in the wilderness. They urged Congress to take heed of the need for physical and emotional refreshment, to remember the value of places far removed from urban crowds, and to protect Dinosaur and other preserves as laboratories vital to modern science.[7] Such rhetoric revealed major elements of preservationist thought in the middle of the twentieth century, striking a chord with increasing numbers of Americans who were ambivalent about the postwar prosperity and population boom. After fifteen years of depression and war, millions of Americans fled from the inner cities and flocked to bright new suburbs, where they developed a culture based on the automobile, television, credit cards, and consumer goods. Prosperity and population growth, while inducing a sense of well being and progress, also created anxieties about complacency, conformity, and the scramble for material wealth.[8]

Advocates of wilderness preservation rebelled against suburban complacency and citified consciousness in the heady days of economic "progress" after the war. They stressed how Americans, lost in their daily striving for material gain, had also lost touch with the natural world, and ultimately with their own inner lives. An outing in the wilderness offered emotional uplift, spiritual insight, and escape from urban sprawl.[9] As Olaus Murie said, Dinosaur must be preserved "for our happiness, our spiritual welfare, [and] for our success in dealing with the confusions of a materialistic and sophisticated civilization."[10] Such sentiments appealed greatly to growing numbers of Americans in the postwar years who yearned to experience natural beauty and escape from urban and suburban daily life.

Nevertheless, while conservationists succeeded in identifying Dinosaur National Monument as a test of the nation's dedication to wilderness preservation, most all the organizations who first came to contest the Bureau's plan did not have such aims in mind. Historians have tended to portray the defenders of Dinosaur as a groundswell of preservationists who came to life after World War II, coalesced around the threat to Echo Park, and fought tenaciously to denounce the dam and dedicate a new generation to wilderness. But not many Americans in the early 1950s thought much about such exalted concerns, nor did most conservationists think primarily in terms of wilderness protection. Instead, groups like the Izaak Walton League and the Audubon Society focused on a particular corner of the natural world. Their concerns lay with protecting certain wildlife species and ensuring good hunting and fishing, and that kind of specific focus had long dominated the mainline conservation movement in the United States. Interest in wilderness, by comparison, was quite new.[11]

The majority of groups who took part in the Echo Park campaign did not do so because of a driving interest in wilderness. Instead, they came together to defend the national monument and, by extension, the rest of the national park system from similar intrusions. The Bureau of Reclamation's proposed dams in Dinosaur became a symbol of growing threats to the national park system following the war, and leading conservation groups believed that guarding the integrity of the national parks was of cardinal importance. They considered the park system to be at the foundation of the nation's conservation tradition, and so long as parks or monuments remained vulnerable to intrusions there could be little hope of protecting other types of preserves, such as wildlife refuges or wilderness areas. Howard Zahniser of the Wilderness Society made this clear when he told Pennsylvania Congressman John Saylor that "a threat to the national park system is a threat to any hope we have of preserving any of our land in its natural unspoiled beauty. It is a threat to our wildlife refuges, a threat to all our state and city parks, a threat to our wilderness areas, our national forests, and to all the areas that we have sought to set aside for special preservation purposes by the device of land classification. It is a threat to all these because it endangers what has been deemed to be the safest of all such systems of land classification, the National Park System."[12]

Zahniser, of course, did have wilderness protection foremost in mind, but he realized that first things came first. Before any national wilderness system could be established, the nation must be fully dedicated to the national parks and guarantee their protection from commercial intrusions. With the society engaged in a major effort to protect landscapes of great beauty and ecological significance, Zahniser knew that past accomplishments in preservation must provide the basis for any further preservation efforts, and that unless the national park system was safe from intrusions, no such efforts could be expected to succeed. Only if existing preserves were invulnerable to encroachment could the wilderness campaign move forward. "We owe it to both ourselves and posterity," he wrote, "once we have set aside an area for protection, to respect its dedication."[13] At its heart, the Echo Park controversy, he later remembered, concerned "the sanctity of dedicated areas."[14]

Here then, was the vital link between the emerging wilderness movement and the Echo Park controversy, and the point deserves emphasis. In the first place, it explains why conservationists had more concern with guarding Dinosaur Monument than with Glen Canyon, a source of great confusion in historical accounts of this controversy. As some of the participants of the Echo Park battle discovered, Glen Canyon contained a maze of scenic wonders that they deeply regretted would soon be lost under one of the Bureau's big dams. In order to prevent Echo Park Dam from being built in

Dinosaur, conservationists agreed to accept the dam at Glen Canyon, making it appear that they had sacrificed the latter to save the former. The notion of a trade-off cannot wholly be denied, but it simplifies the situation greatly and likens the controversy to a simple board game where the choice of moves is always clear. It overlooks the political pressures that defenders of Dinosaur had to confront, and it obscures the choices they felt compelled to make. Committed to guarding the national park system, they saw little need to question Glen Canyon Dam, except for trying to ensure that it would not harm Rainbow Bridge. This part of the story must be fully told in a later chapter. It is enough to say here that conservationists placed less importance on an unknown canyon in southern Utah that was public domain than an unknown national monument in northern Utah and northwestern Colorado.

With the threat to the park system considered to be the critical issue, it becomes clear why conservation leaders accepted much of what the Bureau of Reclamation sought to undertake along the upper Colorado River. With a few exceptions (to be taken up later), they did not question the fundamental purposes of the CRSP, did not challenge the desirability of large dams or hydropower plants, and certainly did not hold the Bureau in the same dark light as a later generation of environmentalists often did. They pressed the Bureau to leave the park system alone and they agreed to support dams and power plants outside of Dinosaur in exchange. While such tactics may appear naive or traitorous to modern environmentalists, care must be taken not to impose present-day standards on the past. While it has been easy for critics of the Bureau's large projects in our own time to denounce those involved in the Echo Park controversy for accepting much of the Bureau's plans (agreeing to Glen Canyon Dam, for example), it must be remembered that they considered it critically important to reaffirm the sanctity of the national park system, a conviction born of the barrage of threats that suddenly emerged after the war. Moreover, that goal united a very diverse group of conservationists. Only by forging a coalition did the opponents of Echo Park Dam have a chance to win, and such a coalition could not have been formed with an all-out attack against the Bureau's upper-basin project.

In seeking to defend Dinosaur and the national park system, conservationists could not help noticing parallels with the conflict involving Hetch Hetchy Dam, constructed on the Tuolumne River in Yosemite National Park before World War I. Once again, a dam had been suggested inside a part of the national park system and all of the arguments for and against such a proposal reappeared. Were alternate sites available? Was the water and power really needed? How much scenery might be lost and how could its loss be measured? Harold Bradley, who was instrumental in bringing the Sierra Club into the Dinosaur fight, had childhood memories of Hetch Hetchy; his father had been active in the Club early in the century when John Muir had worked feverishly to block the dam from Yosemite National Park. Bradley thought the debate over Echo Park was strikingly similar to Hetch Hetchy, that "indeed, the two situations are so much alike that the campaign literature on both sides, might be interchanged, with the appropriate names added."[15]

Yet Bradley recognized that the Dinosaur battle took place in a different generation of national park history. Since that contentious episode dozens of parks, monuments, and historic and cultural sites had become part of the park system. Furthermore, rising standards of living and greater mobility had made the parks substantially more popular than in Muir's time. The automobile made them accessible to the middle class rather than merely the elite.[16] Perhaps most importantly, the Hetch Hetchy controversy had helped give birth to the National Park Service in 1916, a bureau in the Department of the Interior with authority to administer the parks and a mandate to "conserve the scenery and the natural and historic objects and the wild life therein and to provide for the enjoyment of the same in such manner and by such means as will leave them unimpaired for the enjoyment of future generations."[17]

These words, the touchstone of the parks' modern history, did not guarantee protection of the preserves from outside threats. In the 1920s and 1930s, numerous threats emerged and culminated with various efforts during World War II to utilize natural resources in several parks. Companies with wartime contracts demanded access to minerals such as copper, tungsten, and manganese within the Grand Canyon, Shenandoah, and Yosemite national parks, and Park Service Director Newton Drury felt he had little choice but to allow them access. Timber firms in the Northwest sought to harvest the sitka spruce in Olympic National Park, although they were prevented from doing so by Secretary of the Interior Harold Ickes. But despite vigorous protests from the National Parks Association, President Roosevelt approved construction of a tunnel underneath Rocky Mountain National Park, and the project was completed during the war.[18]

The end of the war provided a sense of relief to the NPA and other supporters of the national park system who felt there should no longer be reason to intrude into the preserves in the name of wartime economic needs. Newton Drury now tried to put the best possible face on the war years, indicating how "gratifying" it had been that army and navy officers had tried to look elsewhere for war materials in short supply. Drury thought that the parks had emerged from the war fairly intact, and looked forward to increased travel that would revive public support for the preserves.[19] In December 1947, Secretary of the Interior Julius Krug pledged his support, promising to oppose any effort "to reduce the national park or monument areas." Krug made special reference to reclamation projects, too, announcing that he would not allow dams to be constructed inside any part of the national park system unless the demands of national security should make it necessary.[20] In general, conservationists held high hopes as the nation emerged from the war.

There appeared to be good reason for the optimism. With rationing of gas and rubber lifted, millions took to the highways, ready to celebrate the end of the war with a long-delayed vacation. Americans now could afford to travel to places distant from their homes, and they made the national parks a primary destination, descending on them in record numbers. In 1946, the number of visitors to the park system soared to almost twenty-three million, an increase of nearly 40 percent over the figure for fiscal year 1940.[21] The travel boom signified the growing popularity of outdoor recreation and an appreciation "of natural values by city dwellers," which Samuel Hays claims was part of a broad social revolution in the United States in the postwar era.[22] Organizations like the NPA and the Wilderness Society thought the trend boded well and felt eager to trumpet the national parks for their magnificent scenery. After all, the NPA had been a strong promoter of national parks since its inception in 1919, and had taken every opportunity to advertise them as prime tourist destinations.

At the same time, the NPA had recently begun to pay special attention to large scenic parks in the West, out of concern that the

Park Service was paying too much attention to the many historic sites, presidential birthplaces, cemeteries, and battlefields in the eastern United States which came into the park system in the 1930s. With the NPA fearing that the agency now favored the "cannonball parks," the bi-monthly *National Parks Magazine* began to highlight so-called "primeval parks" in the West to distinguish them from the gamut of eastern sites. After World War II the NPA lauded special scenic qualities of Mount Rainier National Primeval Park, Glacier National Primeval Park, and others, primarily in the West.[23] Protecting "primeval" parks was likened to preserving great works of art, and offered proof that Americans could protect the significant parts of their natural heritage. Thus the western parks took on growing importance to the NPA following the war, and the emphasis on them coincided with the growing activism of the Wilderness Society in the region.

Alfred Runte has pointed out that the national parks were established in order to protect the nation's most dramatic scenic wonders, what he terms monuments of nature, and that "the difference between all parks and national parks lay in the one feature that the latter had had from the beginning—primitive conditions."[24] By definition, *primeval* suggests antiquity and a timeless quality thought to be inherent in wild land. Such a conception of unchanging and pristine landscapes is now less tenable for ecologists and environmentalists who recognize its static formulation of the natural world. At the time, though, the NPA, Wilderness Society and other preservationists regarded "primeval" beauty as a fundamental characteristic of wilderness and a cardinal principle underlying their efforts. By preserving landscapes that supposedly had not changed since time immemorial, moderns could remain connected to their ancestors while hiking in the mountains or forests. Sigurd Olson of the Wilderness Society, a leader in the campaign to establish a national park around the Boundary Waters of Minnesota and Ontario, put it most clearly. Looking inside a dry and dusty museum, he wrote, the historical imagination lies dormant:

> How different to travel down the lakes and portages of such an area as the Quetico-Superior country, knowing one is actually following the unchanged trail of the voyageurs, seeing the shoreline they saw 200 years ago. . . . To camp where they camped, using the very same rocks for our fireplaces, to follow the ancient portages that felt the

tread of their moccasined feet, to read the old diaries while listening to the sounds of the wild they knew, gives one a feeling for early America that can come in no other way.[25]

At the war's end the NPA and Wilderness Society found common ground in a desire to protect "primeval" parks, and the two organizations sought to remind the public that national parks were not simply playgrounds but also prime "samples of original America," to use Olaus Murie's phrase.[26]

The postwar years proved disappointing. In the first place, despite their growing popularity, Congress proved stingy in appropriating funds for improving roads, campgrounds, and public facilities in the parks.[27] NPS Director Drury, who understood the need for adequate facilities to win public support, became convinced that such neglect made the park system vulnerable to encroachment by commercial interests. "With no possibility of developing park lands for public enjoyment," he complained to Bernard DeVoto,"—particularly the newer parks such as Olympic, Kings Canyon, Big Bend . . . we are constantly having to defend them against all kinds of proposals for competing uses. A program which is largely defensive and negative, imposed upon us [by Congress] puts the whole national park conservation program on very thin ice. In public land management, we can maintain a vacuum for a while but not for long."[28]

Drury did not exaggerate. A new series of threats to national parks came to the fore that made the dark days of the war seem bright by comparison. With the demand for housing, food, minerals, and hydroelectric power escalating in the postwar economic boom, timber companies, dam builders, stockmen, and other interests sought access into several of the western parks. Loggers in Washington state set their sights on thousands of acres of forest land in Olympic National Park, and fought to make them available for harvesting. Eventually their effort foundered, but the controversy over Olympic Park awakened conservationists to the gamut of local interests that sought access into several national parks and monuments.[29]

In northwest Wyoming, ranchers railed against the newly established Jackson Hole National Monument and supported legislation to abolish it, setting off one of the hardest-fought battles of the 1940s. Pitting groups like the NPA and the Izaak Walton

League against stockmen and business interests in Jackson, Wyoming, near the Grand Tetons, this bitter controversy was marked by fervent cries about states rights and a heavy-handed federal government. The national monument came under sharp attack, with local interests denouncing it as a symbol of federal intrusion. Kenneth Reid of the Izaak Walton League saw a connection between the bill to abolish Jackson Hole National Monument and what *Harper's* columnist Bernard DeVoto called a "landgrab," a grassroots movement of ranchers from Wyoming, Nevada, and other western states, demanding that millions of acres of public lands be transferred from the federal government to the states. DeVoto attacked the "landgrab" in his columns, convinced that the effort masked a move by oil companies to gain access to Jackson Hole National Monument and other parts of the national park system.[30] The maverick *Harper's* writer soon became a prominent figure in the early stages of the Echo Park controversy.

Yet to DeVoto and many conservationists the most insidious threat to parks and primitive areas came not from private interests but from the Army Corps of Engineers and the Bureau of Reclamation. Following the war both federal agencies introduced plans to construct dams in or near several national parks and national forests, again primarily in the West. (The Bureau, of course, only operated in the West.) The Corps of Engineers, responding to heavy flooding along the Columbia River, proposed a dam along an upstream tributary, the north fork of Montana's Flathead River, which formed the western boundary of Glacier National Park. Glacier View Dam, according to a special commission studying the national parks, threatened to inundate "20,000 acres of the most primitive portion" of the park, including winter range for white-tailed deer and elk. After much wrangling, the Corps agreed to a dam at Hungry Horse, outside Glacier Park, but soon afterward, the Corps began to talk about a dam along the Echo River in Kentucky, which would pose a threat to Mammoth Cave National Park.[31]

Similar proposals by the Bureau of Reclamation followed, threatening other parts of the park system as well as Forest Service "primitive" areas. In 1948, the Bureau announced its intent to dam the outlet stream of Lake Solitude in the Cloud Peak Primitive Area in Wyoming's Big Horn National Forest, sparking a fiercely fought dispute with the Wilderness Society and the National Parks Association.[32] That same year, the Bureau indicated an interest in building dams on the Colorado River at Marble Canyon and Bridge Canyon, locations within the Grand Canyon and close to the national park and monument.[33] Becoming dismayed with this pattern, conservationists concluded that the federal government's dam builders stood ready to violate scenic preserves with the same impunity as loggers, cattlemen, and other interests.[34] In New York state, the Panther Mountain Dam—a power project—threatened the Moose River region in the southwestern portion of the Adirondack State Park, which Paul Schaefer called "the finest remnant of primitive woodland left in eastern America."[35] Leaving aside Panther Mountain Dam and the threat that it posed to Adirondack State Park, most of the dams were slated along rivers in the West and threatened national parks, monuments, or primitive areas in the region beyond the Mississippi. It was the West that provided the setting for the most furious conflict of all, the one over Dinosaur National Monument.

Against this backdrop, the national conservation organizations learned of the Colorado River Storage Project, with two dams slated for Dinosaur, one in Whirlpool Canyon below Echo Park and another in Split Mountain Canyon. They immediately viewed the proposal as the gravest threat to date to the integrity of the national park system, in part because the dams would drastically change the scenery. Echo Park Dam alone would obliterate part of the monument by submerging the lower portions of the Green and Yampa rivers for dozens of miles upstream from their confluence. The heart of Dinosaur, the NPA and its friends maintained, would be transformed from a "primeval" wilderness into a huge artificial lake.[36] Fred Packard, president of the NPA, contended that "many of the outstanding geological and scenic features of the monument, including Pat's Hole, Echo Park, Castle Park, Harding's Hole, and the famous Canyon of Lodore would be destroyed."[37]

The NPA asserted that the dams would establish a precedent of utilizing the national parks for commercial purposes and provide an opening wedge into other national parks and monuments. It warned that both dams must be deleted from the Bureau of Reclamation's plans for the upper Colorado River or the beauty of Dinosaur Monument would be lost, and along with it, the foundation of the national park system.[38] When taken alongside other plans for dams in the Grand Canyon, Glacier, and Mammoth Cave national parks, conservationists viewed the plan for Dinosaur as obstreperous, overbearing, and insensitive to a tradition of preservation as embodied in the national park system. By 1950, the Bureau's proposal had drawn the attention of the Wilderness Society, the Izaak Walton League, the Wildlife Management Institute, and several other groups, who began to coalesce into a united force.

In coming to the defense of Dinosaur, conservationists sought to preserve its "primeval wilderness," and in so doing made the case that the protection of scenic beauty lay at the foundation of the national park system. In one way that argument proved difficult to sustain. National parks had been created for public enjoyment as well as to protect grand scenery, and this dual purpose of the preserves, embodied in the National Park Service Act of 1916, had made for a tricky balancing act. Park superintendents and NPS administrators in Washington had long struggled over how to balance these conflicting aims. How many roads and trails should be allowed to accommodate public access? How much should parks be preserved as wilderness sanctuaries?

The proposal for dams in Dinosaur Monument sparked similar questions. Should dams and reservoirs be permitted in the monument in order to encourage visitation? With the public descending on nearby parks in record numbers, the Bureau and its supporters asserted that the dams would provide added attractions in the intermountain West and help to ease the burden on parks like Grand Canyon, Mesa Verde, and Rocky Mountain, which were already becoming overcrowded. Dinosaur would offer recreational opportunities while also satisfying the upper-basin demands for capturing the Colorado River. Here was a plan in keeping with the interagency agreements signed during the war, and it even appealed to a certain element in the Park Service itself who agreed that additional reservoirs could ease the strains on an overburdened national park system.[39] After all, the first generation of Park Service leaders, Stephen Mather and Horace Albright to name the most obvious (along with the NPA), had heavily advertised the parks and won public support through road building and visitor facilities. They understood that the parks could not flourish without adequate roads, trails, and concessions to attract the public; now, a generation later, the Bureau sought to justify dams in Dinosaur with the same rationale, maintaining that Echo Park and Split Mountain Reservoirs would allow visitors to see the inner reaches of this inaccessible preserve. In a sense, the Bureau coopted a time-honored public relations tactic long used by the Park Service, and this put the latter in a rather awkward position.

The Park Service had trouble defending itself, too, because it lacked funds to build roads into the canyon section of Dinosaur, a fact which had long irritated local residents. G. E. Untermann, curator of the new Utah State Natural History Museum in downtown Vernal, spoke for many in his community when he told NPS Chief Naturalist John Doerr that those living near the preserve deeply resented the National Park Service administration of the monument because it "ain't doin' nothin' with it."[40] So long as Dinosaur remained underfunded by the Park Service the economic benefits to the local tourist economy could only be imagined in a distant future. Until the NPS could promise that adequate roads and public access would be forthcoming, it became vulnerable to the charge that it merely sought to preserve Dinosaur for a handful of wilderness adventurers who did not mind running the rivers in flimsy boats or driving their cars on the rocky and dusty roads.

This broader argument about public access underlay the mounting conflict between the Park Service and the Bureau on reservoirs under NPS jurisdiction, and this issue proved to be the focal point of debate within the Department of the Interior where the battle first erupted. We have already noted that the Park Service had come to administer reservoirs as National Recreation Areas and that Lake Mead and others had been placed under Park Service care. From the Bureau's point of view, the NPS had been a willing partner in managing such areas, as indicated by its consent to study the recreational potential of the upper Colorado River and consider the possibility of dams. The Bureau continued to believe in the merits of the Park Service administration of such areas.

Unfortunately for the Bureau, attitudes toward reservoirs under Park Service care changed swiftly after the war, coinciding with growing interest in "primeval" parks and wilderness. The NPA had just emerged from a bitter controversy involving Jackson Lake Reservoir in Jackson Hole National Monument, and it had failed to convince the Park Service that an artificial lake did not properly belong in a national park or monument. Embittered that the reservoir had ultimately been included in the preserve and convinced that doing so flagrantly violated "national park standards," the NPA had resolved not to accept any more reservoirs within the national park system.[41] That pledge hardened once other proposals emerged for dams in Glacier, Grand Canyon, Kings Canyon, and Dinosaur. The NPA took the view that the preservation of great scenic spectacles should take precedence over recreational demands.

The NPA and the Wilderness Society now took as dim a view of reservoirs under Park Service care as they did of other kinds of intrusions into the parks by private interests. Reservoirs just as clearly thwarted the effort to maintain the "primeval" parks as any other commercial intrusion. Believing that national parks had a cultural value that surpassed their entertainment value, preservationists sought to uphold a rarefied air in the national park system, and encourage a type of recreation that went beyond the pleasures of boating and fishing. They operated on the assumption that national parks served a higher purpose—that the breathtaking scenery of Zion, Yosemite, Grand Teton, and other parks inspired the mind and uplifted the soul. Accordingly, they considered reservoirs such as Lake Mead inappropriate in parks and monuments.[42] As Harold Martin explained in regard to the controversy over motorized access to the Boundary Waters region, "the spirit of the wilderness is too fragile to survive for long in a resort atmosphere. The howling of the airplane engines and the sputter of outboards not only rend the veil of silence that gives to the wilderness its cathedral calm; they cause the moose, the bear, and the beaver to high-tail for distant parts unknown, depriving the canoe camper of the pleasure of their company."[43]

Obviously much truth remained in the notion that "parks were for people" and the NPA hardly abandoned its traditional support of that dictum. Still, with mounting threats to several parks, the NPA allied itself with the Wilderness Society in arguing that parks should serve more noble purposes than public recreation. Instead, like a great work of art, their preservation should stand as a sign of the nation's cultural maturity. National parks should be a showcase of scenic marvels, and stand as proof that the nation had the will to curb its own economic appetite in order to protect remnants of "primeval America." In short, the new wave of threatened intrusions into areas of the national park system stirred a number of organizations to defend the parks as wilderness sanctuaries, as remnants of the "original" American landscape. Adopting such a stance held the most promise for galvanizing the public to ward off such threats. In this way the "primeval" parks became linked to unspoiled wilderness, and that link became a fundamental premise in the campaign to guard Dinosaur from the Bureau of Reclamation and its dams.

For the time being, the changing attitudes toward reservoirs within the park system greatly influenced NPS Director Drury, who soon found himself under considerable pressure to reject the further administration of them by his agency. Drury, to be sure, found himself in a delicate position, wanting to protect his own agency's interests while also recognizing the legitimate aims of the Bureau of Reclamation and the political demands on the Secretary of the Interior. As we have seen, Drury had been willing to cooperate with the Bureau during the war in developing its plans for the Colorado River, and he had accepted the possibility of Dinosaur Monument becoming a National Recreation Area when and if Congress consented to an Echo Park Dam. He had signed agreements with the Bureau indicating that the Park Service and the Bureau could coordinate their efforts, and his approach had created an understanding between the two agencies that encouraged the Bureau to proceed with its plans. He had felt obligated to give the Bureau leeway in building dams that might provide much needed hydropower for the war effort.

Yet the postwar atmosphere surrounding the park system caused Drury's attitude to change considerably. The Director now received a steady stream of correspondence from Bernard DeVoto, Olaus Murie, Fred Packard, and others who constantly reminded him of mounting pressures on parks and monuments and on primitive areas in national forests. Drury knew that, under the leadership of Murie and Howard Zahniser, the Wilderness Society had begun to campaign for a national wilderness system and that it would look to the Park Service as a natural ally in that campaign. Taken together, the NPA, Wilderness Society, and Sierra Club represented a growing interest in preservation, and so Drury found himself pressed to guard the "primeval parks" from intrusive economic interests. The fact that Drury had been under fire among some conservationists for compromising in the latest dispute over Olympic National Park, added to his interest in meeting their demands. By 1949, with a number of park areas in the West threatened by dam builders, stockmen, and other interests, Drury felt he must take a stand.[44]

Drury increasingly indicated that he wished the Park Service to be known for its sensitive management of areas of great natural beauty and for promoting a brand of recreation associated with *wilderness.* He held that the parks should serve a national constituency rather than local recreational interests. As a result, he began to express opposition to the Park Service's management of artificial lakes and to reject any expansion of such administration. He said in a speech that reservoirs like Lake Mead inevitably became dominated by local sportsmen and by business interests, who tended to regard the lakes as their own domain. "Too often," he remarked, "the popularity of [recreation] areas with nearby communities tends to make them local romping grounds."[45] Although Drury had been frustrated with Congress for inadequately funding the Park Service, he did not think that the answer to an overburdened park system lay in acquiring more reservoirs—and certainly not in building dams inside existing parks or monuments.[46] Not surprisingly, Drury now began to exhibit a markedly different attitude toward the prospect of dams in Dinosaur Monument—a shift that greatly surprised the Bureau of Reclamation and proved instrumental in opening the public debate over Echo Park.

Drury's growing resentment of the Bureau grew also from his awareness of the tactics it used to publicize the upper-basin project. With a close-up view of the climate inside Interior during the postwar years, he became increasingly disturbed at the Bureau's habit of pressuring the secretary and flexing its muscles in departmental affairs. He received a number of complaints about the Bureau's tactics in letters from Park Service personnel at Dinosaur Monument and the NPS regional office in Omaha. Their correspondence documented how surveyors, geologists, and engineers from Bureau offices in Vernal and Salt Lake City gained access to Dinosaur Monument without proper clearance from the Park Service and how they promoted the Echo Park and Split Mountain dams among local residents. Jesse Lombard, superintendent at Dinosaur, constantly complained about the Bureau's personnel, who repeatedly entered the monument without his permission. In 1945, the Bureau agreed to keep NPS officials informed of their work, but Lombard continued to be frustrated with the surveyors in subsequent years. The Bureau had also built its own roads into the monument and had drilled test holes in Whirlpool Canyon at the Echo Park dam site.[47]

Taken in its entirety, the Park Service correspondence revealed a pattern of actions suggesting that Bureau officials had taken the lead in promoting and publicizing the CRSP, and had done so well in advance of its approval by the Interior Department. Some Park Service officials speculated that Bureau spokesmen deliberately advertised the package of dams and power plants in order to gain public support and force the Department of the Interior to approve the project.[48] It was clear, in any event, that several officials from the Bureau's offices in Vernal and Salt Lake City had been circulating throughout the upper basin speaking before business and community groups and lauding the many benefits of Echo Park Dam, the CUP, and the rest of the CRSP. Park Service employees at the monument felt that the Bureau's efforts amounted to a public-relations spree, an expenditure of energy that went beyond the agency's proper sphere. Of course, the Bureau's spokesmen felt that they were doing nothing of the kind, that their meetings were intended only to provide basic information about an upcoming project. Still, with growing numbers throughout the upper basin hearing about Echo Park and Split Mountain dams in public hearings and in the newspapers, the Park Service began to fear that the Bureau might go to Congress and present the CRSP as a fait accompli.[49] NPS officials feared that their interest in protecting Dinosaur National Monument was being overlooked by the Bureau and its supporters across the upper basin. As early as 1946, Lombard indicated to David Canfield, superintendent of Rocky Mountain National Park, who shared in administrative responsibilities for Dinosaur, that Vernal and the nearby area "is seething with 'pro Echo Park Project' sentiment . . . and any objections or criticisms on our part would be looked on with disfavor."[50]

Suspicions among Park Service employees continued to grow. In February 1949, the Park Service regional office in Omaha prepared a reconnaissance report of Dinosaur Monument for the Bureau's regional office in Salt Lake City. This report recommended against Echo Park and Split Mountain dams, since "highly important scientific, wilderness, and recreational values would be adversely and seriously affected." The report went on to challenge the Bureau to prove that the value of the dams to the nation outweighed that of preserving the monument.[51] Hoping to obtain a response from the Bureau before the issue became public, Drury soon became dismayed by the appearance of a preliminary report on the CRSP issued by the Bureau. On the front cover of this document appeared the words "For Review Only, Not for Public Release."[52] Drury and other NPS officials learned that the report had been circulated to George Clyde and the Utah Water and Power Board and state water engineers, and they considered this a violation of the trust between the two agencies within the Department of the Interior.

Drury also informed the secretary that the Bureau ignored alternate dam sites outside the monument.[53] The Bureau's own "Natural Menace" report of 1946 listed dozens of possible sites, so it appeared to Drury rather strange for so much emphasis to be given to the Echo Park site. In 1949, the Park Service sent a representative to the Bureau of Reclamation in Vernal to express misgivings about the dams slated for Dinosaur and to suggest alternate sites outside the monument. Yet suggesting alternate sites, as became more apparent throughout the controversy, was difficult and risky. In the first place, the Bureau and the Utah Water and Power Board considered Echo Park Dam a key component of the CUP; Utah had deliberately acquired rights to the Yampa River in the Upper Colorado Basin Compact of 1948, and without a dam below Echo Park the state could not capture and use Yampa River water to purify the Green. Moreover, with its location just below the Green and Yampa rivers at their confluence, the Echo Park site had a decided advantage from the point of view of "river regulation" and hydropower production.

Near the end of 1949, Director Drury realized that his efforts to suggest alternate sites had gone nowhere and that the Bureau had hardened its determination to gain approval of the Dinosaur dams. Losing patience, he began sniping at the agency through several confidential memorandums to Oscar Chapman, who had recently succeeded Julius Krug as secretary of the interior. Pointing out the Bureau's well-orchestrated campaign on behalf of Echo Park and Split Mountain dams, Drury complained that the Bureau had engaged in public relations before the secretary of the interior approved of the project and had gone far beyond its authorized duties of building dams. The Park Service, Drury told the Secretary, "has constantly been at a disadvantage in connection with dam-building projects . . . almost always the knowledge of the projects is given out locally prior to any nation-wide information being given to those who are interested in conservation of all natural resources."[54]

Drury also began to backtrack on agreements that the two agencies had made in prior years about the use of the monument. He now claimed that the Park Service had never meant to commit itself to dams in Dinosaur Monument, and he put the burden of proof for building them squarely on the Bureau's shoulders. Remembering Secretary Chapman's pragmatic outlook, Drury suggested that an unimpaired national monument would benefit local communities through dollars from tourism. Once the monument received adequate funds and the canyons made accessible, its economic value "will increase, whereas Reclamation's anticipated values necessarily decline with the years."[55]

Just what values did Drury hope to preserve? While the director had not been to Dinosaur himself, he had testimony from Superintendent Jesse Lombard about its unique scenery. Lombard made clear that while the fossil bones in the old Carnegie Quarry were not endangered, the dams threatened scenic features along large portions of the Green and Yampa rivers. Split Mountain Reservoir would inundate Island and Rainbow parks in a lovely, expansive section of the Green River east of Split Mountain Canyon. Lombard regretted that prospect because it threatened winter range for mule deer and a dramatic display of geologic faulting at the northeast end of the canyon. As for Echo Park Dam, Lombard informed Drury that it would "destroy, obscure, or lessen the scientific value of many of the geologic features for which Dinosaur National Monument was created . . . its long and narrow reservoir submerging miles of both the Green and Yampa rivers, and with them, innumerable geologic formations, petroglyphs, and unique ecological and scientific characteristics." The superintendent agreed with the Bureau that reservoirs might increase the number of visitors who used the monument, but "it will be an entirely different type of use than was contemplated when the area was established."[56] Here was testimony that cut to the heart of the issue for Newton Drury: reservoirs were incompatible with "primeval parks."

Armed with such information, and under strong pressure from Olaus Murie, Ira Gabrielson, Fred Packard, and other conservationists, Drury decided to make the Park Service position public. In 1950, the Department of the Interior published the results of recreational surveys along the Colorado River, and Drury made certain that the Bureau and the public understood his point of view. Echo Park dam, he said, would be totally alien to the geology and landscape of the monument . . . power lines would have a widely damaging effect upon wilderness and scenic values . . . . Even more unfortunate, would be the submersion of the floor of Echo Park, the great reduction in the visible height of the nearly vertical sandstone walls of Steamboat Rock which rise several hundred feet above Echo Park and partial obliteration of noteworthy folds and faults so strikingly revealed in this area.[57]

Although it was somewhat official and stiff, the passage made clear that conservationists considered Dinosaur a premier example of a primeval landscape, and should be considered in the same category as the national primeval parks. Dinosaur did not have national park status, a fact which bothered the conservationists greatly, and they worked throughout the controversy to alter its designation from national monument to park. Such efforts continued throughout the 1950s.

Of course, Dinosaur was hardly a well-known part of the national park system, lacking the name recognition of Yellowstone, Yosemite, or the Grand Canyon, and it took tremendous effort during the next several years to publicize and promote its scenery. Despite the number of groups poised to defend the monument, its very obscurity weakened their case that the dams threatened to undermine the integrity of the national park system. Building a dam or two in Dinosaur National Monument did not seem analogous to plugging up Old Faithful with a giant stopper, nor did it seem even as serious as placing a dam on the Colorado River in the Grand Canyon. This was Dinosaur, and what did it matter if water covered over some old bones anyhow? Since the monument lacked a clear identity, it would take time to "advertise" this area and make its wilderness values known to the public.

The obscurity of the preserve goes far to explain why the dams had such appeal to residents of northeast Utah and northwest Colorado. To them the dams held special attraction for they had been disappointed for many years with the contributions of Dinosaur Monument to their economy. Prospects seemed considerably brighter, though, with the building of dams and recreation areas in this remote part of the West. The raging Green and Yampa rivers would become large, placid lakes, while boating, fishing, and outdoor activities would beckon many tourists. Having a new playground in the desert seemed a far better way to make use of the remote canyons than holding them in perpetuity for only a few to see. The dams promised to bring forth the great potential that Dinosaur Monument had always offered. Writing to Devereux Butcher of the National Parks Association, G. E. Untermann hammered the point home. "Let the Bureau of Reclamation build the access road for which the Park Service has no funds," he argued. "The Bureau of Reclamation will put Dinosaur on the map and make it important enough so that it can compete with other Park Service areas on a fairer basis for its share of Congressional attention."[58]

Untermann's comments indicate the degree to which the Bureau of Reclamation's proposed dams went far to fill the vacuum of this underdeveloped part of the national park system. The Bureau's plan, in other words, offered to define Dinosaur Monument in a way that those living nearby could understand, and this had been something the National Park Service had never been able to do. At the time, few conservationists knew about Untermann's comments, yet the sentiment which he expressed proved to be an important obstacle to their campaign to stop the dams. Eventually they found they had to counter Untermann's point and demonstrate that the monument not only had great scenic wonders but that its raging rivers could offer a safe and enjoyable experience in the wilderness. This was not an easy task given the fact that river running as a serious sport had barely begun to develop.

The challenge of forcing the Bureau to scrap the two dams had barely begun and was undeniably great. Conservationists could not ask the Bureau to prepare an environmental impact statement because the National Environmental Policy Act was more than two decades in the future. Without the Endangered Species Act (1973), they could not hope to defend Dinosaur by seeking to protect particular species, a tactic successfully used to delay Tellico Dam in the 1970s.[59] With no legal weapons they could only appeal to the public to help uphold the terms of the National Park Service Act of 1916, to help keep the park system "unimpaired for the enjoyment of future generations." But although millions of Americans now enjoyed the national parks, political support for the system, as Drury knew only too well, had been meager after the war.

On the other hand the Bureau of Reclamation had a far more concentrated base of support because of its growing constituency in the American West. The agency had built a strong reputation and become a powerful force in the West, its dams and power plants part of the fabric of many a state's economy. Population growth in the West during and after the war further strengthened the Bureau; by one estimate, eight million people moved west of the Mississippi River between 1940 and 1950, swelling the size of Denver, Las Vegas, Los Angeles, Phoenix, and Tucson. Backed by steadily increased appropriations from Congress, the Bureau expanded across the West throughout the 1940s with major projects along the Columbia, Missouri, Sacramento, and San Joaquin rivers. One historian has pointed out that during the Truman administration the Bureau enjoyed the "most significant period of growth in its first fifty years of existence."[60] Certainly, the Bureau had greater standing in the Department of Interior than the NPS, greatly overshadowing its sister agency.

Then, too, the Bureau had a major presence on the Colorado River with Hoover Dam, the Imperial Canal, and other projects completed in the lower basin. Its interest in the Colorado River preceded that of conservationists by decades, and this, coupled with the lack of legal weapons available to the opposition, allowed the Bureau to set the terms of the debate over Echo Park and Split Mountain dams. Throughout the controversy conservationists had to learn how the Bureau's engineers thought, how such large projects were planned and assembled, and how to challenge their facts and figures. That proved more than difficult as the Bureau had all manner of experts in hydrology, geology, economics, and other fields.

Furthermore, the climate of the times favored large water and power projects. This generation came of age in the heyday of the Tennessee Valley Authority, which had taken dramatic steps to transform a backward region in the South. Most Americans assumed that dams served a useful social purpose, and they took for granted the benefits of flood control, power production, and water for towns, cities, and farms. The Bureau's ambitious project had been designed to help expand a regional economy in the mountain West, a fitting goal to many Americans during a time of economic growth and a growing Cold War. The careful use of water and power seemed necessary to national interests.

In confronting the Bureau of Reclamation, the national conservation groups took on an agency with which they had contended before. In the early 1920s the Bureau had supported a plan of Idaho potato farmers to erect a dam along the Bechler-Falls River in the southwest corner of Yellowstone National Park—though the plan was later scrapped.[61] A decade later, with the Colorado–Big Thompson Project underway, the Bureau sought permission to drill a tunnel beneath Rocky Mountain National Park to carry water from the western part of Colorado to the eastern plains, and this had been done during the war. Yet these episodes, though contentious at the time, did not match the confrontation over the proposed Dinosaur dams. In the first place, neither the dam proposed for Yellowstone nor the Rocky Mountain tunnel posed such a threat to the scenery; in the second place, neither one of these earlier proposals aroused opposition from the public to nearly the same extent.

As Samuel Hays has suggested, World War II and the postwar period proved to be a watershed in the evolution of popular attitudes toward nature. Rising living standards, increasing income levels, and a marked growth in the numbers of Americans attaining a higher education all worked to usher in a post-industrial society and an "advanced consumer economy."[62] These trends underpinned a shift in public attitudes toward the environment. With the middle class increasingly less worried about fulfilling basic living needs and with greater amounts of leisure time at their disposal, this growing segment of the populace expressed a desire to enjoy what Hays has called "amenities" of life, including vacations to parks, forests, and wilderness areas. As Hays makes clear, these changes established the social bases of the postwar environmental movement with stronger public support for preservation of wilderness and wildlife. This, in turn, worked to discredit the traditional conservation agencies like the Bureau of Reclamation which were now perceived as narrowly focused on "resource management" and commodity production. In short, the war gave rise to a new generation of preservationists and planted the seeds for repeated clashes with the Bureau, Forest Service, and other resource agencies. The conflict between the Park Service and Bureau of Reclamation over Dinosaur Monument provides an excellent example.[63]

By 1950, then, two major forces were coming into conflict in the American West, forces which had each been amplified by economic and social changes instigated by the war. The Bureau represented a newly emerging region of the West, which based its outlook for the future on a traditional conception of conservation—in this case, the wise use and management of the Colorado River. Conservationists and wilderness enthusiasts were increasingly concerned about solidifying protection of various public land preserves in the West, and especially determined to reaffirm the mandate in the National Park Service Act of 1916 to conserve the parks "unimpaired." In the early months of 1950, these two forces clashed most directly in the Department of the Interior in Washington, where both the NPS and Bureau of Reclamation had their central offices. For the next three years, the Echo Park conflict was centered in the Interior Department, and Secretary of the Interior Oscar Chapman made some crucial decisions on the dams that brought the controversy into full public view.

FOUR

"A Mere Millpond"

. . . unfortunately, the same natural processes which have carved out great canyons and created magnificent scenery . . . also have produced dam and reservoir sites which are veritable engineers' dreams.

—Oscar Chapman to Charles Sauers[1]

Surely the construction of Echo Park Dam can conceivably be classed as a defense project and a tower of strength in our national economy.

—Bryant Stringham to Oscar Chapman[2]

The battle over Dinosaur erupted in the Department of the Interior in the latter part of 1949, when the National Park Service and the Bureau of Reclamation, each representing strong currents in the American West after the war, clashed over the merits of dams in the national monument. The Park Service stiffly resisted the dams as well as its own administration of reservoirs as national recreation areas, and the NPS and the Bureau soon became irrevocably divided. The directors of both agencies, Newton Drury of the Park Service and Michael Straus of the Bureau, pressured Secretary of the Interior Oscar Chapman to resolve the dispute. Chapman did so by calling for a public hearing, which led to his approval of the dams in June 1950. In the following months, Park Service Director Drury opposed the decision and expressed his clear disappointment with Chapman's leadership at the Interior Department. Drury eventually caused so much turmoil within the Department that Chapman asked for his resignation early in 1951. Drury's removal caused an uproar among major conservation groups in the United States and brought the Dinosaur controversy into the public arena. Newton Drury, in effect, became a martyr of the conservationist position on the national monument and, more importantly, on the nation's threatened national park system.

As the decade of the 1940s drew to a close, Newton Drury knew full well how much pressure was being exerted on the secretary of the interior to approve the Colorado River Storage Project by the Bureau of Reclamation and by most of the public in the upper basin. He was well aware that Bureau Commissioner Michael Straus had been busy lobbying the new secretary of the interior, Oscar Chapman, pressing him to approve the entire CRSP and the controversial Dinosaur dams.[3] Drury also knew that Straus provided a formidable opponent and would not like criticism about either Split Mountain or Echo Park dams. He was right.

Selected by former Secretary of the Interior Harold Ickes as commissioner of the Bureau, Straus brought a strong voice to his corner of the Department of the Interior. According to Marc Reisner, Ickes picked a man in his own image, "a newspaperman, a liberal, a fighter, a curmudgeon."[4] Straus lacked the affinity for paperwork and bureaucratic politics that Ickes seemed to relish, but he had a hard-nosed, aggressive approach that worked well in the Bureau and suited its western constituents. An easterner, born and raised, as well as a New Deal liberal, Straus had unbounded faith in the dams and power plants that the Bureau of Reclamation was building across the American West. The upper Colorado River Basin had yet to benefit from the Bureau's efforts, and Straus thought of the region much as Franklin Roosevelt had regarded the Tennessee River Valley in the 1930s: backward, undeveloped, and poor. He meant to change that, and viewed the Colorado River Storage Project as a great instrument of economic development. Straus intended to build a prosperous economy for Utah and neighboring states.

Michael Straus was in no mood to have the project challenged, and when he received Newton Drury's internal memos, funneled to him from Secretary Chapman with questions about Echo Park and Split Mountain dams, he lost no time responding. He reminded the secretary that the Park Service had cooperated with the Bureau for more than a decade in formulating water management and recreational plans for the upper Colorado River, including the Green and Yampa tributaries. As early as 1941 the Park Service had agreed to the possibility of dams in Dinosaur Monument, and it did not seem fair to Straus that Newton Drury should repudiate the agreements now. Straus also resented the charge that spokesmen for the Bureau improperly promoted Echo Park Dam.

It seemed to him just as inappropriate for Park Service employees at Dinosaur to speak out against the dam, as some had begun to do.[5] Straus dismissed Drury's charge that the Bureau ignored alternate sites outside the monument, and he stated that such sites were inferior to the Echo Park location since they would increase water loss from evaporation. While this point did not strike Chapman as particularly significant at the time, the evaporation rate of Echo Park Dam would later become a major point of contention. Straus also reminded Chapman of the upper basin's interest in making use of its allotted half of the Colorado River, necessitating storage dams in order to comply with the Compact of 1922. Finally, he reiterated the importance of Echo Park Dam as a power source, "the returns from which are essential to the payout plan for the Upper Colorado Storage Project."[6]

Straus and Drury argued with one another for several months, using Chapman as their sounding board. The Echo Park battle was now joined, with the directors of the NPS and the Bureau of Reclamation contesting one another's claims through the channels at the Department of the Interior. With Drury angry at the Bureau's heavy-handedness, and Straus irritated that Drury seemed to be reneging on earlier agreements, Chapman felt pressure growing from both sides.

By early 1950, as a result of their growing conflict, Chapman became the central figure in determining the future of the Colorado River Storage Project and Dinosaur Monument. He had to decide whether to allow the Bureau to proceed with the CRSP and the two controversial dams, though eventually, of course, Congress would have to consider the necessary legislation. With the project in Chapman's hands for the time being, political considerations came to the forefront. A quiet, unassuming sort, Chapman had the reputation of being a loyal bureaucrat and party man, and as a member of Harry S. Truman's cabinet, he had an obligation to carry out political responsibilities to the administration and the Democratic party. He knew that the Republicans were fighting hard to capture the "water and power vote" in the West, and he sought to solidify his own party's reputation as the deliverer of dams and power plants. President Truman, in the spirit of Franklin Roosevelt and the New Deal, even promoted additional valley authorities in the West, although this approach to river basin development—modeled on the TVA—proved unworkable in the region, sparking fears about federal encroachment into states' rights. Still, so long as a Democrat sat in the White House, the dam builders could rest easy, for water and power supplied by the Bureau of Reclamation and the Army Corps of Engineers had become a vital part of the New Deal tradition. Dams, reclamation projects, and power plants all helped to provide rural communities with irrigation and electricity, and in the Colorado basin, dams such as Echo Park and Glen Canyon would also supply recreation, a powerful selling point in a part of the West without many lakes.[7]

Secretary Chapman not only sympathized with development-oriented agencies like the Bureau of Reclamation, he valued a strong water and power program at a time of economic expansion and rising international tensions. He maintained that the Department of the Interior was vital to the country's "prosperity and national security." For Chapman, conservation did not mean "the locking up of some resource in order to keep people from touching or using it. It means to develop the resource in a wise way."[8] A Colorado native from Denver, Chapman appreciated the political power of agricultural, water, and power interests throughout the American West, and he brought to the department a keen interest in their welfare. Historian Clayton Koppes has written that "as an adopted westerner, Chapman understood the region's hunger for economic development; as a liberal, he rejected reliance on state and private solutions in favor of federal programs; and, as a politician, he appreciated the strength the federal development projects provided the Democratic party."[9]

While it seemed likely that Chapman would approve the dams slated for Dinosaur, he did not do so fast enough to satisfy the upper-basin states. Before lawmakers could introduce legislation for the CRSP, Chapman had to approve the Bureau's "interim" report, and to accomplish that task, he had to circulate the report among the upper-basin states as well as among federal agencies with jurisdiction over the Colorado River. Then, with everyone's comments in hand, he could make a final decision on the overall project. Until he released the report from his office, no legislation could be introduced in Congress. In the final part of 1949 and the early months of 1950, Chapman held onto the report because of the squabble between Drury and Straus.[10] Upper-basin politicians grew frustrated, and badgered him to circulate the report. By the early part of 1950, the Denver Post, the Deseret News, and the

Salt Lake Tribune had been buzzing about Split Mountain and Echo Park dams for years, and the Bureau's supporters throughout the upper basin had become restless and intolerant of delay. Residents of Utah had been especially eager to see funds appropriated for the CUP and the entire CRSP. In February, Colorado Governor Lee Knous sent a telegram to Secretary Chapman on behalf of the upper-basin governors, pressing him for a decision. The Upper Colorado River Commission, official representative of the upper-basin states, resoundingly endorsed the controversial Dinosaur dams at its March meeting. Pressure on Chapman grew.[11]

Those who knew the secretary understood that his sympathies lay with the Bureau. Although he had spent part of his career with the National Park Service, he was not the type to challenge political heavyweights at Interior like the Bureau of Reclamation. Some years later, Newton Drury spoke critically about Chapman's unwillingness to stand up to powerful agencies, remembering that he was "very much in the position of the mahout who rides the elephant and thinks he's guiding it but is really being carried along. That wasn't true of men like Ickes, but it surely was true of Chapman as Secretary of the Interior."[12]

But Chapman delayed any decision until he held a public hearing at the Interior building in Washington. Since Drury had made such a fuss about the dams slated for Dinosaur, Chapman thought he ought to give the opponents a chance to speak. Drury reminded Chapman that the Park Service had kept quiet while the Bureau had made plans for the upper Colorado River, and consequently, conservationists had not had an opportunity to express their opinion. Chapman thought no harm could come from letting them have their say, and it might satisfy Drury. He scheduled the hearing for April 3, 1950, and invited a number of individuals to testify on the merits of Echo Park and Split Mountain dams, including Bureau spokesmen, members of Congress, and representatives of conservation organizations. When the day arrived, Chapman provided introductory remarks before announcing that he would base his decision according to "the largest number of people who can be served and the greatest good that can be served."[13]

The hearing, which took place in the Department of the Interior in the nation's capital on April 3, 1950, symbolized the way in which Americans had reached a critical point in their relationship with the national park system. In all likelihood, such a hearing would not have been held five years before, but the two forces of preservation and growth in the American West had grown rapidly since 1945 and brought these issues into high relief. Should the scenic beauty of national parks or monuments be sacrificed for economic gain? Should "primeval parks" be altered in order to satisfy the nation's insatiable appetite for recreation? How important was a wilderness preserve like Dinosaur? These questions were not altogether new, of course, but many conservationists found them particularly relevant after the war, with pressures mounting on national parks and wilderness areas.

Conservationists felt that Chapman had called the hearing rather hastily, and some were unable to make arrangements to fly to Washington to testify.[14] Still, representatives from several organizations did appear, including Bestor Robinson of the Sierra Club; William Voigt, Jr., of the Izaak Walton League; Charles Sauers, superintendent of a forest preserve in Illinois and a member of the Advisory Board on national parks; and Ira Gabrielson, former director of the Fish and Wildlife Service in the Department of the Interior. Gabrielson had since left the government and become president of the Wildlife Management Institute. A veteran of the conservation scene in Washington for many years, "Gabe's" opinions could not easily be brushed aside. When it came his turn to speak, he told Chapman he disliked the way the hearing had been handled, with supporters of the Dinosaur dams given the lion's share of the time to present their case.

Gabrielson and his colleagues sought to take advantage of a growing opposition to expensive water projects in Congress. With the Colorado River Storage Project estimated to cost one and a half billion dollars (in the first phase alone), and with lawmakers increasingly skeptical about the need for large dams and power projects, they hoped to persuade Chapman to delay its approval and scrap its controversial features. While President Truman supported the Bureau and the development of large river-basin projects as part of his "Fair Deal," some Republicans in Congress (with the exception of most from the West) had become more stingy about funding such plans and reluctant to support additional New Deal programs.[15] Les Miller, the former governor of Wyoming, denounced the Army Corps and the Bureau in a 1949 article in the *Saturday Evening Post,* with the apt title of "The Battle that Squanders Billions," arguing that their rivalry caused duplication and encouraged "pork barrel" projects.[16] Irving Brant, historian and former adviser to President Roosevelt on conservation issues, told Chapman that he believed the rivalry between the Bureau and the Corps had given birth to the CRSP, implying that the Dinosaur dams had arisen needlessly. Brant thought the Bureau did not bear "malice toward the park system," but he maintained that the ambitious plan for the upper Colorado River had been the Bureau's way of keeping a step ahead of the Corps of Engineers.[17] Brant and other conservationists felt that the time had come for careful scrutiny of many projects, and they did not want to see Chapman rush to approve any dam of questionable merit.

Howard Zahniser of the Wilderness Society suggested that approval of both dams was being rushed through the bureaucratic process, and he urged the secretary to delay any decision until President Truman's Water Policy Commission issued its final report. The commission, chaired by Herbert Hoover and formed to study the possible reorganization of the executive branch, included a Natural Resources Task Force that had been examining dozens of water projects across the country. Zahniser and his colleagues knew that the Task Force had been sharply critical of the Bureau and Army Corps and had recommended that water projects like the CRSP be scrutinized by some body independent of Congress. Opponents of the Dinosaur dams held out hope that the Hoover Commission's final report would incorporate those recommendations and perhaps single out the Bureau's plans for the upper Colorado River.[18]

While some of the testimony took aim at alleged waste in federal water programs, more of it focused on Dinosaur Monument itself and how the dams might affect the scenery. Unfortunately for the opponents, few of them had actually been there, a fact which did not strengthen their pleas that it be preserved. Chapman had to rely on just two individuals who had floated down the rivers: Frederick Law Olmsted, Jr., the architect, and Dr. Frank Setzler, a member of the National Parks Advisory Board. Setzler compared the scenery in Dinosaur Monument to the Grand Canyon, saying that it offered a dramatic but more intimate view of rivers sculpting through rock. As for the Bureau's proposed dams, they "would submerge the older chapters of earth history now exposed in the Yampa and Green River canyons."[19] Olmsted, for his part, had investigated the monument for the NPS and Bureau of Reclamation in 1943, as part of the jointly sponsored recreational survey. Later, after the Bureau announced its interest in the two dam sites, Olmsted returned to Dinosaur on behalf of the Wilderness Society, the American Planning and Civic Association, and the American Society of Landscape Architects. Like Setzler, he pointed to the dramatic geological history on display in the Green and Yampa canyons, and he left no doubt that he deplored the prospect of dams. Choosing his words carefully, Olmsted pointed out that the canyons of Dinosaur had much to offer:

> Individually, they are highly spectacular, beautiful, and of great variety. Collectively . . . they offer an extraordinarily favorable opportunity for ordinary tourists with little or no previous geological knowledge to [see] . . . the tremendous significance of the geological processes that have shaped the characteristic landscape features of the Colorado River Basin. The scenic and inspirational values obtainable by the public from this National Monument if preserved in its natural condition . . . would not be wholly destroyed by the construction of either or both of the storage reservoirs now proposed; but they would be greatly damaged and reduced.[20]

Ira Gabrielson had not been to the monument but he agreed that the Bureau had given little thought to the loss of such a public resource. After expressing doubt that sufficient attention had been paid to alternate dam sites, he told Chapman that "conservationists do not subscribe and probably never will subscribe to this specious argument that the 'stinking debris' left by a fluctuating reservoir adds to the beauty of any national scenery. I have never seen one that did, and I do not expect to live long enough to see one that so qualifies." Gabrielson regretted that insufficient funds had been appropriated for Dinosaur Monument, and recognized how this had resulted in an underdeveloped preserve that was vulnerable to the Bureau's proposals. His opinion carried weight, and his presence at the hearing caused Chapman to take seriously appeals for preserving the monument.[21]

Newton Drury appeared as well. He asserted that the Bureau exaggerated the need for Echo Park Dam and that it had not seriously considered alternate sites outside the monument. Drury sought to take advantage of the fact that the Bureau had altered course on its earlier plan to utilize Echo Park Reservoir as a source of replacement water for the CUP. The Utah Water and Power

Board had originally rejected Flaming Gorge Reservoir north of Dinosaur for that replacement water because Uinta Basin ranchers feared that the Green River might be contaminated, but in subsequent years, the Bureau had reversed course and decided that Flaming Gorge should be utilized.[22] Other sites, such as Cross Mountain on the Yampa River, were available, too. Drury also argued that the Bureau had misled the public in claiming that Echo Park Dam was vital for storage, because Glen Canyon Dam would hold most of the upper basin's water. Drury tried to ask questions about the logic of the Bureau's proposals. How much storage did the upper basin actually need, and why did it need to have a dam below Echo Park to build it up?[23] For the time being, Chapman did not take such questions seriously. Coming from the director of the Park Service, the suggestion appeared like so much special pleading. Chapman doubted that Drury and the others had sufficient expertise to suggest alternate sites, and he gave little credence to their proposals.

Drury, Gabrielson, Zahniser, and the gallery of conservationists warned that the dam threatened a magnificent national monument, and would set a dangerous precedent for intruding into the national park system. Chapman knew this argument had been made before in defense of national parks. John Muir had uttered a similar warning against the dam at Hetch Hetchy. More recently, the National Parks Association had expressed similar concerns about the Bureau of Reclamation's "Rocky Mountain Tunnel" and about the inclusion of Jackson Reservoir inside Grand Teton National Park. Still, while warnings about "precedents" against the national park system had been made before, conservationists tried to suggest that threats in the past had been qualitatively different. When Muir warned of a dangerous precedent at Hetch Hetchy, the National Park Service did not yet exist. Jackson Lake Dam and Reservoir, though offensive to the NPA, had been built many years before anyone dreamed of a park. And no one could say for certain that the Rocky Mountain Tunnel would damage the park above it. But conservationists had no doubt that the Echo Park and Split Mountain dams would be a direct assault on Dinosaur National Monument and a major challenge to the principles underlying the national park system. Each would alter the scenic character of Dinosaur by flooding out miles of the lower portions of the river canyons with large, fluctuating reservoirs. Charles Sauers claimed that this "would forever preclude full development of this area for enjoyment of unimpaired, superlative natural features that make it distinctive and particularly suitable for wilderness recreation."[24]

Conservationists reminded Chapman of his responsibility to protect the park system, to ensure that it would be "unimpaired for future generations." They appealed to his sense of stewardship for the national heritage, warning that a great American tradition of preserving scenic landscapes could be at stake. Finally, they spoke of the growing popularity of parks and wilderness, and argued that appreciation of nature could not be measured only in dollars and cents. Irving Brant implored Chapman to uphold the scientific and scenic values of the Green and Yampa canyons, and urged the secretary to reject "the specious claim that these dams will not injure the monument and will add to its recreational value." The burden should be placed on the Bureau to justify the dams in the national interest.[25]

Supporters of the dams sought to counter these arguments by emphasizing interagency planning for use of Dinosaur Monument, which began in the 1930s. Just moments before the hearing began, Bureau Commissioner Michael Straus approached Secretary Chapman and showed him a copy of the agreement that Newton Drury had signed in 1944, indicating that the Park Service would not object to a dam at Echo Park, when and if Congress approved it. Arthur Carhart later recalled how Straus's intervention proved to be critical, because it eased Chapman's fears that a favorable decision for the dam would set a precedent detrimental to the park system.[26] John Will, legal counsel for the Upper Colorado River Commission, presented evidence of the prior agreements, citing the presidential proclamation that enlarged the monument in 1938 as evidence that the necessary "reclamation withdrawals" had been made. Will concluded that President Roosevelt, the Federal Power Commission, and Congress understood "that nothing was being done that would adversely affect the eventual use of the area for purposes such as those to which the Echo Park Dam will be dedicated." In expanding the monument's boundaries, Will argued, Roosevelt "safeguarded the use of the area concerned for water control and utilization purposes." Will did not mention the 1935 amendment to the Federal Water Power Act, prohibiting the FPC from issuing permits for power development inside national parks or monuments, though of course that applied only to private firms rather than public projects.[27]

While conservationists did their best, the Bureau of Reclamation and its supporters had the upper hand at Chapman's hearing, and they outnumbered and outmaneuvered their opponents. Given the role of Echo Park Dam in the Central Utah Project, it came as no surprise that the entire Congressional delegation from the Beehive State appeared and urged a favorable decision. Senator Elbert Thomas, Representative Reva Beck Bosone, both Democrats, and the powerful Republican Senator Arthur Watkins all pressed Chapman to approve the project. Together with Harold Linke, Utah's state engineer, the delegation stressed that the CUP promised to continue the state's industrial expansion.

Proponents also minimized the harmful effects of Echo Park Dam and reservoir. While not denying that the lake would inundate the whitewater rapids and canyon bottoms, it would not alter the bulk "of the scenic grandeur which is displayed at higher elevations." Besides, as John Will argued, the turbulent waters posed a danger to the uninitiated river runner, and "history ought to record that those few who have entered these areas have been principally concerned about getting out alive." A physician from Salt Lake City, J. E. Broaddus, was one of the few to have seen the Green and Yampa canyons from the river, when he had floated through the canyons on several occasions. A few weeks before Chapman's hearing, Broaddus told members of the Upper Colorado River Commission that the best reason to build Echo Park Dam was in order "to take people down through there in a boat and have them look up to those magnificent heights. They will get an inspiration such as they will never get anywhere else . . . in the United States."[28] Will's and Broaddus's ideas, of course, disturbed the conservationists, who detested the idea that because a wilderness preserve was inaccessible it should be changed with a dam. Still, their arguments touched on a point that the defenders of the monument had to find a way to counter.

Proponents also emphasized the importance of the CRSP for storing sufficient water so that the upper basin could properly regulate the Colorado River to its own benefit as well as to that of downstream states. In order to ensure a consistent flow to the lower basin, the upper basin had to have storage dams like Echo

Park to build up water reserves for dry years. Such reserves allowed the upper basin to build its own reclamation and power projects like the CUP, while meeting the requirements to deliver water downstream. Along with provisions in the Colorado River Compact of 1922, the United States had also agreed in 1944 to supply to Mexico 1.5 million acre-feet annually. To the Bureau of Reclamation, CRSP was a giant mechanism for regulating the river and for providing water and power to individual states.[29]

Eager to bolster their case, spokesmen of the Bureau informed Chapman that storage dams had become even more critical in the upper basin because of a recent and startling discovery about flows of the Colorado River. Previous studies indicated that the annual flow in the entire river system amounted to sixteen to eighteen million acre-feet, and these figures had provided the basis for the apportionment of water in the 1922 Compact. Since that time, the Bureau had come to realize that annual flow was somewhat less; during some dry years in the 1930s, as little as five and a half million acre-feet passed Lee's Ferry, the dividing point between the upper and lower basin. No longer able to count on an annual flow of sixteen to eighteen million acre-feet, the Bureau argued that storage reservoirs had become even more critical to successful water development in upstream states. N. B. Bennett, the Bureau's chief engineer of the Colorado River Storage Project in the Salt Lake office, said that the upper basin must store forty-eight million acre-feet of water, in order to carry out its own projects while meeting its obligations to downstream users. From the Bureau's standpoint the value of Echo Park Dam had increased since planning for the upper-basin project had begun.[30]

A low rate of evaporation made the Echo Park site the most efficient storage site in the upper basin. Evaporation not only concerned engineers, but also worried the upper-basin states, who had a special stake in the matter. The original Colorado River Compact had not taken evaporation into consideration, so delegates who wrote the Upper Basin Compact of 1948 had taken steps to require each state to share in such losses. The amount of water to which each state was entitled would be reduced proportionally by the amount lost to the atmosphere. How much water would evaporate? Clifford Stone, a veteran in water management from Colorado and the director of the Colorado Water Conservation Board, told Chapman that without Echo Park and Split Mountain dams, the upper basin stood to lose 350,000 acre-feet annually; use of an alternate site would evaporate at least that much more than the Dinosaur dams. Bennett said that such an amount could irrigate 230,000 acres or provide water to a city of 1.5 million people.[31] The Bureau rested much of its case for Echo Park Dam on its low evaporative rate, but technical studies in this field had barely begun. U. S. Grant, III, a retired engineer, and David Brower and Richard Bradley of the Sierra Club eventually brought the accuracy of these figures into question and helped to discredit the Bureau. For the moment, however, Secretary Chapman felt satisfied that the Bureau had marshaled the technical evidence to justify its dams. Chapman later claimed to be most impressed with this testimony about low evaporation losses.

But Chapman found himself most convinced with testimony about the capability of Echo Park Dam to generate hydroelectric power, and that proved to be decisive. The upper-basin states had a great yearning for increased power supplies after the Second World War. Clyde Ellis, director of the National Rural Electrification Administration, told Chapman that upper-basin farmers needed more power and welcomed the prospect of the CRSP. Ralph Goodrich, former dean of the engineering college at the University of Wyoming and a member of the Upper Colorado River Commission, urged Chapman's approval by confidently predicting that "if Echo Park was started at once and completed in five years, its [power] output could be entirely absorbed by the rapidly expanding power market of what is coming to be an industrial empire of the west."[32] Private power companies need not worry either, for they would be able to purchase electricity from the large dams and market the power.

Secretary Chapman deliberated for a few months before announcing his approval of Echo Park and Split Mountains dams near the end of June. Chapman said the low evaporation rates of both dams had impressed him, along with the fact that they had been contemplated since enlargement of the monument in 1938. (Although he approved both dams, Split Mountain Dam did not receive much attention, because it had been designed to operate in tandem with the larger structure at Echo Park upstream. Conservationists objected to both dams, but virtually all of the controversy centered on the major dam site in Whirlpool Canyon.) As for the fear of a precedent that might weaken the national park system,

Chapman tried to be reassuring. After all, Dinosaur was not a national park but a national monument, and since it had been established by presidential decree rather than by Congress, the provisions allowing water development must be respected.[33] Such a distinction stirred great debate at the time and in the following years. Should parks and monuments be treated differently? Were monuments any less "sacred" than parks? Since both belonged to the national park system, conservationists regarded Chapman's argument as specious. Nearly five years later this question was debated in the Senate.

But the decision stood, and political considerations certainly had been taken into account. By approving of the dams, Chapman reaffirmed the Democrats' commitment to developing river basins, something which President Truman had felt obligated to do to help his party in the West. Another political debt had been paid to Utah Democratic Senator Elbert Thomas, who faced a difficult challenge in his reelection campaign in the fall of 1950 and who had pressured the administration to approve Echo Park Dam, the CUP, and the CRSP. Irving Brant later recalled that Thomas's political fortunes carried weight with Chapman and the Truman administration, which did not wish to concede an important seat in the West to the Republicans.[34]

Yet political pressures did not influence Chapman's decision as much as other matters, especially the Interior Department's role in keeping the nation secure during the growing Cold War. While Chapman cited the low evaporation rate of Echo Park Dam, the evidence reveals that he was more interested in attaining a reliable source of hydroelectric power in that part of the West. One month prior to the hearing on the dam, President Truman had approved development of the hydrogen bomb, with testing to take place in Utah and Nevada. As a result of that decision, officials from the Atomic Energy Commission approached Chapman and indicated that they required additional power reserves in Utah. That clinched it. With the Cold War escalating, Chapman felt obligated that the Interior Department contribute to military preparedness. Despite all the talk from the upper basin about its need for water and minimizing evaporation, the decisive factor for Chapman was the capability of Echo Park Dam to generate power.[35] Of course, the secretary did not wish to state this publicly since candor might reveal too much about the administration's Cold War policy. Instead, he said only that the arid West required water and excessive evaporation could not be tolerated. Yet the evidence strongly suggests that early in 1950, the growing Cold War climate was the fundamental factor in Chapman's mind. Perhaps it was no coincidence that his formal approval of the Dinosaur dams came on June 27, 1950, the same day that President Truman ordered air and naval forces to support South Korea.[36]

Chapman's approval did not mean that construction of the two dams could begin. Only Congress could authorize the project and appropriate the necessary funding, and that could not happen until the Bureau of the Budget approved the package of dams and power plants that made up the CRSP. Ironically, while the climate of the Cold War influenced the secretary's decision, the onset of the Korean War in the following months put great pressure on the federal government to watch its purse strings, and so proponents of the CRSP found it difficult to find sufficient support in Congress. Federal budgetary restraints remained strong throughout the Truman administration, and eventually Chapman changed his mind about the Dinosaur dams.

For the time being, Utah and her upper-basin neighbors were delighted. Chambers of commerce throughout the region sent telegrams to the secretary, applauding his decision, while newspaper editorials followed suit. The *Salt Lake Tribune* proclaimed that the decision "opens the way for future industrial and agricultural development in the west." A pleased Senator Thomas said that approval of Echo Park Dam marked "the beginning of a new era in the development of Utah in all aspects of her economy."[37] His state's hopes for construction of the Central Utah Project appeared to be reaching fruition.

This initial euphoria faded somewhat in the next few months, as it became apparent that the CRSP faced various bureaucratic obstacles. Chapman still had to complete his own gathering of comments on the project from other federal agencies at Interior and the upper-basin states, and the Bureau of the Budget still had to consent to the entire package. Upper-basin lawmakers continued to press Chapman to keep the process moving, and for the moment at least, the secretary had confidence that the needed approvals from various quarters would be forthcoming. Upper-basin senators and congressional representatives at this time had much less fear about conservation groups than of states on the lower

Colorado River, which many upper–basin residents suspected of harboring designs on the water. The growth of Las Vegas, Los Angeles, and agricultural areas in southern California had accelerated rapidly during World War II, steadily increasing the lower basin's thirst. Along with southern Nevada, California farmers and urban dwellers swallowed ever-increasing amounts of Colorado River water. Upper–basin states had many reasons for moving ahead with the project, though certainly no better one than this. *The Salt Lake Tribune* warned that no one should doubt that the CRSP should be undertaken with all due speed, since "California is a potent rival for this water which rightfully belongs to the Intermountain region." It made them ever more eager for Oscar Chapman to press ahead.[38]

Meanwhile, Chapman's approval of the Dinosaur dams provoked a quick response from leading conservationists. "I'm grieved that you made the decision as you have," wrote Arthur Carhart, a veteran forester and wilderness activist from Colorado. Carhart scolded Chapman for opening the gates of the national park system to economic developers. "Here is an invitation for those who may . . . demand . . . putting waterfalls into power turbines, [or] cutting timber in the parks for lumber and ties, [or] opening them to grazing, opening them to any sort of raiding which may be argued as needed to support any sort of business of the area."[39]

Carhart knew that Chapman faced a challenging task in balancing the volatile blend of agencies at the Department of the Interior, and no secretary had ever performed the balancing act to the satisfaction of all. Yet Chapman's decision caused him and other conservationists to worry about the changing political currents at Interior, and many of them now looked back wistfully to the years when Harold Ickes was at the helm. Strong willed, combative, and stubborn, Ickes had built a reputation of standing up to development-oriented bureaus at Interior, something made possible by the fact that he did not hail from the West.[40] Among supporters of the national parks, Ickes also had an excellent reputation. Indeed, at his urging, Dinosaur Monument had been expanded in 1938. But Ickes's departure in 1946 had helped resource development agencies regain substantial strength at Interior, a fact that seemed to many conservationists starkly clear with Oscar Chapman's approval of dams in Dinosaur National Monument. As for the Bureau's great power at Interior, Newton Drury put it best when he told an interviewer some years later that "the great Bureau of Rec-lamation was . . . like the state of Prussia in the German empire, where everything was weighted in its favor. That's about the es-sence of the situation."[41]

Carhart and other conservationists regarded Chapman's ap-proval of the dams as proof of the danger in having a westerner as secretary of the interior. The decision suggested that western economic interests had surged to the forefront in the department, and with the Bureau interested in building dams in the Grand Can-yon and in Wyoming's Cloud Peak Primitive Area, the future of the parks and wilderness appeared to be dim. An indication of such a future appeared the day after Oscar Chapman's decision on Dino-saur, when Frank Ward, president of the Vernal, Utah, Chamber of Commerce, appeared before the President's Water Resources Policy Commission. Ward said that all the good dam sites in the West rested inside national parks or monuments, and he urged that all of them be investigated and made available to the Bureau for use. In Carhart's words, "that is the real indication of what lies ahead—the raiding of the national park system throughout for irrigation and power dams first, any other uses serving commerce on the heels of such dams."[42]

Chapman heard similar complaints in the following weeks from other conservation and wilderness leaders, and some of the loudest came from the outspoken columnist of *Harper's Magazine,* Ber-nard DeVoto. A Utah native, DeVoto had left the West as a young man and made his career as a writer in Cambridge, Massachusetts, with a brief teaching stint at Harvard. In the 1940s, DeVoto pub-lished major works on the fur trade and westward expansion, and established a reputation as a leading historian of the American West. After the war, he became an important voice in the conserva-tion movement. In the summer of 1946, he traveled throughout the West, and spent several days in a ranger's cabin in Boise National Forest in Idaho. There, he obtained a close-up view of the United States Forest Service, learning how the agency operated and about the political pressures that it confronted. DeVoto saw the conflict between wilderness lovers and recreationists on one side, and stockmen, miners, and timber companies on the other—each wanting access to the public lands. After returning home to Cam-bridge, he read voraciously in the history of conservation, and quickly became a passionate defender of public lands in the West.

Shortly after his conversion to the conservationist cause, DeVoto learned of a grassroots movement among western ranchers, clamoring for greater control over millions of acres of public land. Decrying federal landholding as burdensome and tyrannical, the movement grew into a loud and boisterous effort to force the federal government to "return" lands to the states. DeVoto bristled. He knew the states had never owned such lands from the beginning, and he believed the cry for "return" masked selfish schemes by western stockmen to have their way with grazing ranges in the national forests. DeVoto attacked the "landgrab" in his "Easy Chair" column, and made it a personal crusade to crush the movement in its tracks. He won the applause of conservationists across the country, while earning the wrath of the West. In his native Utah, where ranching interests were strong and a sensitivity to outside criticism had always been present, DeVoto was ostracized.[43]

It did not stop him. With the "landgrab" having subsided for the time being, DeVoto made himself into a student of western water, reading everything he could find on the history of irrigation, water law, and the West's climate. DeVoto firmly believed in the federal management of resources and in the merits of public power, and he realized that only with federal sponsorship could Utah and her neighbors make good on their claims to the Colorado River.[44] At the same time, his reading of John Wesley Powell and Walter Prescott Webb convinced him that the West's semiarid climate placed stringent limits on its population growth and industrial development. As a student of western history, DeVoto understood the region's propensity to ignore the limits of the natural environment.

In 1950, DeVoto witnessed another example of that historic tradition, when he arranged for a tour of the Missouri River, courtesy of the Army Corps of Engineers. Although much impressed with the massive dams the Corps was then building in North and South Dakota, DeVoto suspected that short-sighted vested interests were responsible. He summarized his thinking in an article in *Harper's Magazine* where he asserted that "limits are absolutely fixed by the topography and distances of the West, by its capacity to sustain population, by its soils and climate, by the amount of water, and by the facts of life. We do not know what those limits are. But they are a hell of a long way this side of what the various planning boards and chamber of commerce promoters have in mind, and what the Engineers and the Bureau of Reclamation talk about. . . ."[45]

In less than four years, DeVoto had become a lover of the West's scenic beauty, a passionate defender of its public lands, and a critic of big water projects. By the spring of 1950 he was primed for entering the debate over Echo Park. He did so in a piece entitled "Shall We Let Them Ruin Our National Parks?" which appeared in a July issue of the *Saturday Evening Post*. It was a hard-hitting article that chastised the Bureau of Reclamation and Army Corps of Engineers for their plans to build dams in various areas of the national park system. Compared to private interests that regularly threatened the parks, the two federal agencies, DeVoto argued, presented a new and more insidious threat. "Each of them," he thundered in his distinctively biting prose, "has about a third of a billion dollars of public funds to spend every year, and so can exert incomparably more pressure than any corporation that ever cast a covetous eye on the wilderness." As for the Bureau's big dam in Dinosaur National Monument, DeVoto expressed outrage. For those who knew his works of western history, the prose was vintage DeVoto, and the message moving:

> Echo Park Dam would back water so far that throughout the whole extent of Lodore Canyon the Green River, the tempestuous, pulse-stirring river of John Wesley Powell, would become a mere millpond. The same would happen to Yampa Canyon. Throughout both canyons the deep artificial lakes would engulf the magnificent scenery, would reduce by from a fifth to a third the height of the precipitous walls, and would fearfully degrade the great vistas. Echo Park and its magnificent rock formations would be submerged. Dinosaur National Monument as a scenic spectacle would cease to exist.[46]

The article delivered a tremendous punch. While members of the Wilderness Society and the National Parks Association already knew about the proposed dam, generally conservation-minded Americans did not. It carried a strong sense of alarm that caught the attention of thousands of people who could help: the rank and file members of the Audubon Society, the Sierra Club, the Izaak Walton League and dozens of other organizations not yet aware of the controversy. And that was the intent. DeVoto wanted the piece to appear in a nationally recognized publication like the *Saturday*

Evening Post. He knew that eastern congressmen regularly helped approve the West's big water projects, that they voted "for Western pork as part of the annual bargain—unless their constituents are stirred up." DeVoto intended for the article to stir up those who could bring pressure to bear on Congress. He knew that unless he raised the hackles of people outside the West, there was little hope of stopping Echo Park Dam.[47]

"Shall We Let Them Ruin Our National Parks?" fully launched the crusade against the dam. It brought the threat to Dinosaur and the national park system into living rooms across the country, and shifted the campaign against the dam into a higher gear. Former Park Service Director Horace Albright, delighted with the piece, told DeVoto he thought it "one of the finest national-park articles that has ever appeared."[48] The only trouble was that it appeared almost a month after Secretary Chapman had approved the dam. DeVoto had finished the piece in early June, but the *Post* could not publish it until July, too late to influence the secretary's decision. Still, though regretting its belated appearance, DeVoto hoped that public opinion might be sufficiently aroused to help block the dam in Congress.

Editorial writers from the *Salt Lake Tribune* and the *Denver Post* knew that DeVoto's article might be damaging, and they struck back. Because Echo Park Dam had been contemplated ever since expansion of the monument, no one could claim that the Bureau had suddenly decided to encroach upon the national park system. The *Post* published an editorial chastising DeVoto for failing to see that there was "plenty of scenery left in our west for us natives." Scoffing at the Cambridge-based author for trying to preserve land that no one from the East might ever see, the *Post* denounced "those who want to maintain the West as a colony so that nature-lovers from east of the Alleghenys can come out every five or ten years to sniff the clean air. Despite the 'desecration' of Dinosaur monument, there's still plenty of scenery left for all America."[49] DeVoto responded in kind, crying that "the National Parks and Monuments happen not to be your scenery. They are our scenery. They do not belong to Colorado or to the West, they belong to the people of the United States, including the miserable unfortunates who have to live east of the Allegheny hillocks."[50]

DeVoto had obviously touched a sensitive nerve, and he may well have been the best person connected to the conservation movement to do so. Unattached to an established group or a particular constituency, he was fairly invulnerable to criticisms from the upper basin. From atop his pillar at *Harper's,* DeVoto quickly dismissed remarks like those from the *Post,* and continued to thunder about the "landgrab" and machinations of the Bureau and the Army Corps. For the moment, he found himself the chief spokesman for those trying to defend Dinosaur and the sanctity of the national park system. Inspired by his piece, conservationists took the offensive, and the battle lines of the Echo Park controversy began to harden. DeVoto and his allies felt that they represented a broad public view, mindful of the growing popularity of parks and other scenic preserves. They saw themselves in the grand tradition of American conservation by seeking to uphold the integrity of the national park system.

Most upper-basin residents harbored ill feelings toward such an attitude. To them, DeVoto and his friends were self-indulgent, armchair "bird watchers," who knew nothing about the practical demands of living. They surely knew little about life in the American West, where water was scarce and every drop had to be put to use. This hostility toward outside "interference" surfaced after DeVoto's *Post* article, and reappeared throughout the controversy. When the *Reader's Digest* reprinted DeVoto's piece a few months later, a determined G. E. Untermann, curator of the Utah Field Museum in Vernal, asked the editor to publish a rebuttal. He was turned down on the grounds that the magazine did not have such a policy. Untermann responded anyhow, and persuaded a sympathetic Senator Thomas to introduce the piece in the *Congressional Record*, remembering later that this had been "a terrible fate" for the rejoinder. Untermann wrote that the *Post* article should not be taken seriously since the author hailed from Cambridge, Massachusetts. He harped away at DeVoto and his eastern friends for thinking that they knew anything about the West or Dinosaur Monument.[51]

In the midst of his diatribe against "eastern" nature lovers, Untermann raised the issue of the inaccessibility of the monument, an issue that opponents of the dams realized they must eventually confront. Few people had ever seen Echo Park because of the lack of roads and the dangers of river running. "Probably less than 100 persons have seen this portion of the monument and they have had to risk their lives to do so," Untermann wrote.[52] Since the Park

Service had never thought it important to make the Green and Yampa canyons accessible, why should anyone worry now about preserving them for posterity in their natural condition? Why preserve an area so few people could actually see? Conservationists could always make the case that public access should not alone determine how a park should be used; nevertheless, ever since Stephen Mather, the notion of "parks for people" had been a primary rationale for their existence. If a park or monument could not be seen by most, what purpose did it serve?

Until that question could be answered satisfactorily, those who pointed out the superlative scenery in Dinosaur found themselves branded as elitist "nature lovers," ready to lock away a vitally important resource necessary for the West's survival. The *Denver Post* branded DeVoto and William Voigt as "nature boys." And the *Salt Lake Tribune* sounded a common theme when it railed that "many of the alarmists are easterners who have never seen the monument and have no understanding of the situation. To these misguided 'protectors' of national parks dam is an evil word and they refuse to recognize facts."[53] Never far from the surface of events, there rested irreconcilable differences between vested interests in Utah, Colorado, Wyoming, and New Mexico, determined to harness the upper Colorado River, and wilderness enthusiasts from Cambridge, Madison, and San Francisco.

Reverberations from DeVoto's *Post* article had only just begun. Bureau of Reclamation Commissioner Michael Straus was quite angry, and he had several reasons to be. He quite obviously resented the charge that his Bureau was set on destroying the national park system. It was not the public image he wanted to cultivate of an agency so important to the West and the nation during the postwar boom and the growing Cold War. Nor did he agree with DeVoto's charges. Since the Bureau and the Park Service had discussed the possibility of Echo Park and Split Mountain dams for more than a decade, it was unfair to cry "invasion" now. Straus also disliked DeVoto's muckraking tone, which portrayed the Bureau as an evil entity the likes of which had not been seen since Ida Tarbell's portrait of Standard Oil.

Even more irksome to Straus was that DeVoto had irritated a highly sensitive nerve in Utah. Bernard DeVoto's name had been branded in Utah for years, a result of the author's criticisms of life and culture in his native state, and especially for his public crusade against the "landgrab" beginning in 1947. Many residents of the Beehive State had had their fill of this native son, and considered his traitorous positions that much worse due to his association with Harvard. They were hardly in the mood to listen to him harp at another one of their favorite projects. Michael Straus realized that Utah had been wounded by DeVoto's pen before, and he feared the wounds had been reopened.

Utah had a high stake in Echo Park Dam because of its connection to the Central Utah Project. Straus and the Bureau had been working with Utah's Water and Power Board for several years in order to incorporate the CUP into the rest of the CRSP. More than any of the upper-basin states, Utah looked to the Bureau for guidance and respected its good intentions to build the Dinosaur dams. Residents wanted assurance from the commissioner that he understood how government worked and how to get things done, that he was fully in charge of Echo Park Dam, and that he would never allow a sister agency at Interior to interfere with any of his dams.[54] Public opinion in the state exerted a good deal of pressure on Straus.

Straus felt the most troubling aspect of the situation was that DeVoto and Newton Drury had recently become close associates. The two men had corresponded on various Park Service issues for some years, and DeVoto had recently taken a seat on the Advisory Board on National Parks, Historic Sites, Buildings, and Monuments. Composed of experts in fields like archeology, history, and the natural sciences, the Board considered a range of issues facing the national parks and made recommendations to the secretary of the interior, generally urging protection and preservation.[55] The Bureau of Reclamation had always cast a wary glance toward the Advisory Board, but Michael Straus now had special reason to be concerned because of DeVoto's presence on the Board and his relationship with Drury. For Straus, it was bad enough that DeVoto had published the *Post* article in the first place. But what if Drury himself had inspired the piece, and enlisted DeVoto to write it? If he had, it meant the worst: that Newton Drury and the Park Service had found a way to subvert Echo Park Dam before it could ever be considered by Congress, that the dam was being pummeled from within the Department of the Interior, and that DeVoto, one of Utah's oldest and worst public enemies, had been called on to do the job. This was something that could not be tolerated.

Straus felt compelled to know exactly what Newton Drury's role had been. Certain that the Park Service had provided DeVoto with some information, he wanted to know whether Drury or his subordinates had gone further and actually assisted with writing the article. Straus asked Drury for copies of his correspondence with DeVoto. Drury resented the request, and strongly denied that he had given DeVoto any information not available to anyone else. Before long, an icy exchange developed between the directors, reviving mutual suspicions from the previous year, when the two had argued about the earlier interagency agreements on Echo Park Dam.[56]

Although Michael Straus posed a problem for Newton Drury, Oscar Chapman soon proved to be a bigger thorn in the director's side. When the secretary had first approved the two dams in late June, Drury considered himself obligated to cooperate with the Bureau, and he had pledged to help implement plans for the National Recreation Area that might soon exist in Dinosaur Monument, which the Park Service would administer.[57] Nevertheless, Drury could not disguise his disapproval of Chapman's decision, and he remarked to the secretary in a memorandum that "Shall We Let Them Ruin Our National Parks?" had obviously been "a rhetorical question. . . . The fundamental question raised by Mr. DeVoto is still the number one problem of your National Park Service," he wrote.[58]

As Drury defended DeVoto and sought to minimize the Park Service's role in assembling material for the *Post* article, he found himself increasingly alienated from Chapman, and the secretary, in turn, began to lose confidence in Drury.[59] It had been obvious to some that Drury had been very disappointed with Chapman's approval of Echo Park and Split Mountain dams, for he had been unable to hide his feelings from his underlings. His scorn for Commissioner Straus persisted as well, due to comments made by Straus that Drury took to be gloating. Drury's anger toward the Bureau chief spread through the Washington office of the Park Service, adding to the rift between the two agencies and further complicating the relationship between Drury and Secretary Chapman. Partisanship crept in, too. Chapman, a loyal New Deal Democrat, had always resented having Drury, a lifelong Republican, as head of the Park Service.[60]

At the heart of their worsening relationship rested the issue of national park standards and the role of the Park Service in public recreation. Chapman had devoted himself to helping the Bureau to build its many dams throughout the West, and he agreed with the Bureau that the task of administering reservoirs could best be handled by the Park Service. For his part, Drury was distressed that the Park Service had drifted from its original mission as guardian of grand scenic parks. He had never liked his agency's administration of reservoirs, and he felt Chapman's approval of a dam at Echo Park was one more sign of the Park Service becoming too oriented toward mass recreation. He believed that the Park Service should be committed to higher values than those associated with playgrounds, and ought to devote itself to protect the most beautiful parts of the American landscape for future generations. Having to administer reservoirs only compromised the primary mission of guarding the jewels of the park system. Drury agreed with his close friend Olaus Murie, executive director of the Wilderness Society, who scoffed at the attitude "that we might just as well do anything we please to a national park, as long as there is some of it left untouched. Millions of people will visit it anyway. It is no answer to say that 'millions' visit the impoundments. Of course. People will always go somewhere. They will accept the next best, if the best is destroyed. The danger is that we will assume that we can't have the best."[61]

In his annual report to the secretary of the interior in 1950, Drury asserted that the parks were under attack by the Bureau of Reclamation, adding that "enlightened and broadgage [*sic*] conservation organizations throughout the country . . . realize, as we do, that the future of [the national parks] is at stake." He wrote that the parks were a unique part of American culture, "essential to fullness of life and not lightly to be destroyed." The director and the secretary had each taken their stand.[62]

As Drury continued to press his views and snipe at the Bureau behind the scenes, Chapman knew a change had to be made. Cooperation of the Park Service director in planning for Echo Park Dam was vital, and Chapman began to realize that he could not count on such cooperation from Drury. He did not want to fire Drury outright because the situation did not call for such a dramatic step, and doing so might needlessly arouse the ire of the NPA and other organizations. Still, he had to get Drury out of the director's chair somehow. In December 1950, Chapman asked Drury to accept a new position as his special assistant. Chapman told Drury that it would be an opportunity to act as liaison between agencies with contending missions, to use his expertise to help the department avoid interagency conflicts in the future. In a sense, Chapman offered an olive branch, suggesting that the disagreement between the Park Service and the Bureau offered an opportunity to create a new position in which Drury would be qualified to serve.[63]

Drury knew better and quickly saw through the offer. He understood Chapman to be saying that the squabble could not go on, that the secretary had taken his stand with the Bureau and the recreation-oriented faction in the Park Service, that the decision for Echo Park Dam stood, and that he thought it would be best under the circumstances if Drury resigned his post. Drury knew he could not possibly accept the offer because the position would have no real authority. Besides, as even Chapman had to admit, the new position carried with it a reduced salary.[64]

After Drury asked Chapman for time to think it over, he returned to California for the Christmas holidays and explored his options. His roots were on the West Coast, where he had many old friends in the Save-the-Redwoods-League and the Sierra Club. As it developed, California Governor Earl Warren delighted Drury with an irresistible offer as director of the state division of Beaches and Parks. Drury quickly accepted. Next to the directorship of the National Park Service, the position was perhaps the most prestigious government post in the field of conservation. Drury would take charge of the California state park system, home of his beloved Redwoods. Upon returning to Washington, he offered Chapman his resignation, and the secretary soon announced that it would take effect on April 1, 1951.[65]

The Drury affair threw members of Chapman's Advisory Board into a frenzy. Board members did not know why Drury had been let go, and they deeply resented the way that Chapman had handled the matter. When the secretary announced Drury's departure, he said he planned to appoint Arthur Demaray to succeed him as director. Near retirement, Demaray had been a loyal employee of the Park Service throughout his career with the government. Chapman indicated that he wished to recognize Demaray's distinguished service by appointing him to the top job in the agency. Chairman Charles Sauers and members of the Advisory Board quickly dismissed that rationale, for Demaray was too close to retirement for the promotion to be credible. In Waldo Leland's words, "the intent to honor him is too flimsy a pretext for Drury's dismissal to be taken seriously."[66] Fred Packard of the NPA believed that Chapman wished to have Conrad Wirth in the director's position, since he seemed more sympathetic with the policy that placed reservoirs under Park Service jurisdiction. Packard claimed that Wirth felt that "inviolate protection of the national parks is a wide-eyed dream." Irving Brant thought the issue was not so simple and indicated to Rosalie Edge that Wirth disliked the proposed Echo Park Dam too; indeed, events in future months supported his claim, when it developed that Wirth disliked the prospect of the dam every bit as much as Drury.[67]

Nevertheless, Chapman was not forthcoming in explaining Drury's resignation, and some members of the Board felt compelled to demand that he do so. Why had it happened so suddenly? What had Drury done wrong? What was going on behind the scenes at Interior? Bernard DeVoto had a special incentive to know the circumstances, for his own integrity had recently been challenged by his old enemies in Utah. An editorial in the *Deseret News* had argued that Drury's removal was easily explained. It had all begun the previous year, when the Park Service secretly authored "Shall We Let Them Ruin Our National Parks?" As the editorial confidently asserted, "Drury was fired precisely because he fathered the DeVoto falsification of the facts about the Utah-Colorado dams."[68] DeVoto had no choice but to respond. For if the paper was correct—if Chapman had removed Drury because the director had inspired the article—then the secretary's actions could be justified. On the other hand, if the *News* was mistaken and, as DeVoto suspected, merely sounding off against Drury and himself in a burst of hyperbole, then conservationists stood on solid ground in questioning the propriety of Chapman's move. They could also be confident in criticizing him for cozying up to the Bureau of Reclamation.

DeVoto knew his article had not originated with Drury. After checking with Waldo Leland about events of the previous year, DeVoto confirmed that the two of them had hatched the idea of the article at a meeting of the Advisory Board early in 1950. (Leland at that time was director of the Board.) With his memory clarified and conscience cleared, DeVoto confronted Chapman at the Board meeting in April 1951. But Chapman offered little information, saying only that Drury had not been asked to leave his position because of the *Post* article. The secretary remained vague, later informing New Mexico Senator Clinton Anderson that he had "personal reasons" for removing Drury that could not be made public. After weeks of pressing for an explanation, DeVoto and members of the Advisory Board still had no satisfactory answer.[69]

Most of those involved assured themselves that Drury's removal had stemmed from his opposition to Echo Park Dam, although Chapman denied it.[70] Because neither Chapman nor Drury saw fit to clarify precisely what had happened, the full story of this episode remains impossible to reconstruct, even to the present day. On the other hand, it seems obvious that a clash between Chapman and Drury was inevitable, given the currents in the preservation movement working on Drury and the pressures in the administration on Chapman. Drury's fate may well have been cast the moment that Chapman approved Echo Park Dam in June of 1950.[71]

Yet the reasons behind Drury's resignation are less important than the reaction it generated. A chorus of protests from conservationists arose, further publicizing the Bureau of Reclamation's plans for Dinosaur National Monument. To many, Drury's removal symbolized the newly emerging shape of the Department of the Interior under Oscar Chapman, with a strong Bureau of Reclamation and other utilitarian agencies taking dominance over Fish and Wildlife and the National Park Service. To members of the Wilderness Society and the National Parks Association, Drury's departure signaled a new era in which the national parks would become outlets for popular recreation, playgrounds rather than sanctuaries of "primeval" wilderness. They had no doubt that Chapman had approved the dam by taking advantage of the internal debate in the Park Service about recreation areas.[72] Drury's sudden departure offered proof of the surging power of the Bureau at Interior, and suggested to conservationists that a major purpose of the national park system—to preserve scenic wonders unimpaired—was fast being eroded. For the Wilderness Society and the National Parks Association, the trend was unmistakable. Thereafter, Chapman could be expected to side with the Bureau at every step; other dams might be allowed in national parks, further eroding the integrity of the whole system, and with "the sanctity of dedicated areas" no longer respected, preservation of any lands as wilderness would be impossible.

Chapman soon found himself besieged by letters and petitions from conservationists across the country, including DeVoto, the NPA's Fred Packard, Howard Zahniser, Ira Gabrielson, and Richard Leonard of the Sierra Club. Meanwhile, the Wilderness Society, the Nature Conservancy, the American Nature Association, and a host of other groups sent a message to President Truman, expressing their concern about the national park system.[73] Truman offered no help. Upon hearing of the dispute over Drury and Echo Park from Irving Brant, the president replied that "it has always been my opinion that food for coming generations is much more important than bones of the Mesozoic period."[74] The comment revealed Truman's ignorance of the dispute. The dinosaur remains at Earl Douglass's old quarry were far removed from the Green River, and could not possibly be harmed by Split Mountain or Echo Park dams—although much confusion persisted on this score in the years to come.

The loud protests following Drury's removal had several effects on the dispute over Dinosaur. Coming on the heels of DeVoto's article in the *Post,* Drury's ouster galvanized conservationists to take heed of the state of affairs in the National Park Service and the Department of the Interior. It wasn't every day that a secretary of the interior removed the director of an agency beneath him. Many now began to take notice of the quiet, unassuming Chapman, and they questioned his attitude toward the national park system.

Chapman felt stung by the criticism, which he obviously did not anticipate, and he took steps to appease opponents of Echo Park Dam. In the midst of the uproar over Drury, Chapman agreed to meet with Leland, Gabrielson, Packard, and other conservation leaders. They pressed him to prohibit engineers in Vernal and in Salt Lake City from entering Echo Park and the canyons of Dinosaur to investigate dam sites without obtaining permission. Such actions had recurred repeatedly, making a mockery of Park Service jurisdiction at the monument. Chapman agreed. He issued a directive that prohibited the Bureau from investigating sites in any national park or monument without the secretary's approval. In effect, Chapman sent a message that Congress had yet to pass on the controversial dam, and the Bureau ought not to assume that it was a fait accompli.[75]

By the early part of 1951, conservationists had gained ground in the battle to preserve Echo Park. The publication of Bernard De-Voto's *Saturday Evening Post* article had focused national attention on the remote Dinosaur Monument and on other threatened areas in the national park system. DeVoto had provided the spark that fired the conservation community into action, and the furor surrounding Newton Drury's treatment at the hands of Secretary Chapman raised questions about the Bureau of Reclamation and its place in the Department of the Interior. The national conservation groups now stood determined to stand up to the Bureau and its supporters for their refusal to recognize the importance of protecting Dinosaur Monument as a scenic preserve. With a host of organizations galvanized, the campaign to save Dinosaur was under way in earnest. Members of Congress now began to hear from constituents protesting the proposed dams. Dinosaur National Monument began to draw the attention of a broad spectrum of conservation and wilderness enthusiasts.

The two disputes over DeVoto's article and Drury's removal delayed the Colorado River Storage Project, and the delay greatly aided conservation organizations. With Chapman increasingly under fire throughout 1950 and 1951, he felt unable to release the Bureau's report on the CRSP, and legislation could not be introduced. This gave valuable time to the NPA, the Wilderness Society, the Izaak Walton League, and other groups who informed their members about the threat to Echo Park and the national park system, and began to assemble a campaign. They knew they faced an uphill battle. The Colorado River Storage Project had been in the works for years, and the Bureau of Reclamation had an enormous amount of power. The nation was in a dramatic period of economic growth fueled by the Cold War, and this could only strengthen the Bureau's hand on Capitol Hill. While it could not be denied that conservationists had gained ground, their campaign to defend Echo Park and the national park system had only just begun.

Looking north into the winding Yampa River Canyon from Round Top, in the Colorado portion of Dinosaur National Monument. Photo by Martin Litton in the early 1950s. (Courtesy Martin Litton.)

Looking east from Harper's Corner into Echo Park and Yampa River Canyon. The Green River, after merging with the Yampa, appears at the lower right. (Courtesy National Park Service Photographic Collection, Harpers Ferry, West Virginia.)

The Green River and the Gates of Lodore, at the north end of Dinosaur
National Monument. In the early 1900s the Bureau of Reclamation regarded
the high walls near the entrance into Lodore Canyon as a prime dam site in this
part of the West. (Courtesy National Park Service Photographic Collection,
Harpers Ferry, West Virginia.)

The western edge of the monument is flanked by the sprawling and broken walls of Split Mountain, with the Green River beneath. A few miles upstream from this location was the proposed Split Mountain dam site. (Courtesy National Park Service Photographic Collection, Harpers Ferry, West Virginia.)

Warm Springs Bend in Yampa Canyon. Photo by George Grant, 1935.
(Courtesy National Park Service Photographic Collection, Harpers Ferry, WV.)

Visitors in Sand Canyon walking on the road to Echo Park in the early 1950s. This photograph appeared in *This is Dinosaur,* published by Alfred A. Knopf in 1955, which was presented to members of Congress at the peak of the controversy. (Courtesy Martin Litton.)

The confluence of the Yampa and the Green Rivers at Echo Park. The sheer rock wall on the right is 800 foot high Steamboat Rock. When John Wesley Powell and his party camped across the river from the wall in 1869, they found that their voices rebounded off the massive rock face, giving birth to the name "Echo Park." Photo by George Grant, 1935. (Courtesy National Park Service Photographic Collection, Harpers Ferry, WV.)

A view of the south side of Steamboat Rock from Pool Creek Canyon at the entrance into Echo Park. George Grant photo, 1930s. (Courtesy National Park Service Photographic Collection, Harpers Ferry, WV.)

Above: Secretary of the Interior Oscar Chapman, left, greets National Park Service Director Newton Drury in December, 1949. Their cordiality masks their growing disagreement over the Bureau of Reclamation's proposed dams in Dinosaur National Monument, a disagreement that became more pronounced in the following year, and ultimately led to Drury's resignation in April, 1951. (Courtesy National Park Service Photographic Collection, Harpers Ferry, WV.)

Bernard DeVoto in Portland, Oregon, January 1954. DeVoto's article in the *Saturday Evening Post* in July, 1950, helped launch the campaign against the dam. (Photo by Carl E. Vermilya of *The Oregonian,* courtesy Department of Special Collections, Stanford University, Stanford, CA.)

Above left: Fred Packard, executive secretary of the National Parks Association. (Courtesy National Park Service Photographic Collection, Harpers Ferry, WV.)

Above right: Howard Zahniser, executive secretary of the Wilderness Society, in a photograph taken shortly after the Echo Park battle. Zahniser was instrumental in winning support from key members of Congress for provisions in the upper basin project bill that guaranteed that no dams could be erected in the national park system. (Courtesy Edward Zahniser.)

On a river trip through Dinosaur in July of 1955, Joe Desloge studies Backus
Rapid in the Canyon of Lodore. Photo by Otis Marston. (Courtesy The
Huntington Library, San Marino, CA.)

Boats dwarfed by the Tiger Wall in Yampa Canyon in 1952. Photo by Harold Bradley. (Courtesy Richard Bradley.)

Stephen Bradley floating the Yampa River in 1952. Bradley's father Harold, who took the photograph, shot a film in the canyons of Dinosaur that summer, which helped to inspire interest in the Echo Park controversy among members of the Sierra Club. The Club subsequently arranged for several river trips through the monument in 1953. (Courtesy Richard Bradley.)

Stephen Bradley on the Green River underneath Steamboat Rock in Echo Park in 1952. Photo by Harold Bradley. (Courtesy Richard Bradley.)

A panoramic view of Steamboat Rock, the Green River and Echo Park, from the south. The proposed Echo Park dam site was located one mile downstream from Echo Park in Whirlpool Canyon, just out of view from this vantage point. (Courtesy National Park Service Photograph Collection, Harpers Ferry, WV.)

Eight-hundred foot high Steamboat Rock looms over Echo Park. Approximately two thirds of the monolith would have been under water if the dam had been built. Note the automobile and haystack at lower left. Photo by Park Service ranger Harry Robinson, 1954. (Courtesy National Park Service Photographic Collection Harpers Ferry, WV.)

Opposite, top left: Richard Bradley, who helped David Brower successfully challenge the Bureau of Reclamation's evaporation statistics for Echo Park Dam. Bradley, then a physics professor at Cornell University, was joined in defending Echo Park by six of his brothers and their father, Harold Bradley. (Courtesy Richard Bradley.)

Opposite, top right: Harold Bradley in Madison, Wisconsin, 1943. After his retirement from the University of Wisconsin a few years later, Bradley moved to Berkeley where he became active in the Sierra Club, and was instrumental in bringing the Club into the Echo Park fight. (Courtesy Richard Bradley.)

David Brower, executive director of the Sierra Club at the time of the Echo Park controversy. Brower took a leading role in the national campaign against Echo Park Dam, and became renowned among many conservationists for his energy, conviction, and bold style of leadership. (Courtesy Sierra Club.)

Left, and above: Sierra Club float trips through Dinosaur in the summer of 1953. (Courtesy National Park Service Photographic Collection, Harpers Ferry, WV.)

A river trip through Dinosaur in 1971. (Courtesy National Park Service
Photographic Collection, Harpers Ferry, WV.)

Opposite, above: Wallace Stegner, professor of literature at Stanford University
in 1961. Stegner edited *This Is Dinosaur,* considered the first fighting book of
the postwar wilderness movement. (Photograph by Stanford University, courtesy
Department of Special Collections, Stanford University Libraries.)
Opposite, below: Richard and Dorry Bradley with their two children, Helen and
Richard, Jr., in Split Mountain Canyon. Harold Bradley took the photograph in
1956, on a float trip celebrating the defeat of Echo Park Dam. (Courtesy
Richard Bradley.)

Glen Canyon Dam and Lake Powell in northern Arizona and southern Utah.
(Courtesy National Park Service, Dinosaur National Monument.)

Searching
for an Alternate Site

The furor over Newton Drury's resignation proved to be the opening chapter in the Echo Park controversy. Drury's departure as director of the National Park Service drew attention to the threat to Dinosaur National Monument and various national parks, as well as to the subsidiary question of national recreation areas under NPS care. Though much disturbed by Drury's sudden resignation, conservationists were galvanized by the episode and quickly came together in a show of force to reaffirm the principle that national parks must be held inviolate. A wide variety of organizations joined in denouncing Drury's departure, and looked upon it as a symbol of threats to the national park system as well as of the Bureau's power at the Department of the Interior. They emphasized that by consenting to the dams, Secretary Chapman had opened a breach into national parks that economic developers could fully exploit, taking comfort in the knowledge that parks were not inviolable. They maintained that Chapman had established a dangerous precedent.

Yet if conservation organizations had gained public attention as well as momentum in the campaign to safeguard Dinosaur Monument, they had not yet fully worked out a strategy for the battle ahead, and this task became paramount in the aftermath of Drury's resignation. That western states held considerable power in the Senate made the need for a strategy particularly evident, for the Bureau of Reclamation had its political patrons there with Joseph O'Mahoney, Eugene Millikin, Clinton Anderson, and Arthur Watkins, who were determined to push the Dinosaur dams and the CRSP forward. As Senators from less populated western states with comparatively little interest in preservation issues, they had no reason to be concerned about the public uproar over Drury's resignation. The Park Service had its base of political support in other regions of the country, mostly urban areas in the East and Midwest, and western states eager to develop the Colorado River gave relatively little heed to the departure of the director of the NPS. With the Senate likely to take a favorable view of the dams, conservationists knew that they must gear their campaign toward members of the House of Representatives who represented their interests to a far greater extent. That meant developing a strong case for protecting Dinosaur and the national park system and demonstrating that the upper Colorado basin need not have dams in a national monument in order to establish beneficial use of their water.

Over time, opponents of the dams developed three main approaches to mount their campaign, and although the chapters of this book tend to separate them for purposes of narration, each of them came into play during the first phase of the battle. First, conservationists knew they must promote the scenic beauties of Echo Park and the canyons of Dinosaur and make them known to the public. Beginning in 1950, leaders of the Wilderness Society, the National Parks Association, and other groups began to visit the monument, and pictures and articles soon appeared in conservation journals like *Living Wilderness* and *National Parks Magazine*. In 1951 and 1952, the Bradley family made their first river trips down the Yampa and Green rivers, which helped promote the sport of river running and prompted the Sierra Club to enter the battle. These forays to Dinosaur Monument by the Bradleys, Devereux Butcher of the National Parks Association, Olaus and Margaret Murie of the Wilderness Society, and others, proved critical to awakening the public to this vast and beautiful preserve, and the publications and films that came forth helped transform the largely unknown monument into a symbol of the nation's wilderness.

A second major tactic of the campaign was the forging of partnerships between conservationists and other political forces which opposed some or all of the upper-basin project, and this eventually proved vital to blocking Echo Park Dam. There can be little doubt that without allies not even a fully united conservation movement would have had the power to force the Bureau and western members of Congress to jettison Echo Park Dam. Such allies emerged with the Army Corps of Engineers, which looked askance at the Bureau's big projects like the CRSP; water and power interests in southern California who feared that their own use of the Colorado River would be compromised; and members of Congress from farm states in the Midwest and South, who abhorred the prospect of adding to the nation's agricultural surpluses. A host of issues involving the cost of the CRSP, political and legal interests of lower Colorado basin states, and the concerns of farm states all played an important role throughout the controversy, and debate over them helped defenders of Dinosaur delay the project in Congress.

Finally, conservationists found that they must confront the Bureau of Reclamation on its own turf by suggesting alternate designs of the CRSP that would eliminate the dams in Dinosaur Monument. Undoubtedly the most difficult of the three prongs of their campaign, the risks involved with challenging the Bureau's preferred arrangement of dams soon became clear: few of the opponents had ever confronted a water and power project on this scale and most of them lacked the technical expertise needed to counter the Bureau's arguments. None had ever questioned the Bureau quite in this way or sought to challenge its experts so directly. If their proposed alternate dam sites should prove unworkable or impractical, the entire campaign could be discredited. Luna Leopold, chief hydrologist with the US Geological Survey, advised against challenging the Bureau's proposals and avoid suggesting alternate sites. Leopold's technical background and knowledge of the Bureau's ways of operating made him a respected adviser from inside the government. He thought it might be quite risky to question the details of specific sites, evaporative rates, and other technical issues. The Bureau's staff of experts had years of experience in determining which dams were needed to fulfill certain functions, and if they could show that alternate proposals were ill conceived it might discredit the conservationists' case. Providing advice to conservation leaders from behind the scenes, Leopold urged that they focus on the encroachment of dams and reservoirs into Dinosaur Monument.[1]

Leopold fully realized that in addition to the Bureau's political strength on Capitol Hill, the agency's power derived from its control over factual data through its professional staff of experts, its hydrologists, engineers, economists, and geologists. Trained specialists from the Bureau had studied the Colorado River for three decades and possessed a wealth of statistics on the merits of various dam sites, water flows, power demands, and financing. Their plans for the upper Colorado River had been worked out carefully over many years, taking into consideration the Colorado River compacts of 1922 and 1948, the legal requirements to supply Mexico, and the demands of individual states for irrigation and power. In short, a legal and political framework was firmly in place which set the terms of discussion of the Colorado River.

Perhaps it is too much to say that the Bureau had a monopoly on "the facts," but the engineers and hydrologists held a great advantage throughout the controversy because they had been trained to devise blueprints for efficient management of rivers and they had invested years of study on the upper Colorado River. Given the Bureau's ironclad grip on hydrologic data, it is little wonder that proponents of the dam near Echo Park scorned the opposition as impractical and sentimental "nature lovers," with little understanding of the American West's reliance on water or its need for power. The public in the upper basin looked down on any challenges to the Bureau and scoffed at those who raised questions. Supporters felt confident that well-educated experts had thoroughly studied the upper Colorado watershed and had devised a plan most suited to its topography and climate. Common sense suggested that since the Bureau had already studied all the options, it was ludicrous to consider alternate sites.

Not surprisingly, suggestions for alternate dam sites went nowhere during the first stage of the controversy. In February 1949, a Park Service official report urged that locations outside the monument be utilized, but the Bureau ignored the report and hardened its determination to win approval of Echo Park and Split Mountain dams. A year later, Newton Drury made the same appeal to Secretary Chapman at the public hearing in Washington, but Chapman brushed it aside as so much special pleading. Other suggestions for alternate sites had similar results. In November 1950, William Voigt, Jr., director of the Izaak Walton League, urged Chapman to scrap the dams and encourage power generation from the vast reserves of coal and oil shale available in western Colorado. According to Voigt, energy firms had already begun to investigate fossil fuels in the region.[2] Chapman again resisted, this time with the claim that such sources would not yield power for many years, a span of time unacceptable to the upper basin and the nation.

Other suggestions for alternate sites came from quite different political motives, and they too were easily rejected. Colorado Senator Edwin Johnson proposed replacing Echo Park and Split Mountain dams with a dam at Juniper Mountain on the Yampa River, east of Dinosaur Monument in his home state. Johnson, to be sure, was less worried about the integrity of the national park system than about the fact that Echo Park, although located in Colorado, stood only five miles from the Utah border, which made it likely that Utah would reap most benefits from the dam.[3] Johnson wanted a large storage dam well within Colorado's boundary to satisfy water interests in the western half of his state who had been battling the Front Range cities over water for years. Chapman and the Bureau, aware of Johnson's motives, did not agree that the Juniper Mountain site provided an adequate substitute for Echo Park and Split Mountain.

No suggestion for alternate sites was likely to succeed unless it took into account the multiple purposes and technical advantages of the Dinosaur dams, particularly Echo Park. That meant considering a range of factors, including storage and power capacities of various dams as well as their effectiveness as "river regulators." Only by examining the place of the Dinosaur dams in the CRSP could an effective challenge be mounted, and doing so required knowledge of multiple-purpose water projects. One person with the needed expertise happened to be the president of the American Planning and Civic Association, an organization active in the campaign. He was General U. S. Grant III, grandson of the famous general and president, and a former staff member with the Army Corps of Engineers. Grant had been with the Corps in a number of positions during his forty-year career, had served as consulting engineer on several dams, and had the knowledge of water projects to enable him to look closely at the Bureau's intentions along the upper Colorado.

Besides being an old hand in hydrology and river-basin management, Grant had been schooled by the often fierce rivalry between the Army Corps and the Bureau of Reclamation, competitors on western rivers and on Capitol Hill. The Army Corps had its own powerful constituency interested in dams and levees on navigable waterways, and the Corps competed with the Bureau of Reclamation on several western rivers. Since they also competed for funding from Congress, the two looked skeptically on one another's plans. The two agencies had fought bitterly over plans for damming the Missouri River during World War II, and though finally agreeing to the so-called Pick-Sloan Plan (after being threatened by President Roosevelt's proposal for a Missouri Valley Authority), they continued to eye one another's projects warily.[4] The Army Corps particularly disliked the Bureau's reliance on hydropower sales to subsidize irrigation, and that fact proved to be critical to the Echo Park battle.[5]

A veteran of these conflicts, Grant went to work examining the CRSP, carefully looking at its cost-benefit ratios. As he studied the Bureau's blueprint for the upper Colorado River, he realized that alternate combinations of dams and power plants could be utilized that would eliminate the need for Echo Park and Split Mountain dams. Grant first raised the possibility of alternate sites when he spoke at the hearing before Oscar Chapman in April 1950. Then, at the encouragement of Howard Zahniser, he published a fuller statement of his ideas in *Living Wilderness,* in the fall of 1950.[6] His article, while not as eloquent or dramatic as Bernard DeVoto's piece in the *Saturday Evening Post,* proved equally important in delaying the dams. Secretary Chapman later reversed his decision, largely due to Grant's efforts and criticisms from the Corps of Engineers.

Grant made his case by drawing on the Bureau's own data from its "Natural Menace" report of 1946. Of the eighty-one potential sites indicated in the report, ten had been included in legislation introduced for the Colorado River Storage Project in 1950. Grant knew that the Bureau sought to build Echo Park Dam in the first stage of the CRSP (Split Mountain Dam in the second), along with Flaming Gorge, Glen Canyon, Martinez (later renamed Navajo), and Curecanti dams.[7] Having the biggest dams in the first stage would accumulate maximum storage and ensure the most efficient "river regulation." Upper-basin states could then be certain of meeting their delivery requirements under the Colorado River Compact and could proceed to use their own share by means of participating projects like the CUP. Moreover, full reservoirs at Echo Park and Glen Canyon would provide maximum power production, which the Bureau preferred in order to begin earning revenues for completing some of the participating projects and helping fund the 1.5 billion-dollar CRSP. The Bureau estimated that the elimination of Echo Park and Split Mountain dams from the CRSP would mean an annual loss of 18 million dollars.[8]

As Grant examined the basic workings of the project, he came to believe that adequate storage and power could be provided without the dams in Dinosaur. He suggested eliminating Echo Park Dam from the first phase of the CRSP and substituting dams at Cross Mountain on the Yampa River, east of Dinosaur Monument, and another at Gray Canyon on the Green River to the south. Together, those two dams offered 800,000 more acre-feet of storage than Echo Park, and 153 million kilowatt hours more power per year.[9] By erecting Cross Mountain and Gray Canyon dams, Dinosaur Monument could be spared and storage and power increased in the first phase. Gray Canyon and Cross Mountain dams would increase the cost of the first phase because two separate construction projects would be undertaken, but Grant claimed that the Bureau could recuperate those costs and ultimately save money on the entire project in phase two. In order to accomplish this, he suggested replacing Split Mountain Dam in the second phase with three others, one near Bluff, Utah, on the San Juan River; a second called New Moab, in central Utah along the main stem of the Colorado; and Desolation Dam on the Green River, to the south of Dinosaur. Under Grant's scenario, no dams need exist in Dinosaur, but the CRSP would store 1.13 million acre-feet above and beyond the original plan, generate 181 million more kilowatts annually, and save the Bureau over fifty-nine million dollars in construction costs.[10]

Grant wanted to challenge the notion that the CRSP depended on Echo Park and Split Mountain dams to be successful. "The important fact to demonstrate and emphasize," he wrote, "is that it is not necessary to invade the national park system in order to realize the objectives of the Colorado River storage project." As for its refusal to consider alternate sites, Grant presumed that "the Bureau of Reclamation has already made site surveys in the national monument and has done other preparatory work . . . to such an extent that it does not want to make similar studies of alternative sites."[11] He was irritated by the not so subtle claim that the upper basin could only claim its rightful share of the Colorado River by means of the two dams, although he realized that such rhetoric had great appeal because it played on fears about the lower basin's

Proposed Reservoir Sites of the Upper Colorado River Basin

insatiable appetite for water. General Grant did not suggest that the Bureau consider a higher dam at Glen Canyon as a substitute for Echo Park, though later this idea became the prime alternative and highly controversial in its own right.

Grant may have been correct that other dams could provide equal or better storage and power—that the technical advantages of the Dinosaur dams were not so great that they could not be replaced—but the political interests supporting them could not be easily overcome. From Utah's point of view, there could be no substitute for the Echo Park site because no other location could capture water from two rivers and allow the state to make good on its Yampa River right obtained under the 1948 Compact. Without the CUP, Utah saw no practical way to capture Colorado River water, and water and power interests in the state stoutly resisted any redesign of the CUP that eliminated Echo Park Dam. As for the other states, particularly Wyoming and Colorado, Grant's alternatives were unattractive because Echo Park Dam would be halfway between Salt Lake City and Denver at the center of the upper-basin power market, an advantage that the Bureau did not like to emphasize as much as its low rate of evaporation. New Mexico also disliked Grant's suggestion to replace Navajo Dam, on the San Juan River, with one near Bluff, Utah. Senator Clinton Anderson was determined that Navajo Dam be included in the CRSP.[12]

Here, then, was the major challenge facing conservationists throughout the controversy, of separating the dams at Echo Park and Split Mountain from the overall project. Because the Bureau had made both of them important parts of the CRSP, questioning the need for either of them was analogous to questioning the need for different parts of an automobile motor. As the controversy unfolded, the Bureau liked to call Echo Park Dam "a piston in the engine," a phrase which obviously made it difficult to suggest that it be removed. Furthermore, while opponents had no choice but to consider the place of those two dams within the larger CRSP, they also found that by examining the overall project they risked appearing to be against the upper basin's legitimate water demands; those fighting to protect Dinosaur had no desire to gain such a reputation. Instead, they sought to make clear their support for the upper basin's interest in utilizing its Colorado River water, and said they would go along with the CRSP as long as it did not threaten the national park system.[13]

Utah Senator Arthur Watkins took the lead in denouncing Grant's proposals. In a Senate speech in March 1951, Watkins rejected alternate sites and dismissed Grant as a nature lover with little knowledge of the Colorado River. To his delight, Watkins found Grant's suggestion for utilizing the New Moab site to be flawed, since its reservoir would enter a portion of Arches National Monument, a point Grant discovered a few months after his *Living Wilderness* article appeared.[14] Taking full advantage of the error, Watkins charged that several of Grant's alternate sites had flaws and that the general had insufficient qualifications to speak on the subject. Their exchange proved to be the first of several angry encounters during the next three years.[15]

After Grant realized the problem with the New Moab site, he suggested the Dewey site on the Colorado River in eastern Utah—upstream from New Moab. Defending himself in the American Planning and Civic Association's journal, Grant then accused Watkins of ignorance about the various sites. Watkins had also criticized Grant for suggesting Desolation Dam, on the grounds that its storage capacity was too small. But according to Grant, the Bureau's own figures indicated that Desolation could store 7.7 million acre-feet more than Echo Park and Split Mountain combined.[16]

Grant also questioned the Bureau's reliance on low evaporation rates to justify the Echo Park site, and this gave rise to one of the most famous aspects of the controversy. While David Brower of the Sierra Club has usually been credited with challenging the evaporation figures, Grant did so long before Brower, and Brower later relied on Grant's figures. The Bureau had estimated that any restructuring of the CRSP that eliminated Split Mountain and Echo Park dams would mean an annual loss of 350,000 acre-feet by evaporation, a significant amount in a chronically water-short region. This potential loss had influenced Secretary Chapman's initial decision in favor of the dams in June 1950. For his part, Grant maintained that scientists had barely begun to study evaporation, and he took satisfaction when Chapman eventually conceded the point by referring to it as "a hydrologic mystery."[17] While the figure of 350,000 acre-feet had obvious appeal to supporters of the dams, Grant considered it "just a guess, a nice big round figure to help the sales argument, and apparently not correct even based on the Bureau's own method of estimating." Grant's suspicions grew after he caught an error in subtraction. The Bureau's estimate of

350,000 additional acre-feet of evaporation was based on the alternative plan to replace Echo Park and Split Mountain with Desolation Dam, but Grant pointed out that if this were done the combined evaporation of the two dams in Dinosaur of 102,000 acre-feet must be subtracted from the 350,000, leaving a difference of 249,000 acre-feet. Grant wondered if the failure to subtract had been deliberate in order to justify the Echo Park site. That may have been the case, although it cannot be proven. In any event, more than two years later David Brower found the same mistake.[18]

Grant's criticisms of evaporation and his suggestions for alternate sites brought action from Oscar Chapman, who had been sharply criticized by conservationists for approving the Dinosaur dams and had been yearning to appease them. A few months after Grant's article appeared in *Living Wilderness* in the fall of 1950, Chapman became embroiled in the controversy over Newton Drury's removal. With Grant continuing to press for alternate sites, Chapman sought to appease the opposition by asking the Park Service to undertake a study of possible locations outside of Dinosaur, and he asked Grant to help. The NPS report appeared in June 1951. It criticized the emphasis on evaporation to justify the Dinosaur dams, and suggested that such loss of water would have little effect on the total amount available in the upper basin or on the upper basin's ability to deliver water downstream. The Park Service report also accused the Bureau of being overly optimistic in judging the market for hydropower.[19]

Bureau Commissioner Straus lost no time in denouncing the report. He said the study had been done hastily and exhibited little knowledge of the Colorado River Compact or of the multiple purposes of the CRSP, and it ignored key studies on evaporation by the Federal Power Commission and the USGS. "Much of the analysis and many of the conclusions," he wrote Chapman, "are not only misleading but contain very serious errors." He assured the secretary that the Bureau had studied alternate sites thoroughly and expressed confidence that "there are no alternatives in the Upper Basin above Glen Canyon having merits equal to the Echo Park and Split Mountain units. The question of satisfactory alternatives is a myth," he declared.[20]

Public opinion throughout the upper basin agreed with that assessment. An editorial in the *Denver Post* insisted that "proper development of the river requires all . . . of these dams and others besides. . . . Unless the engineers can be proved wrong—and they have not been proved wrong so far—any opposition to Echo Park Dam amounts to opposition to the whole river development and must be judged on that basis."[21] Here, again, was the difficulty in challenging the Bureau's experts, of questioning the need for a "piston in the engine." But Grant was on the right track. Evaporation from reservoirs was indeed little understood, a fact that became clearer in 1954, when Richard Bradley and David Brower studied the matter at greater length. Grant had touched on one of the weak points of Echo Park and Split Mountain dams. For the moment, Michael Straus and other Bureau spokesmen defended the evaporation rates and succeeded in keeping suggestions for alternate sites at bay. Straus pointed out that the Bureau had no choice but to guard against evaporation because the Upper Basin Compact of 1948 charged such losses to each state's apportionment.[22]

For the time being, Grant's remarks about evaporation meant less than his criticisms of the Bureau's cost-accounting methods, and this proved to be the most important contribution he made to the campaign to save Echo Park. He asserted that the Bureau had not adequately explained its estimate of the entire project's cost, and argued that cost overruns had plagued the Bureau in the past. For example, the Bureau originally estimated the Colorado–Big Thompson Project at forty-four million dollars, but the final cost soared to two hundred million dollars.[23] Another problem with the Bureau's estimates centered on its reliance on hydroelectric power sales to fund irrigation projects. The subsidy of irrigation by hydropower had been a common practice of the Bureau since the early 1900s, but bringing irrigation water to semiarid lands in Wyoming, Utah, Colorado, and New Mexico promised to be extremely costly, so the Bureau had little choice but to establish high power rates at its big dams. Grant questioned the expenditures and the way that the Bureau accounted for its costs. In many cases, the Bureau could charge a certain percentage of the cost of a storage dam to irrigation, if the dam in question provided irrigation water directly. Echo Park did not, prompting Grant to question its costly power.[24]

Grant was not alone. Dale Doty, assistant interior secretary for public land management, also recognized the problem with expensive hydropower. In January 1952, Doty wrote a long memorandum to Chapman which sharply criticized the cost accounting of the CRSP. He pointed out how the Bureau allocated 30 percent of the total cost of the CRSP to irrigation and the consumptive use of water, but Doty said that such an amount could not be justified because "few of the units will provide any direct irrigation facilities." In particular, Glen Canyon and Echo Park dams were not designed for supplying irrigation water, but strictly for holdover storage and power. As for their cost accountability, "these units should stand on their own feet as power projects exclusively."

Doty maintained that if that were done, then Glen Canyon Dam was acceptable because it could pay for itself with its power rate set at 5.5 mills, but Echo Park and Split Mountain dams could not pay for themselves within the required repayment period. Department of the Interior policy called for amortization of its dams— that is, complete payment with 3 percent interest—within fifty years. In order to fulfill this requirement, power generated at Echo Park and Split Mountain must be sold at 7 mills, and Doty considered this unlikely because power from conventional steam plants sold for 6 mills. Therefore, power from the Dinosaur dams might not be marketable. According to Doty, "if the Congress did not authorize [them] the local area and region would lose what appears to be . . . marginal or submarginal power projects and nearby towns would not enjoy a large construction boom, but the power needs of the region probably would be provided for adequately by more economical projects." Doty urged Chapman to approach the Bureau and ask that the CRSP be redesigned so that each separate project could be shown to be cost effective. In effect, he recommended against Echo Park and Split Mountain dams. While recognizing the delivery requirements of the Colorado River Compact and the upper basin's interest in using its own half of the river, he questioned the need for all the proposed storage and maintained that the Bureau was overdeveloping the river. He claimed that upper-basin demands for irrigation and other consumptive uses were not enough to justify huge storage dams.[25] Such remarks from Grant and Doty carried great weight, and they also foreshadowed criticisms about the CRSP from conservationists and members of Congress during the 1954 and 1955 sessions, when the legislation came under careful scrutiny.

The debate over costly power naturally rankled upper-basin supporters of the CRSP. In the first place, it touched on a very sensitive nerve—the charge often made that farmers and ranchers merely wished to "take a free ride at the expense of power users," as John Geoffrey Will of the Upper Colorado River Commission paraphrased it in a speech before the Colorado Association of REA cooperatives. Will insisted that this was not the case and that, under the terms of the legislation, irrigators must pay "to the limit of their ability." Will also denied Grant's charge that the cost of power had been set artificially high in order to cover unusually expensive irrigation costs, maintaining that the price of the power would be the same even if "irrigation was not involved."[26]

Still, raising the matter of hydropower costs provided fuel to other critics who had begun to question expensive river-basin projects and who took special aim at the Bureau's reliance on power sales to subsidize irrigation.[27] That debate, in turn, went to the heart of the growing wrangling between the Bureau and the Army Corps of Engineers. After the war, despite a generally favorable economic and political climate that boded well for both the Army Corps and the Bureau, the Hoover Commission's Natural Resources Task Force had emerged and had begun to take a critical look at expensive river-basin projects of both agencies.[28] The Hoover Commission had accumulated plentiful evidence of cost overruns, noting that in the Bureau's case, power sales could not pay all costs and taxpayers were left to make up the difference.[29] Not surprisingly, then, Grant's charges about costly power of the CRSP dams caught the attention of the Army Corps of Engineers, and that agency had the power to influence the progress of the CRSP through its own channels at Interior and on Capitol Hill. Once the Army Corps began to raise serious questions about the CRSP, Oscar Chapman realized that he must delay the project—and eventually he began to reconsider his earlier approval of the Dinosaur dams.

Perhaps no single issue divided the Bureau from the Corps as much as the Bureau's cost-accounting methods. The Corps detested the Bureau's reliance on power sales to subsidize irrigation projects, primarily because it gave the Bureau an advantage in acquiring appropriations from Congress. The Corps disliked the fact that the Bureau sought to maintain its image of helping small farmers in the American West—which, in and of itself, had become a fiction in some places—and did so by charging the cost of irrigation projects to hydropower. From the point of view of the Army

Corps, the Bureau was intent on attaining an ever-increasing share of monies from Capitol Hill and spending it on projects of questionable economic value.[30]

In this case, the Army Corps felt that the CRSP had plenty of weaknesses. Major General Lewis A. Pick of the Corps questioned the economic justification of the upper Colorado project, and charged that the Bureau relied too much on so-called secondary benefits of the CRSP, such as increased employment resulting from new industry. Further, Pick said that by subsidizing irrigation through power sales from the big dams, the Bureau masked the costly irrigation projects as well as the expensive power. In a letter to Michael Straus, Pick charged that the Bureau provided inadequate data to support the cost/benefit ratios claimed.[31] Straus lost no time in defending the CRSP and in attacking the Corps for being a hog at the public trough. He knew the Army Corps did not fund its dams and flood-control levees through power sales and depended on direct congressional appropriations, so he could denounce the Corps for building an overabundance of projects at the expense of taxpayers. Straus pointed out that the CRSP would be funded through power sales and would be less burdensome on the public than projects of the Corps. "Since every dollar spent on the construction of the storage project as well as the particular irrigation projects will be returned to the Treasury of the United States, as compared to no return from most flood control and navigation projects, your statement is not understandable and appears to be completely without logic," he told General Pick.[32]

But the damage had been done. Charges from Grant, Doty, and the Corps all had their effect on the one person who controlled the fate of the CRSP and the Dinosaur dams: Oscar Chapman. Sometime during the latter part of 1951, Chapman finally concluded that little good had come from his decision to approve those dams. A host of conservation groups had become enraged, and their anger had grown with Drury's departure from the Park Service. Now Grant and the Army Corps had raised critical questions about the dams and the financing of the CRSP, and Chapman knew he could not dismiss them lightly, especially with the Korean War putting tremendous pressure on the national budget.

Senator Watkins had introduced legislation for the CRSP in 1950, but when the nation became embroiled in Korea it became clear that the CRSP could not be enacted because of budgetary restraints.[33] In 1951 Congress refused to fund water projects unless they were already under way. Upper-basin politicians did not like to have to tell their constituents that Echo Park Dam and the CRSP must be delayed, so Watkins and his colleagues asked that CRSP be authorized, a necessary first step in order for the Bureau to proceed with detailed engineering plans. Large projects like the CRSP took years to plan, design, and construct, and once Congress authorized a project, it became a part of the Bureau's long-range planning; design engineers and hydrologists could then assemble the complex blueprints in preparation for the actual construction of power plants and dams. Because of the long time required to plan and implement the projects, the Bureau had to develop its construction accordingly, coordinating its own engineering plans with the time lag certain to come.[34]

While Watkins sought authorization of the bill, he continued to work to gain full approval from Chapman and the Bureau of the Budget. Chapman had yet to clear the CRSP through the Department of the Interior and send it along to the Bureau of the Budget for its approval, and throughout 1950 and 1951, as evidence mounted about financial weaknesses of the CRSP and as Grant continued to harp about alternate sites, Chapman withheld final approval from Interior. With the Democratic party trying to maintain its reputation in the West as provider of federal water projects, Chapman felt he could not risk having a large project like the CRSP shelved by Congress.

Taken together, restraints on the budget and Chapman's delays made upper-basin residents more than a little restless about their dams. By late 1951, speculation grew that Chapman had changed his mind about Echo Park and Split Mountain dams. John Will, secretary of the Upper Colorado River Commission, complained about the delays repeatedly and sent a resolution to Chapman on behalf of the commission, urging him to send the report forward to the Bureau of the Budget. Will later met with the secretary and won assurance that he was not wavering on the two dams.[35] In November, 1951, Chapman told a reporter from Salt Lake City that he would "abide by my decision reached last year . . . unless suitable alternative projects can be worked out." [36] Still, it became clear to supporters of the Dinosaur dams that Chapman was reconsidering his earlier decision. Later that same month, Chapman remarked in a speech to the National Audubon Society that he was launching another search for alternate sites with the hope "that we might work out a solution whereby the Split Mountain and Echo Park dams need not be built in the monument." After his speech, Chapman established a task force to examine alternate sites, and called on Grant to assist with the investigation.[37]

Elmo Richardson points out that Chapman's willingness to support this search also grew out of the altered political situation that year. Utah's Democratic Senator Thomas had been defeated in his reelection bid and had been replaced by a Republican, diminishing President Truman's interest in the CRSP. With the administration appearing to lose enthusiasm for the project, Democrats from the upper basin now became exceedingly anxious. Reva Beck Bosone, a member of the House from Utah, held a meeting in Salt Lake City, late in 1951, to assure residents that Chapman had not changed his mind on Echo Park Dam, and then sent the secretary a reminder that he had an obligation as a good Democrat to follow through with his approval of the project.[38] From the upper basin's standpoint, Chapman's delays made it impossible for Congress to consider any legislation because lawmakers first had to obtain approval from the Department of the Interior and the Bureau of the Budget.[39]

Chapman's primary reason for holding up approval of the CRSP was because of criticisms from Grant and the Army Corps about the project's expense and its questionable cost accounting. In February 1952, Chapman found himself campaigning for presidential hopeful Adlai Stevenson in Salt Lake City. Asked why he had delayed final approval of Echo Park and the CRSP, he assured the local reporter that it had little to do with those worried about the sanctity of the national park system. "I can take care of the Park Service," he said, "because it is in my department. "The opposition I am concerned about is that from the Army Corps of Engineers. I don't want the report to go to Congress [with their opposition] because I don't think it could get through in that condition."[40] Chapman may have been trying to sound certain that his position did not reflect any fear of the conservation forces, but he could not deny that the momentum behind the Echo Park Dam project had slowed down considerably from two years before. Grant's criticisms had hit their mark, and conservationists had used the opportunity offered by Chapman's delays to continue to rally their forces.

Senator Watkins reintroduced legislation for the CRSP in the Senate in April 1952, still aware that it had no chance of passage without approval from Interior and that Congress had other pressing matters before it in any case. However, simply offering the legislation covered him with constituents back home, and also ensured that Echo Park Dam remained an important political issue in the 1952 election in his state. Watkins wanted to be able to blame the Democratic administration for delaying a vitally important water project.[41]

In truth, the holdup on the CRSP could not be attributed to partisan politics or budgetary restraints alone. Remarkably enough, upper-basin states themselves still had not united on the legislation, another reason that Watkins's bill "went nowhere."[42] New Mexico's Senator Clinton Anderson could not support the legislation because of the tangled water situation in his state. Northwest New Mexico belonged to the watershed of the upper Colorado River Basin, and the Navajo people sought irrigation from the San Juan River. Albuquerque, with a growing population attracted by the defense industry, also wanted water from the same river to be sent by transmountain diversion across the Continental Divide. Because their conflict had not yet been worked out, Anderson could not support the Watkins bill, and the legislation stood no chance of passage without his backing. Nor did Wyoming Senator Lester Hunt support the Watkins bill, saying that a majority of his constituents demanded lower federal expenditures.[43] This lack of agreement within the upper basin proved fatal to the bill.

By this time, Oscar Chapman had accumulated numerous memorandums and reports from the Bureau and the Park Service, and he realized that the two sides were no closer to resolving the dispute over Echo Park than they had been when he had first become secretary. After months of contentious infighting at the Department of the Interior, Chapman finally took action. In December 1952, he sent the Bureau's report on the upper-basin project to the Bureau of the Budget and reversed his earlier approval of Echo Park and Split Mountain dams. Because the Bureau had not requested authorization of Split Mountain in the first phase of the CRSP, Chapman urged that it be deleted from the project altogether. Then he added, somewhat ambiguously, that "in lieu of the recommendation of the Echo Park unit, I recommend that the Secretary of the Interior be authorized to construct and operate facilities, at Echo Park or at an alternative site, to serve the purposes intended to be served by the Echo Park unit." The statement was not quite the unequivocal reaffirmation of the sanctity of the national park system that many conservationists had wanted, though, in Fred Packard's words, it was "a step in the right direction." After three years of contentious debate between the Park Service, the Bureau of Reclamation, and a host of upper-basin business groups and national conservation organizations, the dams in Dinosaur had been put on the shelf.[44]

Chapman had done so because of the cost of power at Echo Park, not because he had suddenly developed an aesthetic appreciation of Dinosaur Monument. With the Corps harping about the high cost of the CRSP and questioning the Bureau's cost/benefit ratios, and with Grant suggesting an alternate plan of construction, Chapman decided he had no choice but to alter the CRSP in some way in order to make it more acceptable. In addition, his original justification for Echo Park Dam—to create power for the defense industry—was no longer tenable, as the Defense Department had changed its plans for added power reserves in Utah.[45] Pressure from the Army Corps had especially been crucial, and in this sense, the Echo Park controversy involved more than a clash between preservationists and the forces of water development: it also brought to the forefront the conflict between the two great rivals of water development in the American West after World War II.

By reversing his earlier decision, Chapman removed the Dinosaur dams from the Interior Department's recommended plan, and left it up to the next administration and secretary of the interior to reconsider the issue. In 1953, as the Eisenhower administration took power in Washington and began to formulate its own water policy, upper-basin states, led by Watkins, once again sought initial authorization of the legislation in Congress, still mindful of the inevitable time lag between the design and actual construction. If Echo Park, Glen Canyon, and other main stem dams could be authorized now, the Bureau might be able to begin building them within three or four years.[46] Meanwhile, aware that the new administration felt obligated to study all options for alternate sites, Senator Watkins and his colleagues pressured the Eisenhower administration during its first few months in office to hasten its analysis of the CRSP.[47]

When Dwight Eisenhower became president in January 1953, the political atmosphere surrounding the Echo Park controversy appeared to change considerably, although the essential question remained: Would the Bureau be allowed to build its cherished dam in Dinosaur, and could its high-cost power be justified? Eisenhower, the first Republican to sit in the Oval Office since Herbert Hoover, symbolized the nation's general frame of mind in the 1950s, a contentedness with growing prosperity and conservatism that distrusted "New Deal–Fair Deal" programs. The administration appealed to conservatives and business interests who had long been hostile toward the Democrats' regulation of the economy. Eisenhower did not try to eliminate the welfare state entirely, but he took steps to lower corporate taxes and eliminate wage and price controls in place during the Korean War. In the words of one historian, Eisenhower "was intent on restraining federal regulation of, and competition with, the nation's private economy."[48]

The new administration also promised major changes in how the federal government managed water and power as well as natural resources on public lands. Ever since the New Deal, Republicans had sniped at Democrats for their willingness to exert federal control over natural resources. With so much public land in the American West, and with strong population growth in much of the region during the postwar years, Republicans understood that political capital could be gained by easing regulations on public lands as well as by reducing federal spending. Eisenhower and the Republican party sought to appease those westerners who had long felt burdened by regulations on grazing, logging, and mining on public lands.[49] Eisenhower ushered in a major change in federal resource policy that was part and parcel of greater economy in the federal government, calling for reduced federal regulations of public lands.

To implement such a policy, Eisenhower chose the retiring governor of Oregon, Douglas McKay, as his secretary of the interior, much to the delight of western states. McKay had a reputation of supporting business interests and of being dubious about Forest Service and federal conservation bureaucrats, a record that had won for him strong support in a state dependent on the lumber industry. His appointment won immediate applause from timber and mining interests in the Northwest, who proclaimed that the region's "natural resources may now be developed under American principles, and through the cooperation of the states."[50] In Elmo Richardson's apt phrase, McKay brought with him to Washington "a businessman's view of resource use," an approach that Eisenhower found appropriate for administering the public lands to ensure economic growth.[51] As McKay put it, in his first annual report as secretary of the interior, "we cannot continue to grow as a nation and as a prosperous people if we adopt the narrow view of conservation as requiring the locking up of our resources."[52] Here was an outlook in keeping with the booming growth in the postwar era.

McKay's appointment aroused suspicions among those who feared he would undermine well-established conservation policies. When McKay appeared before a Senate committee considering his nomination, he kept his answers to questions as vague as possible, raising additional fears about his commitment to sound land-use principles. Oregon's maverick Senator Wayne Morse warned that McKay would cater to "the tidelands thieves, the private utility gang and other selfish interests which place materialistic above human values." Bernard DeVoto, who had crusaded against the "landgrab" since 1947, feared that McKay possessed the same current of hostility against federal grazing regulations. DeVoto indicated that McKay's remarks at his confirmation hearings were "some of the most bizarre ideas that any Cabinet member has uttered in my time."[53]

As for the national parks, McKay reaffirmed the policy announced by Secretary Krug five years earlier—that there should be no relaxation of the safeguards protecting them unless it could "be proven unmistakably that it will produce for the nation values that outweigh greatly those which are [to be] changed or destroyed."[54] In some of his first decisions as secretary, McKay suggested that these were not mere words. To the surprise of some conservationists, McKay refused to accede to proposed reductions of Olympic and Everglades national parks, and supported the search for an alternate site outside Glacier National Park for the Bureau's Glacier View Dam.[55] Despite such actions, national park watchdogs like Horace Albright, Howard Zahniser, and Fred Packard doubted McKay's commitment and felt certain that he did not appreciate the relationship between parks and wilderness preservation. They suspected that he would surrender to commercial interests when the pressure became strong enough.

Still, if McKay caused conservationists to be wary, Eisenhower offered them hope with his dramatic shift away from federal support for large water and power projects. The new president quickly signalled his intention to reduce costly river-basin projects, and indicated that he took a dim view of valley authorities. While Eisenhower did not dislike federal involvement in water development altogether, he had an obligation to be economy minded, and he well knew that the electric-power industry had long been unhappy with publicly produced and marketed power, a legacy of the New Deal and its Tennessee Valley Authority. The TVA had been a sore point with private power companies for years, as it had reduced their own market and symbolized to them the worst features of an overly intrusive federal government. Eisenhower understood that he could not do away with the existing TVA, but he made clear that he did not support the establishment of additional valley authorities, something that Roosevelt, Truman, and various members of Congress had called for during much of the 1940s and early 1950s, notably on the Missouri and Columbia rivers.[56]

Above all, Eisenhower sought to encourage greater cooperation between the federal government and private enterprise in natural-resource policy, to arrange for "partnerships" in the damming of rivers for water and power. He wanted to encourage natural-resource development with due regard for "local interests and local knowledge of local needs. . . . We need river basin development to the highest degree," he said, "but not at the expense of accepting super government in which the people in the region have no voice."[57]

For defenders of Dinosaur, the new policy seemed to bode well. So long as the administration sought to create "partnerships" in developing natural resources, why would it consent to an expensive bill like the CRSP? A number of those engaged in the Echo Park campaign saw no reason why it should. With a Republican in the White House, the time of costly water projects might be coming to an end. Their hopes grew when the Department of the Interior began to analyze water and power proposals currently being considered. McKay eventually announced a "no new starts" policy on water and power projects and froze the Bureau of Reclamation's budget for fiscal year 1954.[58] Then, in a move that sparked a heated debate, McKay reversed a decision by the Truman administration that had blocked a plan by private power companies to build three dams in Hells Canyon on the Snake River in Idaho. The decision demonstrated Eisenhower's desire of encouraging private power projects, while it revived the debate from the New Deal over public and private power.[59]

McKay and the administration suddenly found critics in many

quarters, with certain politicians from the West among the strongest. Senator Richard Neuberger of Oregon, a Democrat, was enraged with the decision in favor of the Idaho power companies, and denounced it as an assault on federal reclamation and power policies that had benefited the West since the Newlands Act of 1902. In Neuberger's mind, all the talk about "partnership" masked the administration's interest in serving private power interests in the Northwest. Neuberger, a strong supporter of the Forest Service and the National Park Service, became a major opponent of the Echo Park Dam in the months ahead.[60]

Conservation leaders thought the decision on Hells Canyon boded well for how the administration would contend with the CRSP and the Dinosaur dams. While Eisenhower did not have to worry about the CRSP being a "public power" project, he could not overlook its huge cost—1.5 billion dollars in the initial stage and approximately 3 billion dollars to complete the project. Approval of such an expensive project did not seem likely, given his interest in reducing federal spending and keeping the budget under control during the Korean War. In May 1953, Ira Gabrielson wrote Secretary McKay, reminding him of the dubious financial merits of the upper-basin plan.[61] As for Echo Park Dam, Park Service Director Conrad Wirth made certain that McKay knew of its history and of Chapman's decision to reverse his approval due to its costly power.[62]

Richard Bradley, Sierra Club member and a physicist at Cornell University, thought that the Eisenhower administration would never consent to the costly CRSP, and he felt confident that Echo Park Dam would never be constructed.[63] Bradley had to wait several months to find out Eisenhower's decision, which Secretary McKay did not announce until December. Meanwhile, in the summer of 1953, Bradley joined nearly two hundred other Sierra Club members on a float trip down the Green and Yampa rivers through Dinosaur Monument, trips that proved instrumental in bringing the scenic wonders of Dinosaur into public view. General Grant's challenge of the CRSP and that of the Army Corps, although critically important in delaying the Dinosaur dams, did little to appeal to the public's interest in parks or wilderness. As we have seen, this national monument in the West had had little public exposure and remained virtually unknown within the national park system. During the postwar years, much of the Colorado River and Plateau was one of the least-known regions of the American West, and few conservationists involved with the Dinosaur campaign had ever seen Echo Park or the surrounding region. Most of them were defending a place unknown to them, but they realized that they could not hope to save Dinosaur from dams unless they made the monument known to more of the public. Grant, the Army Corps, and Chapman had all done their part to delay the dams, and proposals for alternate sites had become a major tactic of the conservationists' campaign. Now, with a new administration reconsidering the entire project, defenders of Dinosaur had an opportunity to visit this place that they were so passionately trying to save. As they did, they discovered a new kind of wilderness landscape that had tremendous appeal to the coming generation of outdoor lovers.

Wilderness for a New Generation

... Beyond this last resting place of ancient reptiles, in the heart of the monument, are vast, wildly beautiful river canyons that few people have seen.
                    —Philip Hyde, "Dinosaur Monument: Rocky Land,"
                          Christian Science Monitor, July 13, 1953

Since the Dinosaur monument was enlarged in 1938, no roads have been built, no improvements have been made. The scenery is available only to a few hardy souls willing to take a chance going over the rapids of the treachorous [sic] Green River.
                    —["Echo Dam Project Would Enhance Scenic,
                      Recreational Values of Dinosaur Park,"]
                        Salt Lake Tribune, January 31, 1950

Michael McCloskey, former executive director of the Sierra Club, in an essay published in the early 1970s, reflected on the history of the wilderness movement and argued that the Echo Park Dam controversy belonged to the first of three phases in the postwar years. Primarily defensive in nature, this initial period in the post-war movement was "concerned with defending previous gains by fighting off attacks on areas already reserved, principally in national parks and monuments." McCloskey concluded that, by the end of the controversy, wilderness preservationists had begun to take the offensive. Having gathered sufficient public support to force Congress to back down on the Echo Park Dam, they took a significant step toward protecting wilderness more permanently. The final CRSP legislation, signed by President Eisenhower in April 1956, prohibited dams in any part of the national park system, a move that strengthened the national commitment to the parks and affirmed the value of wilderness. The controversy, then, represented a milestone in postwar wilderness history, bridging the gap between an older, more defensive-oriented movement with a new generation of preservation, as symbolized by passage of the Wilderness Act of 1964.[1]

That transition over the course of the controversy can be seen in various ways, but becomes most evident in the tactics employed by conservationists to argue the case for protecting Dinosaur. When the Bureau's plan first became public, defensive-minded conservationists lost no time in meeting the threat to the monument. The National Parks Association, the Wilderness Society, and their allies immediately denounced the Bureau for daring to seek access to a part of the national park system, and accused it of deliberately attempting to weaken the park system. Bernard Devoto's article in the *Saturday Evening Post* provides an excellent example. The very title of the piece, "Shall We Let Them Ruin Our National Parks?," indicated the defensive strategy. Hammering at the dam builders for assaulting part of the nation's heritage, DeVoto derided the Bureau for its greed and avarice. This strategy had been used with success on many prior occasions; when confronted with a threat to a park or wilderness area, so the thinking went, immediately identify "the enemy," point out its selfish motives, and denounce it accordingly.[2]

Still, while diatribes like DeVoto's had been effective to some degree, such tactics were not wholly adequate for galvanizing the public to join in the crusade to protect Echo Park. Conservationists soon learned that a defensive strategy appealed to a limited audience and that the claims for protecting Dinosaur and other threatened preserves must be made in a more positive manner. Taking such an approach had already occurred to Howard Zahniser and Olaus Murie of the Wilderness Society. The membership of that organization stood at barely one thousand members in 1945, and they believed the most promising way to increase its size was by courting scouting groups, sportsmen, and wildlife lovers. In this way they hoped to appeal to a broad variety of conservationists and link their separate interests in hunting and fishing, birdwatching, and mountaineering with the broader aims of wilderness preservation.[3] As a result, Murie and Zahniser sought to avoid the kind of sharp-edged attacks against corporate developers long used to defend parks and roadless areas in the national forests. Avoiding a defensive posture and promoting wilderness in a positive manner appeared to them the most promising way to increase their membership. Under their leadership, the Wilderness Society portrayed itself as an educational organization, using the *Living Wilderness* magazine and other forums to build its base. As for Dinosaur, Zahniser urged conservationists in the campaign to make their efforts part of "a positive drive" for wilderness legislation.[4] But how exactly should this be done?

Benton MacKaye, president of the Wilderness Society, offered one approach in an article he wrote in 1950. President Truman's Water Resources Policy Commission, chaired by Morris Cooke, had approached the Wilderness Society for its views on how to reconcile federal water policy with the aims of wilderness preservation. MacKaye responded with a thoughtful article, entitled "Dam Site v. Norm Site," published in *Scientific Monthly.*[5] MacKaye had no desire to comment directly on any controversy, and he made only passing reference to the Dinosaur dams and others being proposed within national parks and wilderness areas. Instead he sought to analyze in a broad, scientific manner the compatibility of dams with wilderness. He recognized that dams had many useful purposes like controlling floods and providing water for irrigation, and that they had their place in fostering healthy land.[6] He also recognized how engineers knew how to justify their favorite water projects with geological, hydrological, and other scientific information, and therefore held the upper hand over wilderness enthusiasts who normally relied on spiritual and emotional rhetoric.

MacKaye believed that the time had arrived to place wilderness preservation on a scientific basis. "The ecologist must do for every ecologic province in the country what the hydraulic engineer has done for every water basin in the country, that is, locate a system of sites needed for his purpose."[7] Having been deeply moved by Aldo Leopold's *A Sand County Almanac,* he set forth his argument that wilderness areas deserved protection since they served as "norm sites," or ecological niches that held "the secret of land health." In a key paragraph, he described the scientific case to be made for wilderness:

> Primeval organic land, the aboriginal food chain, such as a soil-oak-deer-panther food chain, possessed, we may say, perfect internal self-renewal, absolute health, 100 per cent normality. We cannot often say that of the soil-corn-cow-farmer food chain. Such is the difference between wild or primeval land, on the one hand, and tamed or domesticated land, on the other. One is a norm, the other a subnorm. The whole object of conservation is, of course, to emulate nature's

successes; this is not to make tamed land wild, but to make subnormal land normal. . . . Wilderness is the 'perfect norm;' it is wild or untouched land as distinguished from domesticated, such as cornfield, pasture, or farm woodlot; it is the product of paleontology. Wilderness is a reservoir of stored experience in the ways of life before man.[8]

MacKaye's article may be taken as part of the transition that McCloskey detected in the postwar wilderness movement, an effort to leave behind defensive tactics and adopt positive claims for wilderness preservation. Susan Schrepfer has pointed out a major shift in scientific thought in the middle of this century, which greatly influenced the Wilderness Society, the Sierra Club, and other organizations. Biologists, geneticists, and other scientists increasingly departed from the "evolutionary synthesis," which emphasized survival of the fittest and accepted "the importance of genetic diversity and the value of all species."[9] Science, in other words, offered a deep well for preservationists, and MacKaye's piece was an effort to draw from it. Still, if MacKaye connected the emerging field of ecology with the wilderness movement, his article did little to inspire the public to help preserve Dinosaur. "Dam Site v. Norm Site" reached a limited audience of resource professionals and scientists, the older generation of wilderness devotees accustomed to such language and sympathetic to the claims of ecology.

The Dinosaur battle could not be won by appeals to science alone. In order to mount a national campaign against Echo Park and Split Mountain dams, defenders of Dinosaur realized that they must appeal to a broader segment of the public and link the preservation of the monument with a new generation's appetite for recreation. Americans were hungry for an experience in the outdoors and eager to taste the wilderness. Dinosaur offered a landscape of breathtaking beauty that had the potential to draw such people. First, it must be made known by advertising its scenic wonders. Wilderness preservation for a new generation required more than carefully written articles in *Scientific Monthly*. It required poetry and pictures and films—in this case, of the winding, twisting canyons and rivers of an obscure national monument in a remote corner of the American West.

This did not promise to be an easy task since Dinosaur National Monument was a largely unknown place, tucked away in the corner of Utah and Colorado, and hardly a prominent part of the national park system. Most travelers to Colorado and Utah were more likely to visit Rocky Mountain National Park, north of Denver, or Mesa Verde National Park in southwest Colorado, which stood at the epicenter of well-known parks including the Grand Canyon in Arizona, and Zion and Bryce in southern Utah. Dinosaur, spanning the border of Utah and northwest Colorado, did not enjoy such an advantageous position. After the war, as millions made their way to these and other well-known parks in the West, few took U.S. Highway 40 between Salt Lake City and Denver, which provided access to the southern end of Dinosaur. Just a few thousand per year saw the quarry of fossil bones, and there was not even much to see there except a few scattered remnants. With the budget of the National Park Service reduced during the war, the monument remained virtually undeveloped for more than a decade after its expansion in 1938. Few people saw the backcountry either. Only the most adventuresome travelers took the rough and winding dirt road down to the spectacular confluence of the Green and Yampa rivers, and experienced the beauty of Echo Park.[10]

Aside from famous national parks like Zion, Bryce, and Mesa Verde, the canyon country of Utah, Arizona, and Colorado remained largely unknown in the 1950s, even to wilderness enthusiasts themselves, who preferred wooded or high alpine regions like the Appalachian Trail, the Adirondacks, the Grand Tetons, and the Sierra Nevadas. These people discovered a new kind of place in the Green and Yampa rivers, a unique blend of canyons, mesas, and whitewater rivers that distinguished the inner gorge of Dinosaur Monument and the surrounding Colorado Plateau. Many were stunned by the scenery. Olaus and Margaret Murie of the Wilderness Society offer an example. Residents of Moose, Wyoming, a one-day drive to the north, they had been to Dinosaur Monument for a brief visit years before, but had not seen Echo Park. In early July 1950, just one week after Oscar Chapman approved of Echo Park and Split Mountain Dams, the Muries made their first visit, following a meeting of the Wilderness Society Council in Twin Springs, Colorado. David Canfield, superintendent of Rocky Mountain National Park, had attended the meeting in order to inform the council members about Dinosaur. He came armed with maps and offered to take anyone interested to Echo Park and other places in the monument. Along with Jesse Lombard, superintendent of Dinosaur, Canfield escorted the Muries, and other Wilderness Society Board members including Harvey and Anne Broome and Richard Leonard, soon to become president of the Sierra Club.

Margaret Murie wrote of her impressions a few months later in *Living Wilderness*.[11] She found herself surprised at the huge scale of the monument and the intimate beauty of rock walls and twisting canyons. After she and her companions drove the winding and dusty road down into Echo Park, they paused amid the shade of tall cottonwood and box elder trees, walked along a small beach near the grand confluence of the Green and Yampa rivers, and marveled at the cliffs looming all around. Steamboat Rock, whose eastern face had echoed John Wesley Powell's cries nearly a century before, dominated this impressive scene. Opposite Steamboat Rock, high on another rock wall, Murie saw a large petroglyph of a human figure taking aim with bow and arrow at a mountain sheep. She could not easily estimate the size of this petroglyph, as it had been drawn on the face of a wall that was now more than a hundred feet off the ground. All she could do was crane her neck high, marvel at its location, and speculate on how many centuries it had been there. Wondering aloud if the petroglyph would be above water from Echo Park Reservoir, Canfield indicated that it would not. The water would cover it over by hundreds of feet. "We hadn't known it was such a big world, this world of the Green and Yampa," Murie wrote.[12]

Other writers shared Margaret Murie's awe of the canyons and rivers. Bernard DeVoto, in his eloquent article in the *Saturday Evening Post,* described the monument as "a landscape of brilliantly colored, fantastically eroded mesas, buttes, mountains, gulches and high basins. . . . A panorama of fantasy, overwhelming to the imagination, this high rock desert has certain resemblances to the Bryce Canyon and Zion Canyon country . . . and to the setbacks and vistas of the Grand Canyon." Devereux Butcher, executive secretary of the NPA, called it "a region of magnificent scenic grandeur comparable to that in our world-famous national parks."[13] Arthur Carhart exclaimed, in the *National Parks Magazine,* that "the canyons within Dinosaur are, in fact, a succession of smashing gorges between amphitheaters, where the serried cliffs sweep back, and bottomlands afford generous locations for tourist accommodations."[14]

Still, as the title of the most famous book published in the controversy suggested, the "magic" of Dinosaur was less noticeable from an overlook far above the canyons than from a trip down the Green and Yampa rivers.[15] To appreciate the monument's true wilderness character, one had to float through the preserve, but not many who visited Dinosaur had an inclination to do so and only a few hundred individuals had. A carpenter from Vernal named Bus Hatch had begun to run trips down the Green and Yampa rivers, making his first commercial trip in 1932. Later, following the war, Hatch used seven- to ten-man inflatable rafts that he bought from an army surplus outlet in Salt Lake City. Yet throughout the 1930s and 1940s Hatch had precious few customers. In 1945, just 30 persons floated the Green or Yampa; in 1947, the number reached 40; and in 1951, there were 106.[16] Just two individuals who spoke to Secretary Chapman in April 1950 had seen the monument from the river's vantage point, a fact that weakened their case and allowed the Bureau and its supporters to have the upper hand.

Bus Hatch knew that river running had little appeal to the average tourist. Although he had an excellent safety record, Hatch found it difficult to overcome the natural anxieties about running whitewater rapids in spindly rubber rafts. Hatch knew that river running was not so dangerous as to make it foolhardy, but he knew as well that supporters of the dams did not make the distinction. Instead, they emphasized the dangers of river running.[17] In February 1952, for example, local business leaders rallied in Vernal, Utah, the small town closest to Dinosaur. They stressed the importance of hydroelectric power for attracting new industries and underscored the upper basin's interest in utilizing its share of the upper Colorado River. Armed with many pictures of the inner gorge and the rushing rivers, they highlighted the hazards of running the rapids, but ignored the long stretches of calm water in between and did not mention the possibility of portaging around them. Park Service Director Conrad Wirth, who attended the meeting, wrote later that "it was not to be expected, of course, that these gentlemen and their friends from Vernal would, or could, take an unbiased view of wilderness values or understand the purpose and value of a national park or monument."[18]

Wirth did realize how their argument appealed to many residents of small towns in Utah, Colorado, and Wyoming who stood to benefit from construction of the dams and from recreational opportunities provided by the reservoirs. Nor was this the last time that Wirth and Park Service personnel at Dinosaur heard such views. William Wallis, publisher of the *Vernal Express,* later supplied testimony to Congress indicating that more than a dozen people had drowned in the Green River in the previous decade. "Boats of experienced rivermen have been overturned . . . and occupants tossed into the rapids [and] pinned in whirlpools under ledges narrowly escaping with their lives. The river is absolutely hazardous to the inexperienced and unguided venturers."[19] The implication was not very subtle: If the Green and Yampa rivers were a death trap, building the two dams offered the only logical course.

The fact that so few roads had been constructed into the canyons strengthened the case for the dams. G. E. Untermann, curator of the Utah Field House Museum in Vernal, customarily warned those who planned to visit the monument about its lack of serviceable roads. In 1952, Untermann spoke to photographer Philip Hyde, who made the trip to Dinosaur in order to take photographs for the Sierra Club.[20] When Hyde inquired at the museum for information about how to see the preserve, he received the standard warnings about the poor roads. Hyde soon discovered that Untermann had misled him; true, the roads into the monument were dirt, but Hyde still managed to drive thirteen miles down the rough, windy road into Echo Park. Hyde later wrote to Untermann, accusing him of using the state museum as "a propaganda headquarters for the Central Utah Project."[21]

Untermann lost no time in responding. He insisted that "the sheep wagon roads in Dinosaur National Monument are unmarked . . . the few trails are unmarked, and for the most part, [they] lead the traveler into a state of confusion instead of to a definite destination. . . . I feel we are justified in warning the traveler, who doesn't know where he is going, that there is no water that he can count on for either himself or his boiling car."[22] Meanwhile, with help from Frank Ward of the Vernal Chamber of Commerce and Bryant Stringham, a former mayor of Vernal, Untermann assisted in producing a color film of Lodore and Yampa canyons. Local business leaders circulated the film to community groups and showed it to a subcommittee of Congress, which held a public hearing on the Dinosaur dams in Salt Lake City in 1951. According to the *Salt Lake Tribune,* the film "graphically illustrated the inaccessibility of Dinosaur Monument and the difficulty of traveling through it."[23]

To what extent Dinosaur was accessible to the traveling public remained a source of conflict throughout the battle. Conservationists generally admitted that the lack of good roads hampered visitation, although some, like Hyde, realized that the roads were not as poor as Untermann and his friends maintained. Nevertheless, as part of their campaign to thwart the dams, they urged Congress to appropriate funds for paved roads. After all, the lack of roads violated an article of faith in the National Park Service: well-maintained highways encouraged public access and political support. That precept had been established by Stephen Mather, the first director of the agency.[24]

So long as the monument offered poor access to the public, conservationists fighting to keep the dams out of the preserve remained vulnerable to the charge that they wanted to protect it merely for their own enjoyment. "Elitist wilderness lovers" has long been a pejorative brand in the public land states of the American West. Such charges appeared more than once during the Dinosaur controversy, and they struck a chord with residents of the upper basin. The Echo Park controversy revealed the fault lines between western economic interests, determined to guard "their" land, and "eastern" and urban-based environmental interests. Historians of the West have written much about "colonialism." In the heat of this battle, Utah, Wyoming, Colorado, and New Mexico expressed their frustration about being controlled by outsiders, in this case, advocates of wilderness preservation. When Philip Hyde's article and impressive picture of Echo Park appeared in the *Christian Science Monitor* in the summer of 1953, John Will of the Upper Colorado River Commission responded by asking the editor whether "the beauties of the Dinosaur National Monument [should] be reserved for the adventurous and wealthy few? Or, shall they be made accessible, as we propose, and thus available to everyone?" With the monument left in its natural state, Untermann warned Utah Congressman William Dawson, it would remain "the playground of an exclusive few."[25] Newspapers in Salt Lake City and Denver constantly emphasized this point, arguing that the dams would "open" a vast canyon preserve heretofore completely inaccessible. It followed that a reservoir guaranteed popular use of the area by the public at large.[26]

Opponents of Echo Park Dam understood they could not ignore this problem. To argue that Dinosaur should be protected for future generations had limited appeal so long as the average visitor feared the primitive roads and swift rivers. The Bureau and its supporters capitalized on such fears, arguing that the only practical course would be to erect a dam and reservoir and create a national recreation area. That argument went to the heart of the controversy and exposed one of the oldest dilemmas in the national park system: Should natural areas be left as wilderness or "improved" to accommodate visitors? Could national recreation areas be built within national parks or monuments? Newton Drury had already given his answer and had been forced to resign as director of the Park Service. But conservationists realized that Drury's martyrdom on behalf of the national park system meant little in the upper basin. They knew that a way must be found to overcome fears about river running and to make this remote monument known to a larger segment of the public.

Stephen Bradley, a young man from Boulder, Colorado, first confronted this issue. Bradley belonged to a large family of wilderness enthusiasts, many of whom took part in the Echo Park campaign. His brothers, David and Richard, later testified before Congress, while their father, Harold Bradley, helped encourage the Sierra Club to enter the campaign. A boating enthusiast, Steve Bradley had recently begun to build his own kayaks and "folboats"—collapsible kayaks commonly seen on rivers in Europe—and to run them on the Poudre and Arkansas rivers in Colorado. In the summer of 1951, a friend asked Bradley if he might like to run the Green and Yampa rivers, and when Bradley agreed, arrangements were made to hire Bus Hatch as a guide. Steve Bradley had heard about the proposed dams and wanted to learn more about the place he had always imagined as "little more than an arid desolate boneyard."[27]

In July 1951, Bus Hatch took Bradley and his friends and family members for an exciting float through Flaming Gorge and Browns Park, a mostly calm, sixty-mile stretch of the Green River north of Dinosaur Monument. After running the exciting Red Creek Rapids north of Lodore Canyon, the party of nine left the Green and drove to the eastern side of the monument, where they launched their boats in the Yampa. From Lilly Park to Echo Park, the Yampa River carried them for more than fifty miles through the twisting canyon, and the towering cliffs lathered with black stripes of desert varnish left them awestruck. "The Yampa has carved an enchanting canyon in solid sandstone," Bradley later recalled, "one of the most remarkable examples of entrenched river action in America. In some places the walls actually overhang the river so far that a man standing on the brink, two thousand feet above, who dropped a stone would see it fall upon the opposite bank, undercutting on an epic scale."[28]

Obviously impressed, Steve Bradley and his friends quickly decided that a dam and reservoir would ruin this vast wilderness preserve, wiping out numerous whitewater rapids and forever erasing what they considered one of the great river runs in this part of the West. Bradley believed that the national monument had the scenic beauty to be designated as a national park, and he urged that it be preserved for future generations "essentially as it is, natural, undisturbed, and spectacular. . . . The American people should have the opportunity to see and evaluate this park property before forced into a staggering, irrevocable decision they may regret forever."[29] His article, "Folboats through Dinosaur," appeared in the *Sierra Club Bulletin* in 1952, and in the same year the Sierra Club announced its support for a bill to transform Dinosaur into a national park.

While Bradley and his friends thoroughly enjoyed the "magic" of the Green and Yampa rivers, they understood that typical visitors would not so easily be drawn into taking their own float trip. They would need encouragement, and as it developed, Bradley's father helped to provide it. Harold Bradley had been an outdoor enthusiast all of his life, a frequent camper, hiker, backpacker, and canoeist who had instilled a passion for the wilderness into his seven sons—Charles, Harold, David, Joe, Ric, Bill, and Steve. In 1947, after retiring as professor of physiological chemistry from the University of Wisconsin–Madison, he moved to Berkeley where he quickly became active in the Sierra Club. His own father had been active in the Club early in its history and had known John Muir. As a boy, Harold Bradley regularly heard serious talk about a place with an unusual name: Hetch Hetchy. Now, almost forty years later, Harold Bradley found himself involved in a remarkably similar controversy. In the fall of 1951, the Bradley family newsletter, affectionately known as the "Hum O' the Hive," carried Steve's account of his trip down the Green and Yampa. The patriarch of the Bradley clan grew intrigued at what appeared to be a markedly different kind of wilderness landscape. Dinosaur had no resemblance to the High Sierra or the north woods of Wisconsin or Minnesota, but Steve obviously had been impressed and Harold wanted a close-up look himself.[30]

In the summer of 1952, joined by sons Steve and Charlie and their families, Harold Bradley made his first trip to Dinosaur. The party spent "six thrilling days and nights" floating the Yampa to Echo Park.[31] The sheer walls captivated them, while twists in the river offered new angles of vision and treasured sights at every turn. "The experience," Bradley later wrote, of "threading our way through this superb gallery of matchless pictures displayed in ever-changing vistas, left us aghast at the thought that Bureau of Reclamation engineers are calmly planning the destruction of the Monument."[32] The senior Bradley had come prepared with a home-movie camera, a tool that could not have been more appropriate for supplying a close-up view of the Green and Yampa rivers. He shot scenes of his family running their kayaks through rapids, eating lunch on the quiet beaches, and gazing up at the massive cliffs of Weber sandstone. Bradley's movie proved vital to the Echo Park campaign. It revealed the excitement of a new sport to a coming generation eager to experience the wilderness.

The film also sparked great interest within the Sierra Club. Up until that time, the Club's involvement with the controversy had been limited and rather formal. Club President Richard Leonard, also a member of the Wilderness Society Council, had been to Dinosaur Monument with the Muries in the summer of 1950 and come away determined to help fight the Echo Park and Split Mountain dams. Leonard and Bestor Robinson—who testified before Chapman at the hearing in Washington—urged the Sierra Club to stand with other groups in denouncing the proposal.[33] The Sierra Club agreed that the dams would establish a precedent that threatened the integrity of the national park system, and had joined in denouncing Newton Drury's resignation in February 1951.

Yet the enthusiasm in the Club for joining the campaign came more from its leaders than from its rank and file. Most Club members resided in northern California and enjoyed the Sierra, province of the "high trips" in the summer and the focus of virtually all of the Club's conservation activities since its founding by John Muir in 1892. David Brower, recently appointed executive director of the Club, was typical. He had spent years backpacking and climbing in the Sierra and had led a number of the "high trips." The mountains of California had long been Brower's "summertime addiction," as they were for most of the Club's members, and few of them thought much about Dinosaur Monument in Colorado and Utah.[34]

Bradley's movie helped to change that provincialism. Over the next year and a half, he showed the film to fellow Club members and friends as well as to various groups in the San Francisco Bay Area, including the California Academy of Sciences and the University of California Conservation Club.[35] He made a copy and loaned it to Sierra Club members in the Pacific Northwest. He began what he later called a "one man campaign" to reveal the colorful canyons and twisting rivers of Dinosaur.[36]

Bradley found his fellow Club members enchanted with his film, responding in much the same way as Margaret Murie had done two years before. As they watched scenes of the rapidly moving rivers and sheer rock walls, they discovered an entirely new kind of wilderness, one obviously much different than the Sierra, but one that was equally compelling and deserving of greater exploration. The Yampa, Green, and Colorado rivers and the canyons of Utah and Colorado offered a landscape of white, pink, and red cliffs, rushing rivers, sparse desert vegetation, and a unique kind of beauty peculiar to the American Southwest. Bradley's film awakened Sierra Club members to a region unknown to most of them and to the majority of wilderness lovers across the country. Few realized it then, but the Colorado River and Plateau was destined to become a focal point of conservation activity in the decades ahead.

Bradley's movie soon gave birth to an exciting idea: a Sierra Club trip down the rivers of Dinosaur during the following summer. Eager to arrange such a trip, Harold Bradley knew that floating the rivers would acquaint Club members with breathtaking scenery and provide them with a taste of the great wilderness experience that he and his two sons had had the summer before. Certainly, it would expand their appreciation of wilderness beyond the Sierra.[37] He also believed that a successful float trip would demonstrate that river running was not a dangerous sport—not as risky as the proponents of Echo Park and Split Mountain dams wanted everyone to think. And by leading the pack of river runners himself, the seventy-four-year-old Bradley could demonstrate that running rapids need not be restricted to the young. With backing from the Sierra Club's Outing Committee, Bradley threw himself into making necessary arrangements with Bus Hatch for river trips through Dinosaur in the summer of 1953.[38] The February *Sierra Club Bulletin* announced the river trips. Originally, the Club planned for one six-day trip limited to fifty persons, but Bradley's movie sparked such interest that two more trips had to be added.[39]

It became a family reunion for the Bradleys and proved to be a summer that altered the history of the Sierra Club forever. Prominent leaders of the Club went along, including Executive Director David Brower, President Richard Leonard, Bestor Robinson, Alex Hildebrand, Lewis Clark, Francis Farquhar, and Nathan Clark.[40] Bus Hatch's river business flourished, as he carried about two hundred Sierra Club members down the Yampa to Echo Park, and from there followed the Green past the Echo Park dam site in Whirlpool Canyon and beyond to Island and Rainbow parks and Split Mountain Canyon. When the boats stopped for the night, Club members had a chance to discuss the prospect of the dams around the campfire.[41]

Participants of the Sierra Club river trips considered them a great success, with Harold Bradley concluding that "practically all came out filled with wonder and enthusiasm."[42] About five hundred people (about two hundred with the Club) took a float trip through the monument in the summer of 1953 without incident, thereby helping to refute the argument that "the river run is exceedingly dangerous, while the lakes will presumably be safe."[43] Bradley received numerous letters throughout the summer from fellow Club members expressing their exhilaration and newfound dedication to help protect the preserve. Having recently been elected chair of the Club's Conservation Committee, Bradley had a good deal of influence in bringing the Echo Park issue to the forefront of the Club's agenda. Considering Echo Park Dam the most potent threat to the national parks since Hetch Hetchy, he urged the committee to make Dinosaur a major priority.[44]

The river trips also gave birth to another film of Dinosaur, the first film produced by the Sierra Club that featured both sound and color.[45] Shot by professional photographer Charles Eggert, it was entitled "Wilderness River Trail." With the water level particularly high that year, Eggert nicely captured the thrill of whitewater rafting. Recording the *sound* of the water proved to be more difficult. Eggert could not take a tape recorder down the river, so he dubbed into the film the sound of an ocean liner going through "a heavy sea." The effect was somewhat peculiar, and the Sierra Club admitted that it was "a sound you'd probably never hear even if you were on a liner." Nor could such sounds be heard on the Green or Yampa rivers! Yet the total effect of Eggert's work proved to be captivating, complemented by organ music composed for the film by Clair Leonard, a professor of music at Bard College. Brower and the Sierra Club promoted the film by noting that "the total result is beautiful, with many haunting themes."[46]

By the end of 1953 the Sierra Club had made copies of Eggert's film available for viewing across the country, courtesy of several conservation groups, including the American Alpine Club in New York, the Appalachian Mountain Club in Boston, the Izaak Walton League in Chicago and Denver, the Mazamas in Portland, and the Wilderness Society and the NPA in Washington. While Bradley's home movie had inspired Sierra Club members in California, "Wilderness River Trail" caught the attention of thousands of conservation and wilderness enthusiasts around the country. Harold Bradley showed it many times in the Bay Area, favoring it over his own movie. His son, Richard, a physics professor at Cornell University, showed the film sixteen times in Ithaca, and recalls that "Wilderness River Trail" was not only a smashing success, but instrumental in bringing the Sierra Club fully into the campaign against Echo Park Dam.[47]

The Sierra Club intended "Wilderness River Trail" to spotlight the wilderness qualities of Dinosaur Monument. Here was a magnificent part of the national park system with unique and unmatched scenery, which sadly had gone unknown to the public. Dinosaur offered the kind of primeval wilderness and recreational opportunities so cherished by this generation of preservationists, the kind of landscape that could inspire the soul by revealing the "primeval," and the kind of place that Americans living in an increasingly urbanized and technologically oriented culture would treasure. And the Club and its allies asserted that the nation could afford to protect it; no jobs need be lost if Echo Park and Split Mountain dams were jettisoned, and, they contended, neither must the national or upper-basin economy suffer.

While the Sierra Club's float trips generated enthusiasm throughout the organization, no one enjoyed them as much or felt as moved by the scenic beauty as David Brower, recently appointed executive director. Brower had first been made aware of the Dinosaur issue from Martin Litton, who wrote articles on the controversy in the *Los Angeles Times,* and told Brower that he thought the Sierra Club was doing too little to fight the dams.[48] Brower had then assured Litton that he meant to bring the Club into the battle. Even then, however, Brower may have been more influenced by Walter Huber, an engineer and the president of the Sierra Club, who visited Dinosaur and "said that it was just canyons and sagebrush."[49] Not until Harold Bradley produced his film and urged Brower to take a closer look did he become enthusiastic. Then, following his float trip in 1953, he threw himself into the Echo Park campaign with a vengeance.

The Echo Park campaign has long been associated with Brower—and the story has usually been told as a variation of "great man" history, with Brower almost singlehandedly stopping the controversial Dinosaur dams and embarrassing the Bureau of Reclamation. Brower did challenge the Bureau's facts and figures, although the campaign against Echo Park Dam was never his alone. Moreover, his tactics of questioning the Bureau at every step did not always sit well with other groups that made up the coalition. Still, there can be no doubt about Brower's leading role. Besides helping to expand the Sierra Club's horizons beyond California, he brought a vigorous energy to the campaign.[50]

Brower had the same fierce intensity and uncompromising manner that John Muir had displayed during the battle for Hetch Hetchy. Having just attained the leading position in the Sierra Club, his sense of Muir and the Club's past loomed large in his mind. Muir's passionate appeals to save Hetch Hetchy had been drowned out by more powerful voices, and as Brower first became aware of the threat to Echo Park he seemed to hear echoes of Muir resounding in the canyons of the Green and Yampa rivers, calling him and the Club to act. Hetch Hetchy was an old and deep wound in the Sierra Club, and in Echo Park Brower saw an opportunity to make it heal. John Muir had called Hetch Hetchy Valley in Yosemite National Park one of nature's most "precious mountain mansions. . . . Dam Hetch Hetchy," he once wrote, "as well dam for water-tanks the people's cathedrals and churches, for no holier temple has ever been consecrated by the heart of man."[51] Fresh from a river trip down the Yampa, Brower had similar feelings about Dinosaur.

Like most Sierra Club members, he had limited experience with the Colorado Plateau. Except for brief visits to the Grand Canyon,

Zion, and Bryce, he had not seen the canyonlands and rivers of the Southwest.[52] The experience of seeing the Tiger Wall, Steamboat Rock, and the sheer cliffs of Whirlpool Canyon stirred Brower in a way that he did not soon forget. In his family's newsletter, Harold Bradley wrote that Brower emerged "in a state of ecstasy that beats even my own."[53] Brower's river trip proved to be the start of a long love affair with the Colorado River and Plateau. His passion for this unique landscape drove him for another two decades as he fought against the Bureau of Reclamation in Dinosaur and in later controversies at Glen Canyon and the Grand Canyon.

Brower's ability to captivate the public with images and descriptions of wilderness areas turned out to be his greatest gift. Influenced by Ansel Adams, Brower understood the enormous power of still pictures and films in sparking public awareness of places worthy of being preserved. He had a great knack for using pictures to advertise the unique qualities of the Green and Yampa rivers, and he sensed that the coming generation of wilderness activism would respond to them.[54] The publicity skills that he demonstrated so brilliantly in the Sierra Club Exhibit Format books first appeared during the Dinosaur battle. He saw to it that black and white photographs of Echo Park appeared regularly in the *Sierra Club Bulletin*, a decidedly new twist for readers accustomed to seeing snowy peaks and rock climbs in the High Sierra.[55] He kept up a steady drumbeat of press releases and articles to provide constant updates on the state of the battle, informed readers where matters stood in Congress, and told them where they could view "Wilderness River Trail."

Brower had the right kind of personality for the Dinosaur battle; certainly, his demeanor contrasted with that of his counterparts like Fred Packard of the NPA, Howard Zahniser of the Wilderness Society, and Ira Gabrielson of the Wildlife Management Institute. Each of them had roots in the conservation "establishment" in Washington, with Packard once having been associated with the National Park Service, and Gabrielson and Zahniser with the Fish and Wildlife Service. They were accustomed to operating through channels in Congress and among federal and state agencies. Comparatively speaking, they were methodical, patient, and inclined to advance their programs through tried and trued methods. This is not to label them as weak or vacillating—but certainly, they were less aggressive than Brower, preferring to rely on diplomacy and tactful pestering. Brower had a different approach, one that challenged the quiet, gentlemanly styles of his fellow conservation and wilderness leaders. He was a maverick of sorts—a trait he never surrendered during his long environmental career—and he was willing to be more belligerent. He knew how to ask tough questions that others did not think of asking or never had been inclined to ask.

That aggressive style proved to be instrumental in propelling the campaign against the Dinosaur dams into a higher gear. The Bureau of Reclamation, unlike a mining or logging company, presented a new kind of "enemy" to the conservation movement. After all, the agency was founded upon those classic principles of conservation to eliminate waste and manage natural resources in the public interest, and therefore, it could not so easily be chastised as a logging or mining firm. To ask pointed questions about a project like the CRSP—to challenge the traditional tenets behind conservation of water in the West—seemed at that time almost heretical, but Brower proved more than willing to do so over the course of the controversy. While General Grant, Joe Penfold, and others mostly confined themselves to suggestions that alternate dam sites be utilized outside of Dinosaur, Brower wondered if the entire CRSP, with its high financial and environmental cost could be justified. He did not hesitate to question the Bureau's cost-benefit ratios or to challenge the engineers to explain their need for a water and power project on this scale. Exactly how much water must be stored in the upper basin? How much power did the region really need? What was the life span of large dams along a river carrying as much silt as the Colorado? The Bureau and the upper-basin congressional delegation did not take kindly to such questions, but Brower forced them to respond, and in doing so, they provided technical information that proved useful to opponents of the dams.

Small and fairly homogenous, the Sierra Club was poised to take a more activist role on national conservation issues, and members thereby supported the line of questioning pursued by Brower. His tactics did not always suit other organizations in the coalition fighting the dam, and before this controversy was resolved Brower had raised the hackles of every political representative in the upper Colorado basin. But the Sierra Club in the early 1950s was suited for this kind of highly personal leadership, much more so than in the following years when the Club grew and became more politically sophisticated and diverse.[56]

Brower gave the campaign against Echo Park Dam a boost of energy, and the Sierra Club soon became a major force in the coalition of conservation groups. Harold Bradley proved ever helpful, calling on his sons for needed expertise. Stephen Bradley had already provided a close-up look at Dinosaur and sparked the idea of the river trips; David Bradley provided testimony to Congress in 1954; and Richard Bradley soon became a close associate of Brower in questioning the Bureau's evaporation figures.[57] After the river trips, the campaign against the dam escalated. When Brower returned to Sierra Club headquarters in San Francisco in the summer of 1953, he began to work hard on the campaign, and in the next two years he made Dinosaur an obsession of his own as well as of the Club.

Yet for all of Brower's energy and talents, the Sierra Club could hardly have won this battle alone. By 1953, the threat to Echo Park had become a major item on the agenda of many conservation organizations, some of them substantially more powerful in the political arena than the tiny Sierra Club, the Wilderness Society, or the NPA. Although these groups felt they had the most to gain from a successful defense of Dinosaur, they relied heavily on the strength of larger organizations like the Izaak Walton League and the Audubon Society. The weight of conservation sentiment came from sportsmen and bird watchers, not from national park or wilderness lovers, and that fundamental fact had great bearing on the success of the campaign in Congress, where most lawmakers saw little reason to listen to wilderness groups. Even Bernard DeVoto suspected the Wilderness Society and the NPA, telling Congressman Eugene McCarthy "they tend to be starry-eyed and full of nature's beauty."[58]

As the campaign continued to coalesce, the Izaak Walton League proved to be especially important. Under the direction of William Voigt, Jr., and Joe Penfold, the League anchored the Echo Park campaign in traditional conservation goals, namely, by looking out for the best ways to ensure good fishing.[59] League members in Colorado, Utah, New Mexico, and Wyoming had a natural interest in the controversy, wondering how Echo Park and other storage dams and reservoirs would affect their favorite fishing spots. Supporters of Echo Park Dam argued that fishing in the Green and Yampa was poor, that the rivers only yielded 'trash' fish, partly due to the heavy flow of silt in the unchecked rivers. By contrast, the Colorado River below Hoover Dam offered superb trout fishing, while Lake Mead beckoned anglers with its plentiful supply of bass. Confident that the same conditions could be replicated upstream, the *Salt Lake Tribune* editorialized that Echo Park reservoir "would pave the way for splendid bass and trout fishing in the park and the wildlife situation there could be improved otherwise." [60]

Members of the Izaak Walton League were less sure. They knew that storage dams lowered the water temperature and harmed fish habitat, and most felt dubious about claims that fishing must necessarily improve downstream from a dam. Joe Penfold, in testimony before Congress in January 1954, rejected the argument that a reservoir guaranteed good fishing. He said that bass fishing would not thrive because Dinosaur Monument was at a higher elevation than Lake Mead. As for trout, Penfold explained that they spawned best in free flowing rivers rather than in lakes. Trout would have to be raised in hatcheries and stocked in Echo Park Reservoir, but Penfold doubted that local hatcheries could produce enough trout to make fishing very attractive. In 1953, Colorado hatcheries produced three and one-half million catchable-sized trout. "If we dumped that entire production into Echo Park," Penfold declared, "there would be but one-half a trout per acre-foot of water." With fishing likely to be mediocre, the attractiveness of boating would be reduced. To support his contention, he noted that "if the Denver papers on Friday report poor fishing in Granby [Reservoir], the lake is just about deserted over the weekend." [61]

Penfold had a strong influence among League members in Colorado and in the upper basin. As western representative of the Izaak Walton League, with a base of operations in Denver, he lent credibility to the campaign. He understood the West's keen interest in developing its water, and knew that most people in the upper basin eyed the Colorado River and its tributaries much as a desert traveler viewed an oasis. Well aware of the political power of water interests, Penfold kept abreast of the CRSP and Echo Park in the *Denver Post*. [62] Easygoing and informal, he was less of a wilderness crusader than an old-style conservationist, ever mindful of the pragmatic aims of the Izaak Walton League for sound water management and good fishing. While he did not cultivate the public image of more combative individuals like DeVoto or Brower, he shared their conviction about the threat of Echo Park Dam to the national park system, and he was no less devoted to blocking its approval by Congress. Penfold had been suspicious of the Bureau's upper Colorado project from the start. He had been told informally by an employee of the Bureau that the CRSP had been hastily assembled in the Bureau's regional office in Salt Lake City, primarily to keep the agency at work.[63]

Penfold doubted the need for the Dinosaur dams and believed that alternate sites outside the monument could be utilized. Along with General Grant, he thought Cross Mountain on the Yampa River would be a better dam site than Echo Park. East of Dinosaur Monument in Colorado, a dam here could help to supply irrigation water to Colorado ranchers, unlike Echo Park Dam, which the Bureau did not intend as a source of irrigation water. Cross Mountain could not match Echo Park for storage capacity, but its rate of siltation was about half as much and Penfold argued that the terrain near Cross Mountain was more suitable for constructing power lines.[64]

As the controversy unfolded in the summer and fall of 1953, Penfold, Brower, and their colleagues felt encouraged by the attitude of Conrad Wirth, the new director of the Park Service, whom they had suspected earlier of supporting national recreation areas. Wirth had inherited the director's position after the short tenure of Arthur Demaray, and conservationists had assumed that he had been chosen by Secretary Chapman for his willingness to support the Dinosaur dams. To their surprise, Wirth had a change of heart toward the project, much like Newton Drury in prior years. After a float trip with Bus Hatch through the monument in 1952, he confided to Harold Bradley that all of the eloquent words describing its scenery were on the mark, and admitted that Dinosaur "is far more worth-while saving than I had anticipated. . . . If it is possible, I am more determined than ever that we win this particular scrap."[65]

Like Drury, too, Wirth's change of mind also may have been due to his own experience with the Bureau's high-pressure tactics. Shortly after becoming NPS director, he heard about a meeting of the American Planning and Civic Association in Louisville, which brought together a number of speakers from the Bureau and Army Corps to discuss their work. During the business portion of the meeting, General Grant indicated to members at large that the

Board had passed a resolution opposing the Dinosaur dams. Grant said that although the membership did not need to vote on it, the Board would welcome their approval. N. B. Bennett, a leading designer of the CRSP and in attendance as a guest speaker, thought that this was a "most peculiar and undemocratic procedure for governing an organization." Bennett made his opinion known to Bureau Commissioner Straus, noting "that only General Grant and his Board of Directors are opposing Echo Park, and this without the support of the membership." Wirth soon found himself dueling with Straus via memorandums to Secretary Chapman, and, reminiscent of Newton Drury, remarked that it seemed "a little out of line for the Bureau of Reclamation to criticize the operations of a citizens' organization."[66]

Wirth could not publicly oppose the dams in 1952 because Chapman had already given his approval. In the following year, with Secretary Douglas McKay reconsidering the merits of the dams for the Eisenhower administration, Wirth had greater freedom to express the Park Service opposition. He told McKay that approval of Echo Park and Split Mountain dams "would be inconsistent with the long established policy of national park protection and would set a precedent for the invasion of several other areas in which irrigation and power dams are now proposed."[67]

Wirth's comment revealed the basic tactic of the campaign against the dams. Conservationists continued to emphasize the need to protect Dinosaur and the national park system from such intrusions. Bernard DeVoto wrote in his "Easy Chair" column that the dams, if approved, would "breach the principle which has protected the parks from exploitation ever since the creation of the National Park Service in 1916. . . ."[68] By making that principle the centerpiece of their campaign, opponents of the dams sought to appeal to the widest segment of public opinion, to enlist millions of Americans whose knowledge of wilderness and parks may well have been limited but who understood instinctively that the park system was at the core of America's conservation tradition. In addition, that principle worked to unite all the conservation groups under a single banner, a need that became ever more critical as the campaign progressed. Just as the upper-basin states had to be united in order to hold any hope for passage of the CRSP, so did conservationists have to maintain a united front. Howard Zahniser's phrase, "the sanctity of dedicated areas," perfectly captures the central thrust of the campaign, and few of those engaged in it were inclined to lose sight of that central goal.[69]

Given this emphasis, conservationists supported a bill in Congress to change Dinosaur National Monument into Green River Canyons National Park. Introduced in the House of Representatives by California Congressman Leroy Johnson in 1952, the legislation had no chance of passage without the support of the Department of the Interior, which was an unlikely prospect so long as the Bureau insisted on the two Dinosaur dams. Still, the bill expressed opposition to the dams while drawing attention to "the highly scenic aspects of the area."[70] Dinosaur had the scenic qualities to become one of the "primeval parks," and the NPA, the Wilderness Society, and the Wildlife Management Institute urged its passage.[71] This legislation remained stalled on Capitol Hill through the remainder of the controversy, and conservationists gave it little chance until Congress made a final decision on Echo Park Dam. But their interest in converting Dinosaur from a national monument into a national park continued, even long after the battle ended in 1956.

By the fall of 1953, much ground had been gained in the campaign. Secretary Chapman had reversed course on Echo Park and Split Mountain dams and raised doubts about their costly power; then, he handed the problem over to the Eisenhower administration. Eisenhower seemed to indicate that he disliked expensive river-basin projects like the CRSP. Using the opportunity provided by the change in administrations, defenders of Dinosaur had begun to visit the place and to publish pictures and articles. The Sierra Club float trips had introduced members of that organization to the battle, while Brower and the Club had given the campaign new energy. In one sense, the length of the controversy provided an opportunity to move away from a defensive strategy; there was time to explore the Green and Yampa rivers leisurely, and to take pictures and shoot movies and circulate them around the country.

Yet, if conservationists felt more confident, they also had become increasingly worried about how to proceed, and they were in some disagreement. By 1953, Brower, Zahniser, Packard, and other leaders of the campaign found themselves debating over strategy and tactics. While the goal of maintaining the sanctity of the national park system had worked to unite them and to galvanize the public, it had not yet generated sufficient public pressure to guarantee that the Bureau would ultimately surrender on the dam. Nor had the focus on alternate sites, although Secretary Chapman had changed his thinking thanks to General Grant's exploration of such sites and his critique of the costly power at Echo Park Dam. Still, the Bureau and upper-basin residents continued to reject these suggestions, usually with the familiar refrain that without Echo Park and Split Mountain dams they would lose their water allotment to California. "The most charitable thing we can say in regard to these [alternate] suggestions," John Will wrote to Secretary McKay, "is that they are based upon insufficient knowledge and experience in matters hydrological; that they pay too little attention to human needs in our country for water and for water power; and that they countenance waste of water and power on a scale that our people cannot afford."[72]

Will's remark obviously sought to influence the new administration, which had to reconsider the entire Echo Park matter and establish its own policy on the CRSP. The project stood in limbo during most of Eisenhower's first year in office, along with several other water and power projects inherited from the Truman administration. Not until December 1953 did the Interior Department reach a decision on the CRSP, and not until March of the following year, did the Bureau of the Budget give its approval to the project. During the intervening months, the new president's talk about "partnership" suggested that the CRSP might not be fully endorsed, and some believed this offered a prime opportunity to examine every aspect of the upper-basin project critically.

In fact, some of those involved with defending Echo Park preferred to take a more aggressive approach in looking at the Bureau's plan. Martin Litton, for example, who helped interest David Brower and the Sierra Club in Echo Park with his articles in the *Los Angeles Times,* thought that the Bureau's cost-accounting methods and rationale for storage and power should be critically examined, and he wanted opponents of the dam to go beyond suggesting alternate sites and raise questions about the merits of the entire CRSP. He called on Brower to help expand the campaign beyond the "precedent" issue, saying that "we are giving the enemy undue advantage by sticking strictly to our bird-watching and refusing to meet him on his own ground."[73] Litton's argument sparked a debate over the tactics of the campaign. Litton, Brower, and DeVoto tended to think that more should be done than to laud the beauties of Dinosaur Monument and to sound warnings about the eroding sanctity of the park system. Brower and DeVoto, in particular, began to pick apart the CRSP from various angles.

But their approach did not seem wise to other conservation leaders, who feared that broad critiques of the CRSP were impractical and potentially damaging to their cause. Luna Leopold had already offered his opinion about the risks of challenging the Bureau's preferred arrangement of dams, and Fred Packard of the NPA agreed, fearing that the campaign to protect the national park system should not become too immersed in the technical layers of the CRSP. He thought it unwise to raise questions about how much storage the Bureau required or how much power it genuinely needed, that such matters lay outside their concerns and expertise.[74] Packard also felt strongly that safeguarding Dinosaur should be the main focus since it united a wide spectrum of conservation groups. He knew that more conservative organizations were not amenable to challenging the Bureau's entire plan, and that doing so would only provoke suspicions about conservationists's aims throughout the public land states of the West. If they opposed the entire CRSP, they could be dismissed as impractical sentimentalists and lose credibility, something that could poise tremendous problems for battles in the future. Sierra Club President Richard Leonard concurred. He believed that the threat to the national park system was the fundamental issue, that an attack on the entire CRSP might be dangerous, and that there was no reason "to oppose the entire program."[75] To try to stop the CRSP would neither "be practical nor politically sound. . . ."[76]

This debate continued, and since the ties between conservation groups were informal, several different approaches were undertaken at the same time. In general, however, the defenders of the monument opposed the Dinosaur dams and tried to make clear they had nothing against the upper basin's desire to utilize its half of the Colorado River.[77] Repeatedly, they indicated their willingness to support the rest of the project and to help the upper basin achieve sound water development in the public interest. That stance may seem naive to environmentalists in our time, but for those who had little by way of legal methods for stopping dams in the national park system, this seemed the wisest approach.

This debate over strategy also had to take into account the politics surrounding the CRSP legislation, especially the mounting conflict between the upper and lower basins over the Colorado River. Those tensions had long been present, but they became more intense now that the CRSP was being seriously considered by the Eisenhower administration and might soon be taken up in Congress. Conservationists did not wish to become embroiled in the tug of war between the two basins, and they tried to avoid all appearances of favoring one basin over another. Lower-basin states, notably southern California, had begun to express doubts about the CRSP, and that caused sharp anxiety within the Sierra Club, whose membership was largely based in California. As Brower put it, "there are always a few suspicious people who tend to think a Californian's only interest in the Colorado River is that its lower western bank should always be wet."[78] If the opposition to Echo Park and Split Mountain dams became associated with water and power interests in southern California, their ability to stand up on behalf of the park system might be quickly undermined.

Indeed, the upper basin's wariness toward California emerged in response to Congressman Johnson's bill to create a Green River Canyons National Park. Roscoe Fleming, writing in the *Denver Post,* claimed that Johnson cared little about highlighting scenery in Dinosaur, "but he well knows that denial of the Echo Park Dam site would upset the plans of the Upper-Basin states for another five years at least, and all the time California would be making hay."[79] Fleming had touched a nerve. Ever suspicious of southern California's motives when it came to Colorado River water, upper-basin states scoffed at any politician from the wealthy downstream state who had the gall to aid in guarding Echo Park. Hostility toward the lower basin was mounting throughout the upper basin, and would boil over in a rage during the congressional session in 1954.[80]

All of this revealed the hazards of trying to block one part of a large-scale river project of the Bureau of Reclamation. With a federal water project the size and scope of the CRSP—one involving the fortunes of four states—risks seemed to exist for every avenue the conservationists considered taking. Suggesting alternate sites had seemed a logical approach, but none of them could replicate the advantages of Echo Park, with its prime location below the confluence of the Green and the Yampa rivers. To challenge the Bureau's technical rationale for the dams offered another way of defending Dinosaur, but this ran the risk of confronting engineers and hydrologists whose expertise on such matters was formidable. In 1957, Norman Wengert pointed out that the public could easily be manipulated by the language of the dam builders. When the Bureau or the Army Corps employed terms like *multipurpose, integrated,* or *unified,* it masked "the real issues of costs, benefits, alternatives, and consequences."[81] Finally, to take a strong stand against the entire upper-basin project risked being labeled as radicals or being branded as dupes of lower-basin water and power interests. Some conservationists felt as if they were scaling a sheer wall of granite with no obvious route to follow. Above all, they needed a foothold to begin to make their way, a "hook" in order to grab onto some weakness of the Dinosaur dams.

Ironically, Secretary Douglas McKay provided them with a path to follow. In December 1953, after months of studying the matter, McKay announced his acceptance of Echo Park Dam in a move that foreshadowed the administration's approval of the CRSP. McKay rested his case on the cardinal importance of minimizing evaporation from the Echo Park reservoir. Conservationists quickly reacted. They knew that this rationale could hardly be sufficient for a multimillion-dollar dam; no one would build a dam only because of its low evaporative rate. They suspected that "evaporation" had been fabricated merely to provide a reasonable sounding justification. McKay had now made evaporation the primary issue in the Echo Park controversy, and, in so doing, he provided the foothold that conservationists needed to continue their battle.

SEVEN

The Great Evaporation
Controversy

A noble vision and 80 years of its fulfillment stand at bay amid the scenic grandeur of Dinosaur monument. Are they to be brought down by a disputed calculation in hydraulics?
—Christian Science Monitor, March 17, 1954

All this evaporation study can hope to do is cast doubt on Reclamation rather than cast any light on evaporation.
—Richard Bradley, in family newsletter,
March 2, 1954

The Sierra Club float trips down the Green and Yampa rivers in 1953 revealed a new world of "primeval" wonders to a couple of hundred wilderness enthusiasts, and sparked a fascination with white-water river running and with the Colorado Plateau. By revealing the unique scenery of Dinosaur National Monument to the public with pictures, films, and firsthand accounts, the Sierra Club and other conservation groups engaged in the Echo Park campaign found an opportunity to spotlight a wilderness landscape and make their case for the protection of Dinosaur and the rest of the national park system. At the same time, while the river trips supplied plentiful evidence of the beauties of the Green and Yampa rivers, it was doubtful that such testimony could persuade lawmakers from upper-basin states who held high positions on congressional committees and eventually would consider legislation for the upper-basin project. Because a final decision on the whole matter had to come from Congress, the Bureau's political strength within key committees on Capitol Hill gave it a decided advantage.

Of course, millions of Americans had been flocking their way to national parks in record numbers following the war, so conservationists continued to appeal to lawmakers to help strengthen the "sanctity of dedicated areas," to return to Howard Zahniser's phrase. Still, armed with few legal means to challenge the Bureau's proposed dam, many felt that they must challenge the Bureau more directly, and they searched for some way to confound the experts or for a weakness or flaw with Echo Park Dam or the CRSP. To challenge engineers who had studied the upper Colorado River for years did not seem to be a promising avenue, especially since few defenders of Dinosaur possessed technical knowledge of water and power projects. While it was true that General Grant, along with the Army Corps of Engineers, had been able to delay the Dinosaur dams in the Truman administration, no one could be sure if such a strategy could work again. By 1953 a new administration held office, and Secretary of the Interior Douglas McKay had begun to consider the CRSP in light of new political and budgetary considerations. It remained to be seen whether Grant could persuade another Interior secretary of his contention that Echo Park Dam was overly expensive and that alternate sites should be used. In any event, the problem of contending with the Bureau's experts and its cost-benefit ratios once again had come to the fore.

As it developed, Secretary McKay provided conservationists with the avenue for challenging the Dinosaur dams that they badly needed. Ironically, McKay did not reject the Dinosaur dams at all, but his decision about them in December 1953 provided a way to question the need for them within the larger CRSP. McKay knew that his predecessor at Interior had reversed course on both dams primarily due to their costly power, and McKay himself had already rejected a federal proposal for dams in Hells Canyon for similar reasons. As the secretary reexamined the CRSP during the early months of the Eisenhower administration, he turned for assistance to Ralph Tudor, undersecretary of the Interior Department. Tudor, formerly of the Army Corps of Engineers, had a fine reputation as one of the designers of the Bay Bridge connecting San Francisco with Oakland. Shortly after becoming undersecretary, he discussed the CRSP with the Army Corps of Engineers, trying to find a way to encourage the Bureau's rival to reduce its opposition to the project. Tudor sensed that the Corps might be more amenable to the legislation if its cost-benefit ratios were reformulated, and if the Bureau agreed to delay some of the storage dams to a later construction stage. Tudor soon familiarized himself with all of the issues raised by the Corps and by General Grant, and, of course, with the controversy over Dinosaur. McKay, who relied on Tudor in this instance, asked for an investigation of the matter with a firsthand look at Echo Park. In late September 1953, McKay dispatched Tudor to the monument to examine the two dam sites, to consider alternate locations, and to provide him with a recommendation.[1]

For three lovely days in early fall, 1953, Tudor floated down the Yampa and Green rivers, joining in the parade of those who had done so that year. Park Service Director Conrad Wirth and the new commissioner of the Bureau, Wilbur Dexheimer, successor to Michael Straus, accompanied him. The group inspected Dinosaur Monument by boat and from the air. Tudor found the winding sandstone cliffs and swift rivers truly impressive, and he enjoyed the opportunity to sleep under the stars on a beach along the Yampa River.[2] He later told Secretary McKay that if a dam were to be built, "the alteration will be substantial and if conflicting interests did not exist, I would prefer to see the monument remain in its natural state."[3]

But Tudor put his own feelings aside and recommended that McKay approve the Echo Park site. Compared to the most likely sites outside Dinosaur—New Moab, Dewey, Desolation, or a higher dam at Glen Canyon—he considered the Echo Park location superior because high rock walls in Whirlpool Canyon offered plentiful shade that would minimize evaporation. Split Mountain Dam, less important in the first phase of the CRSP, could be delayed, but the bigger Dinosaur dam should not be postponed and it could not be replaced. Tudor told McKay that the use of any alternate site meant a loss of between 100,000 and 200,000 acrefeet of water annually, and "even the lower figure is enough to provide all of the domestic, commercial and industrial water for a city the size of Denver. . . . In an area where water is so precious," he continued, "this is a matter of very serious consequence."[4] Privately, Tudor did not find himself persuaded by opponents of the dam whom he considered to be governed solely by emotions.[5]

McKay dutifully followed Tudor's recommendation. On December 12, 1953, the secretary announced his approval of the initial phase of the CRSP, recommending Echo Park and Glen Canyon dams. In his announcement, McKay repeated Tudor's words, noting the importance of minimizing evaporation losses in a region where every drop counted. He claimed that the dam sites at New

Moab, Dewey, and Desolation Canyon did not provide acceptable alternates for Echo Park. McKay did not recommend Split Mountain Dam in the first construction phase, although this did not entirely remove the smaller of the two Dinosaur dams from the controversy.[6] McKay's decision, unexpected given the administration's attitude toward such projects, did not surprise many of his detractors, who thought "Giveaway" McKay had lived up to his name.[7]

That the secretary placed such emphasis on evaporation was undoubtedly due to his realization of the high cost of Echo Park's power, and his feeling that he must avoid any claims to the contrary that might irk General Grant or the Army Corps all over again. Of course, the Bureau of Reclamation by no means had lost interest in hydroelectric power, the revenues of which would help fund the participating projects of the CRSP. Nor had rural electrical cooperatives, businesses, or the Utah Water and Power Board. Secretary Chapman, McKay recalled, had originally justified both Dinosaur dams due to pressure from the Defense Department which sought a source of power somewhere in Utah; since then, the Department had acquired power from another source. Still, McKay understood that power generation remained a critical aspect of the CRSP, and, along with President Eisenhower, he recognized that the upper basin's growing economy and its political importance to the Republican party made the CRSP a top priority. Despite these considerations, however, McKay made only passing reference to the power capacity of Echo Park Dam, preferring to stress its low evaporative rate. McKay made evaporation the fundamental issue of the Echo Park debate, and set the framework for the next several months, when the battle moved to Congress.[8]

The *Denver Post,* the *Salt Lake Tribune,* and other papers applauded the Secretary's decision. Editorials called for quick enactment of the project in order to allow "full consumptive use" in the upper basin and to bring hydropower to industry and communities. Wyoming State Engineer L. C. Bishop expressed satisfaction to McKay, saying that the benefits of Echo Park Dam "will much more than offset any primitive area value."[9] McKay's decision cleared the way for Congress to consider the CRSP when it convened in January 1954. Eisenhower's Bureau of the Budget still had to approve the upper-basin project, but McKay's decision strongly suggested that such approval would be forthcoming. Hearings on the CRSP were now scheduled to begin on January 18, 1954. In anticipation of those hearings, Utah Congressman William Dawson introduced the appropriate legislation in the House of Representatives.

As those hearings drew near, both sides of the debate churned out a flood of publicity, including press releases, pamphlets, and letters-to-the-editor. The Sierra Club, the NPA, and the Wilderness Society worked on the issue day and night for nearly a month after McKay's decision to ready themselves for the hearings, while the *Salt Lake Tribune,* the *Deseret News,* and the *Denver Post* published numerous editorials and feature stories on the still largely unknown Dinosaur Monument. Much of the debate centered on familiar subjects like the inaccessibility of Dinosaur, the hazards of river running, and the precedent that conservationists claimed could be established should Congress approve of the dam. Newspapers from Wyoming to New Mexico assured any doubters that the project was sound and that preservationists were simply misinformed, especially with their worries about flooded dinosaur bones. Utah Congressman William Dawson circulated a letter to House members listing all of the "facts" and "fictions" about Echo Park Dam, including the false charge—which he attributed to the opposition—that the dam would harm dinosaur bones.[10] Such claims by Dawson and the *Salt Lake Tribune* angered Olaus Murie, and prompted him to write to Wyoming Senators Frank Barrett and Joseph C. O'Mahoney, insisting that the bones were not in danger and that any such claims merely detracted from the major issue at hand.[11]

McKay's recommendation for Glen Canyon and Echo Park dams galvanized Murie and other leaders of the campaign to save Dinosaur in the same way that Chapman's decision had three years before. Despite numerous indications that Eisenhower shrank from expensive water projects, conservationists now feared that the president's approval of the CRSP would soon be forthcoming. McKay's decision obviously did not bode well, and they lost no time denouncing the secretary's decision. Ansel Adams of the Sierra Club expressed regret to the president, and asserted that "the true measure of a civilization lies not in its mere physical and practical accomplishments, but in the achievements of the spirit and in the protection of spiritual and inspirational values." He added that the national parks possessed a value that could not be measured in economic terms, but only in "the priceless and profound qualities of natural beauty and wilderness moods."[12] Adams found some encouragement when the president mentioned national parks in his State of the Union address in January, although he made no mention of the great controversy brewing in the West. A *New York Times* editorial reminded the president of his party's past and warned that if the Echo Park Dam were allowed, "we might as well look ahead to another dam flooding out part of Glacier National Park, still another one wrecking a chunk of the Grand Canyon, and lumber companies moving in on Olympic National Park. We also will hear Theodore Roosevelt, Gifford Pinchot, and the other great Republican conservationists of a half-century ago turning in their graves."[13]

On January 4, 1954, directors of several conservation organizations held a press conference in the nation's capital, to focus public attention on Echo Park in advance of the hearings. Those attending included Charles Callison, of the National Wildlife Federation; Ira Gabrielson, of the Emergency Committee on Natural Resources; Fred Packard, of the NPA; Michael Hudoba, of the Outdoor Writers Association of America; and General Grant, of the American Planning and Civic Association. Gabrielson informed reporters that the original justification for dams in Dinosaur used by Chapman—to supply power for the defense industry in northeast Utah—no longer existed because the needed source had been found elsewhere. Grant continued to press for alternate sites. He accepted the Bureau's plan for Glen Canyon Dam in the first construction phase, but urged replacement of Echo Park and Navajo Dam in New Mexico with dams at Cross Mountain, Whitewater, and Flaming Gorge. These four dams would cost 54.6 million dollars less than the Glen Canyon–Echo Park–Navajo combination, and would store an additional three million acre-feet.[14] (What usually went unsaid was that Glen Canyon Dam, just north of the dividing line between the upper and lower basins, would store substantially more water and generate far greater power than Echo Park Dam. The Bureau planned for 26 million acre-feet of storage at Glen Canyon, making Echo Park's 6.4 million pale by comparison. Still, engineers in the Bureau had designed the CRSP to be a team of dams, and they considered it critical to have a large storage reservoir near the northern end of the upper basin to aid in river regulation. They considered Echo Park excellent for "river regulation" due to its location below two major tributaries.)[15]

Although angry with McKay's decision, conservationists realized that they had been given an avenue for challenging Echo Park Dam. By justifying the dam for its low evaporation, McKay provided a focal point of attack. Grant distrusted the evaporation rates because the Bureau had altered them more than once since the controversy had begun. Oscar Chapman had said that replacement of Echo Park with any alternate site guaranteed a loss of 350,000 acre-feet per year. But the figure had been revised several times since 1950, and presently stood somewhere between 100,000 and 200,000 acre-feet per year. Grant questioned the accuracy of such figures, for he knew that scientists had barely begun to study evaporation. "This loss of water," he argued, "is very much in the imagination of the reclamation board whose figures on losses also seem to be evaporating."[16] For the first few weeks of January, Grant and others tried to draw attention to the uncertainties of evaporation, but they found it impossible to capture public attention on this highly technical issue until the hearings before a House subcommittee were held later that month.

The scene of the controversy now shifted to Washington, where members of the House Subcommittee on Irrigation and Reclamation prepared for the hearings, set to begin on January 18. An offshoot of the Committee on Interior and Insular Affairs, the subcommittee included Wayne Aspinall of Colorado; William Dawson of Utah; and the chairman, William Henry Harrison of Wyoming. Eager to obtain quick approval of the CRSP, they made the case that their region of the West had waited long enough for a federal reclamation project of this kind; dams had been going up all along the Missouri River since the end of World War II, and the Bureau and the Army had worked their wizardry with rivers all over the West. The upper Colorado River Basin remained one of the few regions still to benefit from this federal largesse. Further delay would only strengthen southern California's already dominant stronghold on the Colorado River and impede the upper basin's economic development. "The water not put to use in the upper basin," intoned the *Salt Lake Tribune,* "naturally flows to the thirsty lower basin. And once captured and put into consumptive use in Southern California it may be impossible to retrieve, irrespective of the 1922 [Compact]. If Echo Park is scuttled the future development of the Intermountain West is jeopardized."[17] The Upper Colorado River Commission, the Utah Water and Power Board, and members of Congress from each upper-basin state continued to press for the dam, claiming that without it water rights could be lost to the lower-basin states.

Supporters of the project also made the case that federal development of water in the upper Colorado was long overdue. Aspinall and Dawson, along with their colleagues in the Senate, reminded the subcommittee that upper-basin states had contributed heavily to the Bureau of Reclamation's initial source of funding, the sale of public lands. Wyoming Senator Barrett, for example, eager to see approval of the LaBarge, Lyman, Seedskadee, and Eden projects, which stood to benefit ranchers in southwestern Wyoming, pointed out that his state had contributed 100 million dollars from public-land sales to the reclamation fund, helping to build dams and irrigation works across the American West. These four projects along the Green River, estimated to cost 43 million dollars, were simply a fair return on years of down payment.[18] Barrett, a conservative Republican and always sensitive to the tough brand of individualism and self-reliance in his home state, argued that the projects did not constitute a federal handout. Utah Senator Watkins won applause from the CRSP backers when he said it was no "giveaway." Unlike flood-control projects in eastern states, which relied on congressional appropriations, the CRSP had its own built-in repayment plan with power sales.[19]

Supporters of the CRSP maintained that the upper basin's economic future depended on Echo Park, Glen Canyon, and all the other storage dams and participating projects. They stressed that their states had great potential for agriculture and industry, limited only by the lack of water and low-cost power. Utah Congressman Stringfellow cited statistics demonstrating how reclamation projects in other states improved agricultural output, increased federal tax revenues, and fostered industrial growth. Stringfellow sounded a common theme when he lamented that "the Colorado Basin States are suffering economic atrophy. We need water. Utah is looking toward the Colorado River as its only hope."[20] Senator Watkins echoed him in saying "the only place we have any water left for Utah is in the Colorado River."[21] Utah especially liked to point to the Colorado River as its "last waterhole." Utah's lawmakers spoke for their constituents who had long believed that water "made the desert blossom like a rose." This phrase recalled the ingenuity and heroic deeds of early Mormon pioneers who had settled in the Salt Lake Valley in the 1840s. Moreover, by framing the issue in this way, Utah lawmakers linked Echo Park Dam with life-giving water.[22]

Finally, supporters of the project felt obligated to stress how the CRSP, in strengthening the economy of the upper basin, stood to benefit the nation as a whole. In his testimony before the House subcommittee, New Mexico Senator Clinton Anderson stressed that Navajo Dam was necessary to help supply water for the White Sands Proving Ground and for Los Alamos, center of the nation's atomic research.[23] Linking the Colorado River dams with national-defense needs may have been the most potent argument of all, and this explains why the CRSP backers constantly referred to the upper basin's mineral wealth waiting to be tapped. Congressman Stringfellow mentioned the availability of coal, oil shale, phosphate ore, potash, and uranium across much of eastern Utah, western Colorado, and parts of New Mexico and Wyoming. Senator Wallace Bennett claimed that Utah possessed rich sources of uranium, a critical raw material in the atomic age, and said that the processing of uranium and other radioactive ores (along with metals like zinc, copper, and silver) necessitated capturing Colorado River water.[24]

Public opinion in the upper basin stood solidly behind the Echo Park Dam, so Senators Anderson, Watkins, Barrett, and Millikin could easily ignore the small number of wilderness activists from their own states. This fact, coupled with the regular horse trading over water projects that occurred regularly in the Senate, made it next to impossible for the Dinosaur dam or the CRSP to falter there. When the Senate subcommittee held hearings on the CRSP in June 1954, the opponents agreed that they would not change many minds.[25] No one doubted that the Senate subcommittee would favor the project or that the full Senate would not follow suit. On the other hand, members of the House from eastern and midwestern states could not afford to ignore pleas from the Izaak Walton League, the Audubon Society, or the National Wildlife Federation, groups with substantial strength in some states; supporters and opponents alike realized that the House of Representatives held the key to resolving the controversy.

For this reason, the House hearings that opened in January 1954 took on great importance. When their turn came to testify, opponents of Echo Park Dam emphasized the importance of safeguarding the national park system, and warned that approval of the dam would merely open a breach into the parks that might never be closed. Fred Packard asserted that Echo Park Dam "would constitute the first invasion of the national park system by an engineering project since the National Park Service was established and would open the door to similar invasion of other national parks. . . . If our national park system is to be of any permanent significance—and it is the model for some 44 nations that have established national parks—it must retain its integrity and continue to remain inviolate." Packard added that they did not oppose the CRSP or wish "to impede orderly development of the water resources of the Western states." [26]

Brothers David and Stephen Bradley brought pictures from the river trips of the previous summer to share with the committee. They showed Sierra Club members camping out along the rivers and indicated how much land and water would be inundated by Echo Park Reservoir. David Bradley reminded the committee that a variety of people had floated the river, including his father, who was in his mid-seventies. His brother Joe, despite a recent brain operation, had also joined the group, and later complained that the six-day trip was too short. Bradley obviously wanted to make the case that white-water rapids were not dangerous when they were treated with respect. "Echo Park," proclaimed David Bradley, "is a temple which has been many millions of years in the building. It belongs to the people of this country and has been reserved to them from all forms of appropriation. . . . We have had money changers in our temples before. We have thrown them out in the past, and with the help of this good committee we shall do it again." [27]

Joe Penfold admitted that Dinosaur had poor roads and that a dam would increase public use. At the same time, he lamented the inundation of the Green and Yampa canyons, saying that "with a reservoir people could get around by boat, but in acquiring that means of transportation they will have lost a major portion of what it was they came to see and enjoy." [28] Secretary McKay had recently announced the availability of twenty-one million dollars for the Park Service to build roads and campgrounds in Dinosaur. Penfold disliked the fact that the Interior Department would provide the Park Service with these funds only after the dam was built, and he asked why the money to build roads could not be offered now.[29]

Undersecretary of the Interior Ralph Tudor provoked sharp criticisms from his testimony claiming "that the choice is simply one of altering the scenery of the Dinosaur National Monument without destroying it."[30] Testifying on behalf of the Federation of Western Outdoor Clubs as well as the Sierra Club, David Brower provided one of the more memorable comments from the hearing. Scoffing at Tudor's comment, Brower said that if Echo Park Dam would not alter Dinosaur Monument, then neither would a dam between El Capitan and Bridal Veil Falls in Yosemite National Park alter that grand scenery; nor would removal of the rain forest from Olympic National Park or the huge trees in Sequoia National Park alter those preserves. "After all," he mocked, "the ground would still be there, and the sky, and the distant views. All you would have done is alter it, that is, take away its reason for being. Maybe 'alter' isn't the right word. Maybe we should just come out with it and say 'cut the heart out.'"[31]

Brower's testimony at the House hearings proved crucial, though not because of his remarks about the scenery, which upper-basin lawmakers found unimpressive. Brower correctly surmised that no amount of eloquence would persuade Aspinall, Stringfellow, or Harrison, nor would dire warnings about the integrity of the national park system. Secretaries Chapman and McKay had already dismissed the "precedent" argument, claiming that the necessary claims on dam sites in Dinosaur had been made prior to the establishment of the monument.[32]

Instead, Brower challenged the estimates of evaporation rates, and in so doing, he discredited the justification for Echo Park Dam that Secretary McKay had utilized. His appearance before the subcommittee has since become a famous episode in the annals of American environmental history, taking on almost legendary proportions. The story has been told with overtones of David and Goliath, and generally, it has been assumed that Brower brought an end to Echo Park Dam almost singlehandedly. Like most legends, it has become larger than life. When Brower challenged the Bureau's figures at the hearing, it was merely the beginning of many weeks of intensive investigation of evaporation computations.

For weeks, Brower had been looking for ways to discredit the "fundamental issue" of evaporation, but he did not know exactly how to do so until he arrived in Washington for the hearings. In fact, he had been discouraged from taking such an approach by Luna Leopold and Walter Huber. Charles Woodbury of the Wilderness Society had suggested to Brower that he ask Huber to testify. A former president of the American Society of Civil Engineers, Huber had the knowledge to question the "seepage hazards" of large storage dams, a potential problem that the USGS had recently uncovered in its studies. Perhaps Huber could also raise questions about the high cost of power at Echo Park Dam. Unfortunately for Brower, Huber did not think that he had sufficient knowledge of the CRSP, and he turned down Brower's request to appear at the hearing. Huber also repeated Leopold's advice and warned against challenging the Bureau on technical matters. If the Bureau found flaws in such criticisms, the campaign against the dam could be badly hurt.[33]

Somewhat discouraged, Brower went to Washington uncertain of how to proceed. On the first day of the hearings, he listened to Ralph Tudor's testimony, and found his plan of attack. After describing the major features of the CRSP, Tudor stressed that Echo Park and Glen Canyon dams would provide holdover storage so that the upper basin could meet its legal obligations under the Colorado River Compact, "while permitting full development and utilization of the waters available to the upper-basin states."[34] Tudor then turned to the controversy surrounding Dinosaur. He informed members of the subcommittee about his own investigation of Echo Park and alternate sites, and of his conclusion that Echo Park minimized evaporation. He then presented the Bureau's estimated evaporation rates, and by means of a chart indicated the storage capacities and surface areas of reservoirs to be included in the first and second phases of the CRSP, including Echo Park, Split Mountain, Flaming Gorge, Glen Canyon, Navajo, Gray Canyon, and Cross Mountain. Tudor explained that because all of the substitute dam sites outside Dinosaur had a greater exposed surface area, they had a greater evaporation rate than Echo Park.[35]

Working with figures supplied by the Bureau, Tudor said that Echo Park Reservoir had an evaporation rate estimated at 87,000 acre-feet annually, and Split Mountain at 8,000 feet, with a combined total of 95,000 acre-feet per year. Any of the alternate reservoirs (such as New Moab, Dewey, or Gray Canyon) would evaporate at least 100,000 acre-feet per year *more* than Echo Park and Split Mountain—and some would evaporate almost 200,000 acre-feet more. "In the final analysis," Tudor told the subcommittee, "the increased losses of water by evaporation from the alternative sites is the fundamental issue upon which the Department has felt it necessary to give any consideration to the Echo Park Dam and Reservoir." [36]

Tudor then mentioned the possibility of eliminating Echo Park and Split Mountain dams from the CRSP and constructing a higher dam at Glen Canyon. Adding thirty-five feet to the height of Glen Canyon Dam would increase its storage capacity by more than 5 million acre-feet (from 26 to 31.7 million acre-feet), almost making up for the storage lost from Echo Park.[37] For various reasons, however, including the political support of Echo Park Dam as well as the importance of hydropower for funding the CRSP, the Bureau opposed a higher dam at Glen Canyon. The engineers preferred a low dam at Glen Canyon combined with the dam at Echo Park (Split Mountain Dam was less critical). Any change in that blueprint would weaken the CRSP, both technically and politically. Not wanting to admit that political considerations had to be taken into account—or even the fact that the Bureau's major interest in building Echo Park Dam was to generate hydropower—Tudor dismissed the idea of eliminating the Dinosaur dam and replacing it with "high" Glen Canyon because, like other alternates, it would mean an unacceptably high rate of evaporation. He explained that a reservoir behind a "high" Glen Canyon Dam would have a surface area of 186,000 acres and an estimated annual evaporation rate of 691,000 acre-feet.[38] The Bureau's preferred option of "low" Glen Canyon–Echo Park–Split Mountain would evaporate 621,000 acre-feet per year. The difference between those two figures was 70,000 acre-feet, but Tudor claimed it to be 165,000 acre-feet, not because he deliberately sought to mislead the committee but because the Bureau had bollixed the figures on which he relied.[39]

As Brower studied these figures later that night in his room at the Cosmos Club in Washington, he realized that the Bureau had repeated an error first caught by General Grant three years earlier: failing to subtract the estimated annual evaporation at Echo Park and Split Mountain dams (95,000 acre-feet) from the 165,000 figure. Ignoring the advice of Leopold and Huber, Brower decided to confront the experts on their own turf. The following day, he told the subcommittee that if "high" Glen Canyon replaced the "low" Glen Canyon–Echo Park–Split Mountain combination of dams, then the difference between the two would be 70,000 rather than 165,000 acre-feet per year. Brower noted that Grant had found the same error, and said it was remarkable that the Bureau had never corrected such a simple mistake.[40] "It is hard to believe, I know, to someone sitting here who has no engineering experience, but if I am wrong," Brower declared, "it must surely be because [Tudor] is wrong, and he is not supposed to be wrong in engineering matters or figures."[41]

Brower then argued that the Bureau had also erred in calculating the annual evaporation of "high" Glen Canyon at 691,000 acre-feet, further undermining its case for Echo Park Dam. By assuming that the rate of evaporation depended on the surface area of a reservoir, Brower could not understand the claim that a "low" Glen Canyon Dam with a surface area of 153,000 acres should evaporate 526,000 acre-feet annually, while "high" Glen Canyon, with a surface area of 186,000 acres, should evaporate 691,000 acre-feet per year. By setting up a straight line ratio and using cross multiplication, Brower calculated an annual evaporative rate for "high" Glen Canyon of 640,000 acre-feet—not 691,000—a difference of 51,000 acre-feet. The discovery of this second error prompted further revision of the figures. Now it appeared that the difference in evaporation between "high" Glen Canyon and the "low" Glen Canyon–Echo Park–Split Mountain combination was neither 165,000 nor 70,000, but 19,000 acre-feet per year.[42] By claiming to catch errors in subtraction and multiplication, Brower reduced the amount of additional evaporation from the "high" Glen Canyon alternate proposal by 146,000 acre-feet annually. Congress would make "a great mistake," he told the committee, "to rely upon the figures presented by the Bureau of Reclamation when they cannot add, subtract, multiply, or divide."[43]

Colorado Congressman Wayne Aspinall responded to Brower by suggesting that he find Tudor, "because most certainly it is a direct criticism of Mr. Tudor's ability which has been challenged here."[44] Congressman Dawson of Utah, already irritated with having to listen to the "nature lovers," fumed. "There are some 10,000 employees in the Bureau of Reclamation, and 400 engineers at Denver alone, who have been investigating these sites and working on them. . . . this is like taking the pistons out of the engine if we delete Echo Park, [and then] we must say that those engineers are all wrong." At Dawson's suggestion, Cecil Jacobsen from the Bureau's regional office in Salt Lake City came before the subcommittee the following morning to respond.[45]

Jacobsen defended the 691,000 figure for the evaporation of "high" Glen Canyon. In making his computation on a straight line ratio, Brower had argued that evaporation was a function of the area of a reservoir, but Jacobsen informed the subcommittee that "evaporation is not a direct relationship of the maximum area of a reservoir."[46] Surface area was merely one factor used in calculating evaporation. Such estimates had to consider a number of factors and required the use of higher mathematics, including algebra, plane geometry, trigonometry, and calculus, and he declared, "these matters enter into such a complicated matter as computing evaporation from reservoirs." Rejecting Brower's ninth-grade arithmetic, Jacobsen said that "had he compared the areas at the centroids of the respective volumes, he would be more nearly correct." And before long, much to the delight of Aspinall, Harrison, and their colleagues, Jacobsen was tossing around phrases like "square root of the radius" and "cube power." Then, in a phrase that came to symbolize the dispute over evaporation, he concluded that "you just cannot use ratios and run the old slide stick and get any answer you want."[47]

Upper-basin members of the subcommittee rejoiced. In their minds, everything had been set aright. Brower had been proven wrong convincingly in his effort to second-guess the experts with simple-minded computations. He had tried to walk into the castle and challenge the king, and he had been duly rebuffed. More importantly, the scientific case for Echo Park Dam had been reaffirmed.

Richard Bradley of the Sierra Club later asserted that Jacobsen had given the committee "a snow job," a charge based partly on the fact that Jacobsen did not admit the subtraction error at all but only the error in multiplication. By doing so, Bradley contended that Jacobsen tried to confuse Brower and impress the subcommittee with terminology from higher mathematics.[48] Whether or not Jacobsen tried to stonewall the alleged error can probably be debated. A case can be made that the figure of 691,000 acre-feet was not exaggerated, that well-meaning engineers from the Bureau's

Salt Lake office provided the best estimates on evaporation that they could possibly provide, based on factors such as the slope of canyon walls and the amount of shade—factors that might well require higher mathematics to compute evaporation. It must also be said that someone—Jacobsen or some other Bureau official—did correct the subtraction error in the written transcript of the hearings.[49] Bradley, who soon became closely involved with the evaporation issue, came to believe that evaporation estimates could not be made with great accuracy; later, he speculated that Jacobsen's mathematical jargon simply masked the Bureau's errors.[50]

The evaporation debate had just begun to unfold. Brower's appearance before the House subcommittee turned out to be merely the opening round in the evaporation battle, a battle that continued for several months after the hearing. For Brower and his colleagues leading the campaign, the hearings had mixed results. They had had an opportunity to tell the subcommittee about Dinosaur and further publicize the remote monument in the national press. Brower, for his part, found himself applauded for boldly taking on the Bureau. No one had ever quite criticized the agency so publicly before. After returning to San Francisco, he won accolades from many of his counterparts. Howard Zahniser said, in a telegram to the Sierra Club, that if the Bureau was considered to be Goliath, then Brower was surely David: "Salute him well. He certainly hit the giant between the eyes with his five smooth stones."[51]

Despite the praise, Brower felt a letdown after the hearing. He continued to believe that the Bureau had been mistaken in its computations and botched its case, yet his triumph had certainly not been as complete as the applause from fellow conservationists indicated. As far as members of the House subcommittee were concerned, Jacobsen had defended the case for Echo Park Dam and effectively closed the subject to further discussion. In their minds, no one need question the experts again. But opponents of the dam felt certain that all of the talk about evaporation merely provided a convenient rationale and masked the Bureau's primary interest in supplying power and satisfying upper-basin desires for large dams. While they could continue to question the high cost of power from Echo Park Dam, the "fundamental issue" of evaporation remained.

Brower wanted to pursue the issue, but he realized that he could not continue to question well-trained engineers with his "ninth grade arithmetic." He needed help from someone who knew higher mathematics. General Grant was a possibility, but someone whose arguments about the CRSP were not known and who did not have ties to the Army Corps of Engineers seemed preferable. Harold Bradley suggested that Brower contact his son Richard, a physics professor at Cornell University. In writing to Ric Bradley a few weeks after the hearing, Brower argued that all the talk about evaporation simply masked the fundamental reasons for Echo Park Dam. "Evaporation is just a lot of chit-chat," he said, "but we do have to demolish that herring, and can gain quite a bit, I think, in doing so."[52] Ric Bradley soon became Brower's chief assistant on the evaporation front. He agreed that no progress could be made toward saving Echo Park until the "fundamental issue" was discredited.[53]

Bradley set to work on evaporation in a methodical way. He knew that the Bureau's estimates had been altered, since Oscar Chapman had first mentioned the evaporation matter in 1950. So he began with a simple question: How did the Bureau—or any scientist, for that matter—estimate evaporation? To find out, he sent a one-page questionnaire to seventy specialists in hydrology and meteorology. He asked them how they went about making such calculations, how evaporation was affected by wind, air temperature, and reservoir size, and how reliable they considered the various formulas for calculating evaporation to be.

What he learned proved to be most enlightening. As he heard from hydrologists around the country and read the technical literature, he began to realize, as he told Zahniser and Brower, that "evaporation is at best an inexact science."[54] To be reliable, calculations for a given body of water had to take into account a variety of factors such as surface temperature, humidity, and wind velocity, and such factors had to be measured carefully over the course of a year before any estimates could be made. Bradley also learned that the Bureau, working with the Geological Survey and Weather Bureau, had studied evaporation on Lake Hefner, a reservoir in Oklahoma. The Lake Hefner report, completed in 1952, had been, to date, the most rigorous and accurate examination of evaporation by public agencies. Not surprisingly, the Bureau relied on the report for estimating evaporation at Echo Park and its other proposed dams along the upper Colorado River. When Bradley learned how much the Bureau relied on the Lake Hefner report, he became suspicious. The climate in the upper Colorado basin was markedly different from that in Oklahoma, and other differences such as altitude and the shape of the reservoir seemed to Bradley to have been overlooked. He thought it unlikely that a report focusing on evaporation in Oklahoma could reliably predict water loss from reservoirs in Colorado or Utah.

Furthermore, the evaporation studies at Lake Hefner had been done with measuring pans on shore, and some hydrologists considered this to be a very crude way to test evaporation rates. Professor Harold Byers, chairman of the Department of Meteorology at the University of Chicago, responded to Bradley's questionnaire by saying that "pan measurements bear so little relation to evaporation from larger bodies that I would be skeptical."[55] Moreover, even considering the unreliability of this method, Bradley knew that the Bureau had not placed pans at Echo Park or near any of the proposed reservoirs. Instead, information had been gathered from "widely scattered Weather Bureau stations of which only six had evaporation pans."[56]

As the uncertainties of measuring evaporation became increasingly clear, Bradley obtained confirmation from Luna Leopold. Until now, Leopold had not encouraged conservationists to question the Bureau on such technical issues, but by the spring of 1954 he became more willing to provide information from behind the scenes. Chief hydrologist with the Geological Survey and one of the nation's premier experts in the field of water, Leopold's knowledge proved invaluable to Bradley and his side. He confirmed to Bradley that "estimates of evaporation from dams not yet constructed are still subject to considerable error," and he indicated that the studies of evaporation done at Lake Hefner revealed the inadequacy of measuring using "ordinary evaporation pans."[57]

Leopold also pointed out that by emphasizing water loss from evaporation, the Bureau conveniently ignored other ways that water could be conserved. As an example, he informed Bradley that the Bureau could save 100,000 acre-feet of water from Lake Mead if it released water from the surface instead of from the depths.[58] He went on to point out how the city of Los Angeles disposed of its untreated sewage by sending it directly into the ocean, unlike most cities in the eastern United States, which treated sewage and reused it for municipal water. Leopold maintained that Los Angeles was "throwing away more potentially usable water than would possibly be saved by the construction of Echo Park [Dam]. Engineering wise it would be possible and probably more economical to obtain additional water by the treatment of Los Angeles sewage effluent rather than by additional storage in the upper Colorado basin."[59] Bradley pursued the matter with officials in Los Angeles and learned that 450,000 acre-feet per year was dumped into the ocean, and later he discovered that Salt Lake City disposed of 35,000 acre-feet of untreated sewage annually.[60] Before long, it became clear to him and others that water could be conserved in any number of ways and that more efficient use of water should be undertaken. The more Bradley learned about wasteful water practices, the more he saw the absurdity of the Bureau's evaporation argument. Joe Penfold urged that the Bureau line its irrigation canals—an excellent way to save water that would also allow the preservation of Dinosaur.[61] Remarks such as these revealed how the close analysis of evaporation led Bradley, Brower, and their friends to begin to ask questions about the CRSP as a whole. How much water really had to be stored in the upper basin to allow for full consumptive use? And did all of these main-stem dams have to be constructed in order to acquire it? Such questions increasingly became a part of conservationists' strategy.

For the moment, the focus remained on evaporation. As Bradley proceeded with his investigation, he concluded that the major problem with the Bureau's estimates sprang from the impossibility of predicting evaporation from reservoirs that did not yet exist. Variables affecting evaporation were such that estimates from *existing* reservoirs were subject to a 10 percent to 15 percent error, but to measure evaporation for a reservoir not yet in existence merely increased the uncertainties. It became that much harder to consider wind velocity, humidity, and water temperature, not to speak of difficulties in foreseeing variables like climatic change and economic growth. Bradley concluded that to accept the Bureau's estimates of evaporation "requires some assumptions regarding the future flow of the erratic Colorado River, east slope diversions, changes in climate and upstream diversions . . . based upon a hypothetical economy of this region seventy five years from now."[62] Bradley did not intend to discredit all of the Bureau's studies or deride the hydrologists. "They undoubtedly are top men in their field and did the best job possible under the circumstances," he wrote. "I question only the soundness of giving overriding importance to these estimates when viewed in the light of the overall uncertainties of hydrologic data."[63]

He concluded that estimates of evaporation were subject to a 25 percent error, and that the Bureau's computations for Echo Park and other dams "are in doubt by an amount of water comparable to that being used to justify the construction of the controversial dams."[64] In a letter to his father, Ric Bradley put it less formally: "evaporation-shmevaporation. The more I study the problem and the more people I contact the more I'm convinced no one, repeat NO ONE knows anything about what goes on on a real lake. Reclamation was pretty smart in selecting a field in which there are no experts and setting up camp and then giving out the word."[65]

In one respect, this foray into methods of estimating evaporation rates proved to be academic. Despite the apparent weaknesses of the Bureau's methods, Bradley had to admit that the engineers knew as much about evaporation as anyone else, and he was not inclined to think that the Bureau had simply made up the figures. Even after learning the shortcomings in the Bureau's studies, he conceded that Echo Park provided a site that would reduce evaporation. Common sense pointed to the advantages of the site, and no amount of rigorous investigation of the matter could thwart the argument. William Romig, retired from the Bureau of Reclamation, suggested that Bradley momentarily discount the math and asked him to think about "where would you logically place a reservoir to conserve water. I'm sure that you too would seek a narrow canyon . . . shielded from winds and shaded by the canyons walls so sun exposure is short."[66]

The proof of that appeared later in 1954, in publicity supplied by the Upper Colorado River Grass Roots, Inc., an entity representing thousands of citizens dedicated to gaining approval of the CRSP. Adopting the nickname "Aqualantes" (meaning "water vigilantes"), it churned out an impressive variety of news releases and pamphlets testifying to the unmatched qualities of the site. "Aqualante kits" began to appear in the mail of key political figures, and they included arguments for the dam presented in a variety of formats (including a "dial a wheel" and note cards for speakers). Echo Park, the Aqualantes argued, was "a natural place for a reservoir—at a high altitude, in a deep, narrow canyon."[67]

Yet Bradley and Brower refused to surrender. Instead, armed with a wealth of information about evaporation, they pestered the

Bureau, Secretary McKay, Undersecretary Tudor, and members of the congressional subcommittees with letters of inquiry. They questioned the inexact nature of evaporation studies, and argued that the case for the dam ought not to rest on evaporation alone.[68] Bradley also asked the Bureau to verify the evaporation rate of "high" Glen Canyon, because he and Brower remained unsatisfied with Jacobsen's response at the House hearings.

Their efforts bore fruit. One day in April, Bradley looked at his mail and was startled to find a letter from Floyd Dominy, then an acting assistant commissioner of the Bureau of Reclamation. Dominy sent along a revised estimate for the evaporation rate of "high" Glen Canyon of 646,000 acre-feet per year, extremely close to the figure of 640,000 that Brower had presented to the House subcommittee three months before.[69] This figure vindicated Brower's claim about the multiplication error, although Dominy did not explain how the new figure of 646,000 had been determined. In any event, the "Dominy" letter, as it came to be known, broke the Bureau's case wide open. With evaporation of "high" Glen Canyon confirmed to be 646,000 acre-feet and "low" Glen Canyon 526,000, the difference was 120,000 acre-feet; subtracting the 95,000 estimate for Echo Park and Split Mountain now yielded a *total* difference in evaporation of 25,000 acre-feet. That was the amount of water that stood to be lost each year if Echo Park Dam (and Split Mountain) were scrapped and the Bureau built "high" Glen Canyon instead. The Bureau could build "high" Glen Canyon and eliminate both dams in Dinosaur and barely notice the water loss. Bradley sent Dominy's letter to Brower, who immediately realized its value to the campaign. He issued a press release from Sierra Club headquarters, and within a few days national newspapers carried the story. Ric Bradley read his name the following Sunday in the *New York Times*.[70] After years of claiming that low evaporation justified Echo Park Dam, the engineers had admitted they were wrong.

The upper basin became exasperated. In an editorial that caught the attention of conservationists everywhere, the *Denver Post* asserted that "the net effect of this fumbling, whether witless or calculated, is to compound and confirm the skepticism and opposition that much of the country, outside the west, has for the reclamation-power program which has contributed so much to western growth and prosperity."[71] The *Post* had long been a champion of

Echo Park Dam and of the CRSP, making its remarks particularly damaging to the Bureau.

A few weeks later, the Bureau admitted its errors to Undersecretary Tudor, whose investigation of Dinosaur had been the basis of McKay's justification of Echo Park Dam. Tudor fumed. He had spent weeks investigating the subject, and he had gone before a congressional subcommittee to make his case. His anger grew when he learned that individuals inside the Bureau had known about the errors for some time, but had kept them to themselves. Tudor scolded that "the failure to advise superiors of this error promptly upon its discovery is a matter of serious embarrassment to the Bureau of Reclamation, to the Department of the Interior and to the Administration in view of the endorsements that have been given." [72]

Tudor had no choice but to inform the House subcommittee of the errors. He sent the revised figures to Wyoming Congressman Harrison, chairman of the subcommittee, stating that a "high" Glen Canyon Dam would evaporate 25,000 acre-feet annually more than the "low" Glen Canyon–Echo Park–Split Mountain combination. Then, to be certain of the accuracy of the latest figures, Tudor ordered that all calculations on evaporation for the CRSP reservoirs be checked thoroughly again. [73] It took several months for the Interior Department's technical-review staff to do so. When staff member John Marr presented the final figures in October, they confirmed Dominy's calculations, and sustained the argument that a higher dam at Glen Canyon offered an appropriate substitute for Echo Park Dam. Marr's conclusions vindicated Grant, Bradley, and Brower once and for all. [74] The outcome of the evaporation controversy thus reinvigorated the Bureau's own suggested alternative of a higher dam at Glen Canyon. That option suddenly became more tenable now that the Bureau's evaporation argument had dissolved, and in the summer and fall of 1954 "high" Glen Canyon became a major part of the controversy. [75]

Brower's and Bradley's triumph in the evaporation dispute discredited what had been the major justification for Echo Park Dam for more than four years, and put to rest what Secretary McKay had called the "fundamental issue" of the dispute. Their efforts (along with Grant's) had shown the Bureau's experts to be wrong, and their spadework proved to be of major significance in the controversy. From the very beginning, opponents of Echo Park Dam had been trying to find out how best to challenge the experts in the Bureau, and with the exception of Grant, they had generally been ignorant of how the Bureau operated, how its engineers and planners went about designing the CRSP, and how its "inner mind" really worked. The evaporation episode broke the Bureau's iron-clad grip on factual information; it provided conservationists with crucial insight into how the Bureau thought, and taught them that the experts could be challenged. Until then, everyone familiar with the Bureau's role in developing the American West considered it to be immune to such criticism, invincible to the kind of attack that Brower, Grant, and Bradley had mounted with their slide rulers and legal pads. The campaign to safeguard Echo Park had suddenly gained legitimacy, in showing that evaporation was truly a surface issue.[76]

The evaporation episode uncovered a technical world of hydrology and "river basin management" that not only proved critical to the success in defeating Echo Park Dam, but was an important legacy of the controversy. Brower later said that he had attended and graduated from the University of the Colorado River.[77] In future battles over dams along the Colorado River—most obviously the conflict over dams in Grand Canyon—this technical knowledge proved invaluable. An important group of wilderness advocates and white-water enthusiasts became familiar with the legal, technical, and political forces governing Colorado River use, and they formed the core of the first generation of environmentalists who defended this region of the West with special fervor.

Conservationists now felt more confident in scrutinizing the Bureau's plans and more sure of themselves in pointing out shortcomings in its public statements. As Congress continued to consider the upper-basin project, it became fully clear that the CRSP was a vast plumbing system—a giant machine of dams, power plants, and irrigation projects, intricately connected. Brower and Bradley, joined by Grant, Penfold, and Bernard DeVoto, increasingly explored the CRSP from a number of angles and asked questions about how much storage and power really was needed. Quite apart from Echo Park Dam, did the Bureau have a realistic sense of what it wanted to do and why? Did the nation really need this Colorado River Storage Project? As the multiple layers of the CRSP were revealed, it appeared to some that the Bureau was overdeveloping the Colorado River with too many dams, storing excessive amounts of water, overestimating demands for power, and overlooking alternate sources such as steam plants and atomic power, which was then regarded as a prime source of energy for the future. Brower, for his part, had concluded, late in the spring of 1954, that the CRSP was a boondoggle and that the whole project should be thought through again and radically reformed.[78]

One fact became clear: The Bureau's immense project was designed to gather together a tremendous amount of water, upwards of 48 million acre-feet above Lee's Ferry, the dividing point of the basins (more than half of it behind Glen Canyon Dam and 6.4 million acre-feet of it at Echo Park). Yet a substantial portion of that storage was not required for river regulation or for meeting requirements of the Colorado River Compact. A Bureau report, made public in March 1949, made clear that the upper basin did not need all of the storage planned under the CRSP for twenty-one years, that the big dams were not needed in the project's first stage *if* their only purpose was storage. With twenty-three million acre-feet of storage, the upper basin could still proceed with a number of irrigation projects and meet the terms of the Compact.[79] At the Senate hearings in June 1954, Brower suggested that the upper basin could make use of more than 70 percent of its allotted Colorado River water without any storage, and he questioned whether the Bureau had paid sufficient attention to the need for soil conservation to prevent its reservoirs from filling too quickly with silt. Some of the proposed reservoirs were to act as giant settling ponds; Glen Canyon Reservoir (later called Lake Powell) was to help keep silt from building up in Lake Mead and thereby enhance the power capability of Hoover Dam—but paying more attention to soil conservation appeared to be an alternate and less costly strategy.[80]

Why, then, all the storage? After the evaporation controversy, the Bureau had no choice but to admit its primary interest in hydropower. The planned storage of forty-eight million acre-feet was critical to generate hydropower, and that had been, of course, a major function of Echo Park, Glen Canyon, and other storage dams from the start. Conservationists and other skeptics of the CRSP now began to raise charges about its cost and marketability.

For the time being, defenders of Dinosaur Monument basked in their triumph over the Bureau's mathematicians and engineers, and they used the opportunity to focus on Echo Park. Leaders of the campaign had been greatly heartened with Floyd Dominy's admission of errors. The national press and the public began to notice the controversy. In 1954, seventy-one thousand people traveled to Dinosaur—a great increase over prior years. When they arrived at the quarry—now being developed into a visitor center—many asked what the controversy was all about. Where would the dam be built, and what harm would it do? Did the Bureau need this particular dam? Where could they go to see Echo Park for themselves? As thousands more people made their way to the monument the controversy entered a new phase. Exposure of the Bureau of Reclamation's errors of evaporation rates had been vitally important for encouraging the diverse and loosely organized coalition of garden clubs, hunting and fishing groups, and organizations devoted to the national parks and wilderness. Conservationists were devoting substantial time and resources to saving the preserve. Dinosaur, once a little-known part of the national park system, was fast becoming a symbol of endangered wilderness, the center of attention in the nation's environmental affairs. Congress had become deeply involved in the debate, too, and as the battle began to peak in that heavily charged political cauldron, conservationists began to make the alliances that they needed to win. The Bureau was by no means ready to surrender, but a coalition of forces began to converge, which did not bode well for the Echo Park Dam.

EIGHT

The Politics of Preservation

... the Sierra Club and other so-called conservation groups in California are not mainly concerned about Echo Park and Dinosaur National Monument. Their main concern is to let that water flow on down the river so California can continue to use it, even though it rightfully belongs to the upper basin States by virtue of the 1922 compact. As the battle continues, the Californians are making it more and more clear that this is their true purpose.

—Theron Liddle, Managing Editor
Deseret News and Salt Lake Telegram[1]

After the evaporation dispute, conservationists took advantage of the Bureau of Reclamation's errors and their newly gained knowledge about the technical and legal aspects of the Colorado River and continued to challenge the proposed dam at Echo Park. Seeking to escalate their campaign, they sharply attacked a project they deemed to be a financial boondoggle and insensitive to the new interest in parks after the Second World War. Yet despite the momentum they gained from showing the Bureau to have been wrong, the controversy was by no means over, and the Bureau and upper basin were hardly prepared to surrender. Echo Park Dam had substantial political support across Wyoming, Utah, New Mexico, and Colorado, and little had happened to change the region's interest in acquiring low-cost power or in capturing its rightful share of the river.

Historians of this controversy have given the impression that the revelation of the evaporation errors quickly brought an end to conflict, that after Brower and his friends exposed the mistakes, the mighty Bureau quickly retreated, and the demise of Echo Park Dam was at hand. Such a scenario has great appeal to admirers of David Brower and the Sierra Club, and to those fond of such David and Goliath encounters in environmental history; yet, Brower's triumph over the Bureau on the evaporation figures did not deliver the fatal blow. Even though some important ground had been gained, conservationists soon found themselves wrestling with entirely different arguments advanced by the Bureau of Reclamation. It soon developed that forcing admission of the evaporation errors alone was not going to be adequate to block approval of the dam. Storage and power had, of course, been the essential purposes of Echo Park Dam from the start, despite what Secretaries Chapman and McKay had said about evaporation, and nothing had occurred to change the importance of either one. Accordingly, the Bureau simply shifted its public stance and emphasized the dam as a centerpiece of the overall project, a necessary part of the complex of dams needed to regulate the river and to generate power.

As a result of the Bureau's new emphasis on hydropower and "river regulation," some decided to question the entire upper-basin project, and David Brower was at the forefront. Fresh from his foray into evaporation rates, Brower had become well-versed on the legal treaty, and he speculated that the main problem with the CRSP—particularly the fact that it called for upwards of forty-eight million acre-feet of storage—could be attributed to the framework established by the Colorado River Compact of 1922. It seemed obvious to him that with population growth in California, Nevada, and Arizona rapidly outpacing that of the upstream states, it no longer made sense to allocate dams and power plants equally among the two basins. If so many people preferred to reside in the lower basin, he reasoned, perhaps most of the water should be allowed to go there. Why not determine how best to use the Colorado River's precious supply in advance, Brower asked, instead of basing all decisions on a compact forged in 1922? Brower concluded that the upper basin would not need its full share of allotted Colorado River water for several decades, and that it did not require all of the big storage dams which the Bureau had sought to erect. The Sierra Club leader summarized much of his thinking in a *Pacific Spectator* piece in April 1954, where he referred to the project "as a costly device to lift Colorado River economy by its bootstraps."[2]

Bernard DeVoto also questioned the Compact, and told Arthur Carhart that he might suggest dissolution of it, adding that "the more divisions we can create among the sons of bitches the better off we will be."[3] For DeVoto, the CRSP exacted exorbitantly high financial and environmental costs, and he charged the Bureau and vested interests behind it with a greed and arrogance that ignored the West's climate in the name of the almighty dollar. Holding as an article of faith that the West's semiarid environment placed rigid limits on its population, he looked with skepticism on the alleged technical wizardry of the Bureau of Reclamation.[4] DeVoto distrusted the Bureau and the Army Corps for bringing forth projects that could not be justified economically and for being "intolerably irresponsible, arrogant, and in their public statements and backstage dealings dishonest." He believed that the CRSP was too costly and had not been adequately scrutinized by the public.[5]

Brower and DeVoto did more to question the Bureau's assumptions than any other leaders of the campaign, and in this respect, they foreshadowed the coming generation of environmentalists. Yet most conservationists at the time did not regard the Bureau of Reclamation with the same kind of suspicion nor did they examine its projects with the same degree of scrutiny. Recent books like Marc Reisner's *Cadillac Desert* and Donald Worster's *Rivers of Empire* sharply critique the dam-building agency and point to its arrogant approach toward manipulating western rivers; and Worster claims that it helped to spearhead the postwar West's interest in "achieving nothing less than total control, total management, total power [over] the western American landscape of scarcity."[6] In their own way, Brower and DeVoto provided similar critiques, but it is important to distinguish between the more flamboyant opponents of the project from the great bulk of those fighting to stop Echo Park Dam.

The all-out attack by Brower and DeVoto did not sit well with most of those engaged in the campaign, and the debate over tactics, which surfaced first in 1953, emerged once again. Most of the other organizations did not wish to take a stand against the entire upper-basin project, for they might then be seen as obstructionists who sought to block any kind of legitimate water "development." That could harm the entire conservation presence in the American West. Nor did they seek to become embroiled in the politics of Colorado River water, especially with the fierce tug of war between the upper and lower basin, for little could be gained by being associated with the powerful water and power interests of southern California. Suggesting alternate dam sites outside Dinosaur still seemed the best tactic (along with appeals for protecting the sanctity of the national park system), but even this proved more than challenging, especially once they began to urge raising the height of Glen Canyon Dam, a suggestion which the Bureau as well as some wilderness lovers in Utah stiffly resisted. Glen Canyon Dam soon became a key part of the controversy, and questions about that dam put Brower and his friends into anguish. Above all the coalition of conservationists had to keep in mind that they began this campaign to stop Echo Park Dam, not to question the upper basin's right to make use of its water.

As the battle unfolded in Congress, it became clear that the controversy over Echo Park would finally be decided not by fundamental changes in the ways that the Bureau did business or by altering the Colorado River Compact, but through politics and political alliances. Opponents of Echo Park Dam now found themselves struggling to be heard amid a variety of voices and contending with issues that went far beyond the matter of a dam in a national monument. Questions about the cost of the CRSP, the Bureau's accounting methods, and the need for more irrigated land were each taken up in Congress, and the campaign against Echo Park Dam now became entwined with these and other issues. The debate in Congress, by pitting eastern and midwestern states against those in the upper Colorado basin, brought forth allies for the defenders of Dinosaur, allies who proved instrumental in the campaign's ultimate success. Sectional tensions over reclamation—tensions that had been building since the 1930s, when the Bureau of Reclamation entered a period of great growth—now erupted, and a powerful coalition emerged that blocked the CRSP and eventually forced upper-basin lawmakers to surrender Echo Park Dam in their desire to gain approval of the remainder of the legislation. During the last two years of the controversy, regional conflicts between the West and other areas of the nation proved crucial to the outcome of the controversy.[7] They demonstrated that the Bureau and its upper-basin constituents did not have as much power as they had thought.

DeVoto tried to warn the West of this fact in March 1954, when he lost his temper with Palmer Hoyt, publisher of the *Denver Post,* saying that he and his staff were ill informed on the Echo Park issue. The *Post* had been irresponsible in linking conservationists to southern California water interests, a charge that DeVoto said was "a cheap slur and altogether false":

For I remind you of something. The development of the Colorado River is a national problem and a national concern. So are the public lands, including the National Parks. So is the U.S. Treasury. The expense of Colorado River development will be borne by the American taxpayer at large, and he lives in far greater numbers outside than inside the states that will directly profit from it. Massachusetts will pay a much larger part of the cost of any dam in the Project than Colorado will.[8]

DeVoto's remarks proved astute. Much of the debate over the CRSP in the following two years centered on its cost and value to the nation at large. From the beginning the upper basin had maintained that the project offered benefits to the entire country by dispersing industry to the West and ensuring a ready supply of food for a hungry nation and world. Proponents had sought to portray Echo Park and other dams as vital for national defense and for fostering industrial and agricultural productivity in a nation with a rapidly growing economy.

President Eisenhower, who fully shared this view, delighted the upper basin in March 1954, when he announced approval of the legislation, after the Bureau of the Budget acceded to the 1.5 billion dollar package. The president referred specifically to Echo Park and Glen Canyon dams, calling them "key units strategically located to provide the necessary storage of water to make the plan work at its maximum efficiency."[9] Eisenhower's approval of the bill led to a full debate in Congress, and brought to a head many of the tensions between the West and other sections of the nation over the merits of reclamation—tensions that had been building up for years.

From one standpoint, the president's support for CRSP seemed curious. Eisenhower had called for reduced expenditures and he had taken steps to stop other costly federal water projects, such as the high dam at Hells Canyon in Idaho. With the CRSP estimated at 1.5 billion dollars in the first phase, Ike's approval made many members of Congress highly skeptical. Even members of the Hoover Commission Task Force on Water Resources and Power had to ask how the president could have made such a decision, especially because the Task Force had not yet made any recommendations on water projects.[10]

Politics provides part of the answer. Eisenhower's rejection of Hells Canyon Dam and the curtailment of other projects had provoked sharp criticism from various quarters in the American West, and the administration had sought a way to appease some of these critics. In approving the CRSP, the president found a way to demonstrate that he did not dislike every big water project. Nor did he worry about the project competing with the power industry, for it bore little resemblance to the Tennessee Valley Authority. While the Bureau of Reclamation would build power plants at Echo Park, Flaming Gorge, Glen Canyon, and elsewhere, it had no intention of *marketing* the power; private companies would do so, and indeed they supported the legislation fervently on Capitol Hill.[11] Finally, Eisenhower wanted to encourage the Republican vote in the intermountain region. As DeVoto put it in his "Easy Chair" column, the administration would "not risk losing the West—and failure to support reclamation development on an enormous scale would lose the West. As regards dams there are no party lines."[12]

Eisenhower also may have supported the project due to his political troubles with Senator Joseph McCarthy. By 1954, the Wisconsin Republican had developed a national reputation for battling communist influence in the federal government, but he had also become a growing liability to the Republican party and to the administration. Eisenhower did not wish to confront McCarthy publicly, but he did want to reduce his influence and prestige, and he could do so by helping certain Republicans in Congress who also disliked McCarthy and might be able to bring pressure to bear against him.[13] One of the president's most prominent allies was Utah Senator Arthur Watkins, and although it cannot be proven, Eisenhower's support of the CRSP may have been linked to his need for Watkins's aid in the Senate to thwart further attacks from McCarthy. Senator Watkins did eventually lead the Senate's move to censure McCarthy. When asked about the possible link between Echo Park Dam and McCarthy, Sherman Adams denied it, but the defenders of Dinosaur remained unconvinced.[14]

In any event, the president's announcement opened the final phase of the controversy, and conservationists quickly expressed their disappointment. Most doubted that he knew any more about the Echo Park dispute than Truman, whose ignorance had been demonstrated with his remark about old fossil bones being less important than increasing the nation's food supply.[15] During Eisenhower's first year in office, controversies raged over Olympic

National Park and Dinosaur Monument, but according to Rosalie Edge, of the Emergency Conservation Committee, "our dear child-like President probably knows nothing of either project." Irving Brant agreed, noting that "Eisenhower *reads nothing* except one-page memos handed him by his aides. He sees no newspapers, no magazines, and has not the slightest understanding of conservation (or any other nonmilitary and nonforeign) issues. His inclination will be to do whatever McKay asks him to do."[16]

Their disappointment partly reflected how Ike had given Echo Park Dam and the CRSP a significant boost, and strengthened the upper basin's case that it was no mere regional project. By placing the prestige of his own office behind the project, the president gave comfort to upper-basin residents struggling against the flourish of publicity by conservation groups. Utah Governor J. Bracken Lee said that the president's approval would help "to counteract the propaganda against the project that certain conservation groups have been spreading."[17] His approval also aided the Bureau of Reclamation, which now decisively shifted its public stance on Echo Park Dam. With the evaporative rate of the structure no longer able to serve as its primary justification, the Bureau had little choice but to emphasize its fundamental purpose: to regulate the river and produce hydropower.

Buoyed by the administration, the upper basin pushed to gain approval of the project from Congress. With southern California continuing to grow at a rapid rate, residents understood that Colorado River water would become increasingly valuable to both basins, and that fact alone provided ample justification for the project. Bills appeared in the House and the Senate, additional hearings were called, and the debate began to crest; as it did, opponents of Echo Park Dam found allies among a host of political forces, some in the Midwest and East, and others in California. Lawmakers analyzed the CRSP from many angles, including its overall cost, the expense of its hydroelectric power, and the need for more irrigated land. Conservationists found themselves asking a wide range of questions about hydropower, storage, and cost, and while most did not wish to block the entire project, helping to delay it aided their campaign to stop the Dinosaur dam and reaffirm the sanctity of the national park system.

To some members of Congress, in particular Oregon Senator Richard Neuberger, Eisenhower had committed a grave error by approving the bill. Many residents of Idaho and Oregon had seen their hopes for a dam at Hells Canyon dashed when the administration had decided to scrap the plan, and they could not understand how that move could be reconciled with the approval of dams for the upper Colorado. A disappointed Senator Neuberger soon became an important ally to conservationists fighting to save Echo Park in the Senate.[18]

Neuberger was not alone in denouncing the CRSP. Illinois Senator Paul Douglas, also active in the effort to protect the Indiana Dunes as a national park, likewise denounced the upper-basin bill. In addition to expressing his fondness for the national parks, Senator Douglas spoke on behalf of the farm states of the Midwest and the East, deeply worried about adding to the nation's agricultural surplus and exacerbating the problem of low commodity prices. When Eisenhower took office, the decline in farm prices was a burdensome political and economic issue, and eventually, the president and his agriculture secretary urged "flexible price supports." With farmers worried about low commodity prices, it is not surprising that they questioned the merits of an expensive water project in the West designed to place even more land under cultivation.[19] Supporters of the CRSP argued that population growth in the nation and in the world required increased food production, but detractors argued that such increases could be accomplished more efficiently through improved plant breeding, pesticides, and mechanization than with additional land under cultivation.[20] Senator Douglas took note of the expense of irrigating semiarid lands in the upper Colorado River Basin, saying "it is too bad that these fine people live in [such a dry region] with a river running through deep canyons. We are sorry for them, but I do not think that creates for them a perpetual claim on the public treasury."[21] Douglas went on to lament the fact that although the monies Congress appropriated for the CRSP would be repaid from power sales, such funds would not be used to reimburse the General Treasury but to undertake additional irrigation projects. "To paraphrase the old miner's song," Douglas told fellow senators, "Oh, My Darling Clementine, the irrigation money once appropriated is lost and gone forever; no matter how dreadful sorry we may be, it will avail us nothing."[22]

While the issue of agricultural surpluses caused many members of Congress to look skeptically upon the bill, others questioned the

CRSP due to its price tag, estimated at 1.5 billion dollars in the first stage, with an ultimate cost of 3 billion dollars. This, too, gave some of the lawmakers reason to pause, especially newly elected Republicans who had arrived in Congress in large numbers in 1952, determined to trim the federal budget. After the president's approval of the bill, members of his party came under great pressure to go along, and Eisenhower, of course, could count on many to do so. But some Republicans simply could not sanction such a costly project. They applauded the columns of Raymond Moley in *Newsweek* and the *Los Angeles Times*. Moley, once a member of Franklin Roosevelt's "brain trust" and an avid New Dealer, had since changed parties and had become a leading spokesman for conservatives. Moley regularly harped about the CRSP, arguing that it was too costly and offered little benefit to the nation. Republicans, he charged, "when their sectional interests are involved, are no more for economy than was Truman at his most reckless period." Moley criticized Republicans from the West who backed the project, and he suggested that their support weakened the Republican party in the same way that McCarthy did. According to De-Voto, Moley did much to arouse the ire of eastern taxpayers against western reclamation, and was instrumental in turning the controversy over the CRSP into a sectional battle.[23]

Perhaps the most galling aspect of the legislation for many critics was its costly hydropower, which Neuberger, Douglas, Moley, and others constantly assailed. "It is extraordinary," Senator Douglas said, "that this administration, which has frowned upon additional dams on the Columbia, which has turned down a high dam at Hells Canyon, and which has certainly frowned upon power projects in the Niagara, should choose the upper Colorado as the place where it intends to launch a public power program."[24] The cost of hydropower in the project had also been controversial during the Truman administration, and had been the primary cause of Oscar Chapman's decision to reverse course on Echo Park Dam in 1952. Charges about CRSP's expensive power now came to the forefront again, with many of these complaints designed to show how Eisenhower's talk about trimming costly water and power projects amounted to nothing but lip service.

Hydropower had been a primary purpose of the CRSP from the start, for the Bureau wanted to sell power to subsidize irrigation projects—projects too expensive for individual irrigators. Yet, as

Chapman and Dale Doty had recognized two years earlier, the Bureau relied heavily on power revenues to pay for irrigation projects. Despite the fact that Echo Park, Glen Canyon, and other storage dams were not designed to irrigate any land, the Bureau still charged some of their cost "off" to irrigation, and by doing so, other irrigation projects too costly for individual irrigators to pay could be justified. As DeVoto put it in a letter to a friend, "the Bureau wants power rates as high as possible so that power income will pay for uneconomic reclamation . . . where the public power interests, REA cooperatives . . . want power rates as low as possible."[25] In short, high-cost power helped to fund expensive irrigation.

Expensive indeed. According to information supplied to the Wilderness Society and other groups by the Second Hoover Commission's Task Force on Water Resources and Power, chaired by Ben Moreell, the cost of irrigation in the various participating projects was exorbitant. For instance, in arriving at the expense of the Central Utah Project, the Bureau charged 127.3 million dollars of its cost (the initial phase of the CUP was estimated at just over 231 million dollars) to irrigation that it would provide. Yet the CUP was designed to irrigate just 160,000 acres, meaning that the cost *per acre* reached 124,000 dollars. Other participating projects followed suit: the Hammond project in New Mexico supplied irrigation water at a cost of 100,000 dollars per acre, and the figure for the Paonia project in Colorado approached 64,000 dollars per acre. Costs like these sparked charges that such irrigation was "wholly unwarranted."[26]

Critics also pointed out that the Bureau, by using power revenues to fund irrigation projects, accounted for the expense of the entire project, but did not justify the cost of each dam, power plant, or irrigation project on its own. Groups such as the Engineers Joint Council maintained that the government should not allow a policy of tying together a package of dams and more than a dozen participating projects into a single bill, for none were made subject to individual scrutiny. The council, a federation of eight engineering societies, including the American Society of Civil Engineers and the American Institute of Electrical Engineers, had been closely viewing federal water projects since 1947.[27] While the group did not try to influence particular legislation, it was part of a reform movement after the war that was concerned with federal water policy, and it looked down upon the Bureau's cost accounting in the case of the CRSP. In a congressional hearing, the council stated that the Bureau's method of "basin accounting" allowed it to use "the actual or presumed favorable margins of financial justification in the cases of better projects to make up for the deficiencies of inferior projects."[28] The council urged the Bureau to justify its irrigation projects as cost effective and to authorize each project separately.

Still another problem with costly CRSP hydropower was its marketability. As Dale Doty had pointed out to Oscar Chapman in 1952, some of the power rates at CRSP generating plants were too high to attract buyers. Power at Echo Park, for instance, would be sold at a rate of 6 mills, but whether or not the Bureau could sell sufficient power at such a rate and over a long enough period of time to pay for the project remained unclear. With alternate sources of power becoming a possibility (including nuclear power), many doubted that such costly hydropower could be justified. David Bradley, son of Harold and brother of Richard and Stephen, had recently authored *No Place To Hide,* a chronicle of atomic testing after the Second World War. Speaking to a congressional subcommittee, Bradley urged that members look ahead to a day when nuclear plants might provide energy to the nation. Bradley spoke at a time when the future of nuclear power appeared bright, and he argued that plans for erecting a hydropower dam in a national monument were exceedingly shortsighted.[29] Alex Hildebrand and David Brower of the Sierra Club studied the power issue extensively, and concluded that substantial savings could be accomplished with federally subsidized steam plants, pointing out that alternate sources of power lay amongst the rich sources of coal, oil shale, and uranium embedded throughout the upper-basin states. Les Miller, a member of the Hoover Commission, advised upper-basin senators that steam plants could provide ample power at less cost.[30] Miller's remarks, it might be said, particularly irritated upper-basin states, especially Wyoming, where he had once served as governor. Upper-basin states badly needed to be united on the project, and criticisms from leading political figures within the region could not help but damage their cause.

For defenders of Dinosaur, the problems with Echo Park power greatly aided their campaign by effectively holding up the legislation. When the House and Senate began to assemble versions of the CRSP bill, pressure from fiscal conservatives to delete the dams with costly hydropower helped to delay the bill's progress, particularly in the House. The Senate, of course, clung to Echo Park Dam for a much longer time, though eventually it too realized that it had to dispose of the dam.

As for the Sierra Club and its friends, many were irritated with the Bureau for choosing not to emphasize power until it was forced to do so—until the evaporation rationale for Echo Park Dam had been completely discredited. At that point, the Bureau had shifted its stance on the dam to emphasize hydropower and river regulation, and it had been able to do so almost effortlessly because of the multiple purposes of the CRSP dams. But the way the Bureau had altered its course only hardened the conviction of conservationists that Echo Park Dam was not needed and that adequate substitutes could be found. Fred Packard told a Senate subcommittee in 1955 that the Bureau had changed its mind about the need for the dam more than once: Oscar Chapman had initially approved of the project due to pressure from the Defense Department; later, when the needed defense facility was built elsewhere, evaporation became the "fundamental issue," and still later when that rationale was shown to be wrong, the Bureau suddenly claimed that it had to have the dam "to firm up power produced at other sites." Packard insisted that the Bureau had constantly refused to study the possibility of producing power from other means.[31]

The Bureau was able to emphasize the role of Echo Park Dam for generating power so easily because of the strong support for cheap power across the upper basin. Since the end of the war, industries had been clamoring for low-cost power, and entities like Utah's Water and Power Board had worked feverishly to lobby Congress for the project to provide it, bringing forth yearly reports testifying to that need and based on projections of population growth and new industries, increased per capita income, and the expansion of Rural Electric Cooperatives. It was hoped that large blocks of power from Echo Park and other dams would draw to the region industries that were eager to mine its coal, phosphate, gilsonite, iron ore, and other resources. Continued industrial growth and expanding urban centers would thereby absorb the hydropower and furnish adequate funds in the Upper Division Development Account to pay for the participating projects.[32] The Bureau also claimed that building steam plants was more expensive than operating hydroelectric-power plants, and it insisted that obtaining power from coal or oil shale did not offer long-term energy solutions because they were finite sources. The Bureau even predicted how much coal, oil, and natural gas would be saved with the use of hydroelectric power.[33] Finally, private-power firms supported the CRSP legislation, which also aided the bill on Capitol Hill. Power companies such as Utah Power and Light, Southern Colorado Power, and the Public Service Company of New Mexico agreed to back the legislation because the government would construct power plants and the primary transmission lines from the dams, thus saving them substantial costs of their own. True, the cost of power from Echo Park Dam would be high, but the Glen Canyon power promised to be quite reasonable, and power companies knew that Glen Canyon powerhouse would be a major supplier. Support for the CRSP from private-power firms helped keep the legislation alive in the face of so much opposition.[34]

As these kinds of discussions over power and other aspects of the CRSP continued, the legislation met its first test vote in April 1954 in the House Subcommittee on Irrigation and Reclamation, a body of nearly two dozen members, more than half of them from western states. A favorable vote on the CRSP seemed likely. Just a few days before Floyd Dominy admitted the evaporation errors, the subcommittee considered an amendment to the bill offered by Pennsylvania Congressman John Saylor, who had become the primary spokesman for conservationists in the House.[35] Saylor's amendment deleted Echo Park Dam from the legislation and offered substitute sites. The amendment failed 12 to 5, and the subcommittee soon after approved the CRSP with Echo Park Dam by a vote of 12 to 9 and sent the legislation to the full Committee on Interior and Insular Affairs.[36]

The full committee did not have a chance to vote on the legislation until May. By that time critics of hydropower and agricultural surpluses had had time to make their case, and Dominy's letter admitting the evaporation errors had been made public by the Sierra Club. The atmosphere had changed. Thousands of letters opposing Echo Park Dam reached the desks of House members, and thousands more came in from farmers and other opponents. Members of the Interior and Insular Affairs Committee became wary of the bill, some of them because of Echo Park Dam and others because of obvious problems with cost accounting. Publicity over the Bureau's errors lent weight to the hundreds of letters pouring into Congress pleading that lawmakers do their part to protect the national park system. On May 13, Saylor again offered his amendment deleting Echo Park Dam, and this time the full Interior and Insular Affairs committee turned it down by a vote of 13 to 10.[37] Though Saylor lost for a second time, the vote was much closer than it had been the month before. Congressmen who joined Saylor in asking that Echo Park Dam be deleted came from large states in the Midwest and the East—New York, Florida, and Ohio—where the campaign against the dam had reached members of bigger conservation groups. Considering the number of westerners on the committee, the closeness of the vote seemed quite remarkable and an indication of the gathering strength of conservationists. Finally, on May 18, the committee voted 13 to 12 to approve the initial phase of the CRSP (including Echo Park Dam) 13–12.[38] CRSP now had to go to the Rules Committee before it could reach the floor.

Defenders of Dinosaur took such a close vote as a clear sign of their efforts, and they knew that such a division in the committee did not bode well for the legislation on the House floor. Upper-basin Congressmen Wayne Aspinall, William Dawson, and William Harrison now sensed that they had further work to do before any concrete could be poured in Whirlpool Canyon. By June 1954, it became clear that the full House would not vote on the CRSP during that session. Opposition to Echo Park Dam had been too great, and together with doubts about the bill's cost accounting, expensive hydropower, and the problem of additional agricultural surpluses, had given many House members reason to be wary. By summer's end, the House had still not voted on the bill and it was dead for the session.

As for the Senate, the Interior committee reported favorably on the bill, but the administration opposed it on budgetary grounds after the committee inserted additional projects including the 68 million dollar Navajo Dam in New Mexico and participating projects for Colorado and New Mexico that cost 318 million dollars. Furthermore, there remained disagreement among the upper-basin states about some of these projects. Coloradans could not agree on which participating projects to include in order to benefit both the western slope and the Front Range near Denver; and Senator Millikin, who chaired the Senate committee, feared that until his state agreed on the projects it would be vulnerable to losing its water to New Mexico. Millikin thereby proved reluctant to push for a vote by the full Senate. Senator Clinton Anderson of New Mexico, who badly wanted Navajo Dam included, had to be content with promises that the CRSP would be taken up in the following session of Congress.[39]

Leaders of the Echo Park campaign expressed surprise with the Senate's inaction on the bill, though they were grateful that they had gained valuable time. They were also pleasantly surprised when Senators Hubert Humphrey and John Kennedy spoke against Echo Park Dam.[40] Still, it was a foregone conclusion that the full Senate would approve of the bill when the opportunity arose in the following session, and conservationists knew that the House offered them their best hope. During the recent session, dozens of House members had become uncertain about the CRSP, and many had become determined to block the dam in Dinosaur, thanks especially to larger organizations like the Audubon Society and the Izaak Walton League. Combined pressure from conservationists, farm states, and fiscal conservatives from other regions effectively blocked the bill and delayed it until the Eighty-fourth Congress. As we shall see in the next chapter, conservationists took full advantage of delays in the legislation to prosecute their campaign and bring forth a flurry of publicity about the remote monument in Colorado and Utah. With articles appearing in major newspapers and in all of the major conservation publications, Dinosaur became a symbol of the nation's endangered wilderness.

Having gained important allies in the Congress, the search went on for ways to persuade the Bureau to jettison Echo Park Dam. General Grant continued to promote the substitution of sites outside Dinosaur, such as Flaming Gorge and Dewey Dam; and Joe Penfold joined with Grant in urging the use of Cross Mountain. Penfold stressed methods of water conservation as well. He asked the Bureau to line its irrigation canals, claiming that doing so would save twenty million acre-feet annually, enough to supply water to two hundred Denvers![41] Brower suggested reducing the number of dams from ten to four by eliminating Glen Canyon and Echo Park and keeping Flaming Gorge, Cross Mountain, Navajo, and Curecanti. Doing so would supply adequate storage to allow for full consumptive use and to meet requirements of the Compact.[42]

But most of the debate over alternate plans through the remainder of 1954 and into 1955 centered on the possibility of a higher dam at Glen Canyon as a replacement for Echo Park. This idea had been mentioned first by the Bureau, in one of its early reports, and by Undersecretary Tudor, who raised the option at the House hearings in January, only to reject it because of the higher rate of evaporation. But with the Bureau's calculations discredited, the "high" Glen Canyon substitute seemed the most logical means of redesigning the CRSP and eliminating Echo Park Dam. Brower and Richard Bradley urged that the "high" Glen Canyon option be taken, and this led to months of debate that proved to be one of the most difficult, complex, and, in hindsight, bittersweet aspects of the whole controversy.

The mere mention of Glen Canyon Dam causes anguish and anger among environmentalists today. In the aftermath of the Echo Park controversy, members of the Sierra Club, the Wilderness Society, and other groups began to float through Glen Canyon—a remote and seldom visited area on the Colorado Plateau in southern Utah—and found, to their utter astonishment, some of the most awesome beauty that any of them had ever experienced. There were sandy beaches, cliffs draped with maidenhair fern, petroglyphs and other prehistoric remains, and a maze of narrow canyons and colorful walls. In 1963, the Sierra Club published *The Place No One Knew,* a memorial to Glen Canyon, which was just then being inundated by the dam. A generation of environmentalists became so enchanted with Glen Canyon and the surrounding landscape of the Colorado Plateau that much has been forgotten about its obscurity in the early 1950s.[43] At the time of the controversy over Echo Park, Glen Canyon remained largely unknown but for a handful of river runners in Utah. Furthermore, because it did not belong to the national park system, it lay outside the realm of concern of those engaged in the Dinosaur battle.

As for Glen Canyon Dam itself, the Bureau considered it a linchpin of the upper-basin project. Its reservoir capacity was slated at 26 million acre-feet, compared to 6.4 million at Echo Park, and Glen Canyon Dam would have by far the largest power plant of all the upper-basin dams. (Together, the two dams would capture more than 32 million acre-feet, more than half of the total storage that the Bureau planned to acquire.)[44] For most of the controversy, no one much questioned the merits of Glen Canyon Dam. In May

1954, David Brower appeared at a hearing before the Water Resources and Power Task Force of the Hoover Commission in San Francisco. Brower told the Task Force he did not think "there will ever be any alternate found for the Glen Canyon reservoir. That is such an important part of the old Upper Colorado project I don't see how even the nature-lovingest person of all . . . could find a way to save that."[45]

Discussion about Glen Canyon Dam escalated after the evaporation debate. In April 1954, the Bureau admitted that the "high" Glen Canyon substitute for Echo Park Dam would only increase the total CRSP evaporation by twenty-five thousand acre-feet, and so this substitution appeared to be a logical way to resolve the Echo Park controversy. Brower formally made the suggestion in a letter to Congressman John Saylor, immediately after Floyd Dominy admitted the evaporation errors.[46] Brower pushed for "high" Glen Canyon at the time, not only because he found it a logical suggestion after the Bureau's admitted errors but because he had come to believe that the Bureau was overdeveloping the river with too many dams, and he thought that the upper basin could acquire all the storage that it needed behind "high" Glen Canyon. As time went on, Brower decided that Glen Canyon Dam was not needed either, particularly because he became convinced that with atomic power coming into its own, the nation did not require such an expensive hydropower project.[47] (So Brower eventually disliked Glen Canyon Dam—but not *primarily* because he knew about the scenery that stood to be lost beneath it.) In any event, Brower was preoccupied with Glen Canyon Dam for much of 1954 and into 1955, and this was more than a mere sideshow to Echo Park; it was inextricably connected to the Dinosaur battle, and all the attention on Glen Canyon caused upper-basin lawmakers to be greatly suspicious of the motives of Brower and his friends.

It must be understood that the Bureau preferred to construct a "low" dam at Glen Canyon, combined with the dam at Echo Park; and it regarded the two as anchor points of the CRSP. Engineers envisioned these two as the workhorse dams of the project, each providing critical storage, power, and "river regulation." What the Bureau did not wish to admit was that this preferred arrangement of dams had a political rationale as well as a technical one. While the argument might be made that a "high" Glen Canyon Dam could supply all of the power and storage needed—eliminating the need for Echo Park on technical grounds—the dam in Dinosaur was dear to the hearts of residents in Utah, Wyoming, and Colorado. Those three states preferred that a large storage reservoir be located in the northern part of the upper basin—and Echo Park was much larger than Flaming Gorge, Cross Mountain, or any other site. They disliked the prospect of concentrating storage at Glen Canyon Dam in the southern part of the upper basin. More importantly, Utah wanted a source of power near the Wasatch Front, and Echo Park could provide it; the Echo Park location lay at the center of the upper-basin power market and offered substantially more power than Flaming Gorge. Then, too, without Echo Park Dam, Utah could not make good on its water rights to the Yampa, which it had acquired in the Upper Basin Compact of 1948. Perhaps most critically, upper-basin lawmakers recognized the political importance of a major power dam in the northern end of the upper basin. If Echo Park Dam were eliminated from the CRSP, Glen Canyon Dam would then become the main source of hydropower, but much of that power would have to be sold to the lower-basin states. Despite all of the talk about a growing upper-basin power market, the CRSP would provide a surplus of power and would be sold to the huge market down the river. In that event, with Echo Park Dam out of the bill, Congress might view the CRSP as a benefit to southern California rather than to the upper basin.[48]

All of this went unspoken, for the Bureau did not wish anyone to think that it assembled its projects on any basis other than the technical—and certainly it could not admit that the Dinosaur dam had a political as well as a technical use. Besides, it was easy enough to emphasize the technical advantages of Echo Park Dam—its location at the center of the upper-basin power market and below the confluence of the Green and Yampa—and by doing so the Bureau could avoid admitting the political ones.[49] In short, the Bureau clung tightly to its preferred arrangement of "low" Glen Canyon Dam combined with Echo Park Dam. But the "high" Glen Canyon option still had to be rejected, and when the Bureau did so, conservationists once again discovered the difficulty of second-guessing the experts.

Why was a "high" Glen Canyon Dam not feasible? In the first place, Rainbow Bridge stood in the way. This gigantic sandstone arch rested in a side canyon off Glen Canyon, was considered sacred by the Navajo Indians, and had been designated a national monument in 1910.[50] If a "high" Glen Canyon Dam were built, the reservoir might come close to the base of Rainbow Bridge, so the Bureau claimed that it would have to build a dike below the arch to prevent water from lapping at its base, thereby adding to the expense of the CRSP. Brower, closely following the Bureau's every move, thought it significant that the Bureau had not paid much attention to Rainbow Bridge until it had been forced to, until the original justification for rejecting "high" Glen Canyon because of excessive evaporation had been shown to be false. He thought it revealing that suddenly Bureau engineers started to sound like concerned members of the Sierra Club, fretting about "high" Glen Canyon damaging a national monument. Brower also knew that the Bureau had already admitted the need to build the protective dike to protect Rainbow Bridge from a "low" Glen Canyon Dam, but the Bureau wanted to give the impression that only a "high" Glen Canyon would bother the monument.[51] (In fairness, the Bureau had recognized the problem of Rainbow Bridge earlier, but the matter had not come up for discussion until Brower urged that "high" Glen Canyon be built.) By raising the issue of Rainbow Bridge, conservationists now demanded that it be protected, regardless of the height of Glen Canyon Dam. In a sense, the Bureau had reminded conservationists that a second national monument was involved in the debate, by pointing out the threat that Glen Canyon Dam posed to the arch. Two national monuments rather than one, now hung in the balance of the CRSP controversy.

But Bureau spokesmen realized that they could not reject "high" Glen Canyon solely because of Rainbow Bridge, and they soon began to stress another problem with a higher dam: the rock structure at the Glen Canyon site. In essence, the Bureau admitted that only a "low" Glen Canyon could safely be built in the Navajo sandstone. In June 1954, Bureau Commissioner W. A. Dexheimer told the *Denver Post* that "the greater evaporation loss from the higher dam was not the reason we proposed the lower one. The reason is geological. Our proposed dam 580 feet high ["low" Glen] is the maximum that can be built on that site geologically."[52] From Brower's point of view, this concern also had not emerged in any of the Bureau's previous statements, and he thought that such fears merely amounted to a convenient way of sidestepping the "high" Glen Canyon option. Brower became convinced that the Bureau had shown little regard for weaknesses in the foundation of Glen Canyon Dam until it became politically necessary to do so.

There followed an intense debate between the Bureau and its congressional spokesmen and Brower, who hammered away at the dam builders for continuing to change their story. Throughout the fall and winter of 1954 and well into the spring of 1955, controversy raged over the strength of the Glen Canyon site. How did the Bureau know that it could not build "high" Glen Canyon? What studies had it done? How many test holes had been drilled? Brower fed questions such as these to Pennsylvania Representative John Saylor, the leading opponent of Echo Park Dam on the House Interior committee. The questions became tougher after Commissioner Dexheimer got himself caught in a series of contradictions. In October 1954, he wrote to Richard Bradley, stating that the Bureau was worried about the rock structure even with a low dam at Glen Canyon.[53] Then, six months later, Dexheimer appeared before Saylor in the House hearings and expressed certainty that a low dam was perfectly safe, while a high one was not. Since "high" Glen Canyon would be only thirty-five feet higher than a low one, his reasoning seemed flawed. Besides, Saylor pestered the commissioner, what studies had the Bureau done in the six months since October which had convinced it that a low dam could now be built? Dexheimer could not explain the change, and he had to call for Bureau geologists to appear before a dubious Saylor to explain the matter.[54] The Bureau had been damaged again. In the Senate hearings in March 1955, Brower pointed out how the Bureau's rejection of "high" Glen Canyon had shifted ground over time. He argued that the agency would do anything to justify its preferred CRSP plan of "low" Glen Canyon combined with Echo Park Dam.[55]

All but lost in this contentious debate was a small coterie of voices in Utah, who made urgent appeals for recognizing Glen Canyon's great beauty. In June 1954, a committee representing two hundred people appeared before a Senate subcommittee and urged creation of a Glen Canyon National Park.[56] However, until the Echo Park issue was resolved, those voices in Utah largely went unheard because a higher Glen Canyon Dam appeared to be the best replacement for the dam at Echo Park. Not until the later part of 1955 did Brower, Packard, Zahniser, and other leaders of the

Echo Park campaign begin to realize that saving Glen Canyon had great merit, but at that point they continued to feel hamstrung because Echo Park Dam remained a threat, and they did not see how they could simultaneously demand that Congress turn back a threat to an existing national monument while also expanding the park system to include Glen Canyon. It has sometimes been said that Glen Canyon was sacrificed; and while such an interpretation contains a measure of truth, it must be remembered that Glen Canyon remained outside the basic agenda of the campaign to reaffirm the integrity of the national park system. In December 1954, the Sierra Club Board of Directors passed a resolution saying that it did not oppose any of the CRSP dams that lay outside the boundaries of national parks or monuments.[57]

Here, then, was the best illustration of the Bureau of Reclamation's ability to set the terms of the debate. The fact that conservationists had to choose between protecting Echo Park and Glen Canyon illustrates their subservience to the Bureau. If they tried to block Glen Canyon Dam along with Echo Park, it would be tantamount to blocking the entire project, and that could place their own credibility at stake, something that appeared less than wise, especially in the postwar West. The tale of Glen Canyon illustrates, perhaps better than any other aspect of the story, the challenge facing those who questioned the Bureau's arrangement of dams and the powerlessness of conservationists without any legal weapons to fight the project.

More than that, raising all these questions about Glen Canyon Dam did not help conservationists in courting the upper-basin lawmakers, who, ultimately, must be persuaded to delete Echo Park Dam from the CRSP legislation. Discussion about Glen Canyon Dam was extremely tricky politically because it added to upper-basin suspicions of conservationists' motives. So much exertion had gone into raising questions about "high" Glen Canyon and "low" Glen Canyon that it appeared to the CRSP supporters that the opponents disliked any sort of Glen Canyon Dam; and perhaps they even disliked any reclamation or hydropower development along the upper Colorado River. Brower was not alone responsible, but all of the searching questions about the Bureau's engineering capabilities did not sit well with Senators Watkins, Barrett, Millikin, or Anderson. The more questions that conservationists raised about Glen Canyon—the more they doubted that the rock structure could hold the dam and that the Bureau would protect Rainbow Bridge—the more difficult it was to prove that their primary concern lay with Echo Park. Upper-basin lawmakers increasingly suspected that conservationists really hoped to block the entire legislation, and although this was untrue, the mere perception hardened their determination to win approval of the controversial dam in Dinosaur. And ultimately, the upper basin would have to consent to deleting the dam from the legislation because it was not possible to force its deletion through any other legal means.

But there was another important source of upper-basin suspicions toward conservationists, which arose from the political tug of war along the Colorado River basin. After Congress began to consider the legislation in 1954, most of the public in the upper basin became convinced that the strongest opponent of the CRSP was not the wilderness crowd but the far more powerful interest groups grasping for water and power in the lower-basin states, especially California. Southern California political figures, lawyers, and water experts were denouncing the upper-basin project as stridently as eastern and midwestern states were, and thus contributing to its delay in Congress. While the upper basin resented opposition to the CRSP from other regions of the country, residents found California's position especially troubling, for it undermined the unity among western states that was critical to gaining passage of big-water projects. Tensions between the upper and lower basins became central to the politics of the Dinosaur controversy in 1954, and remained crucial until the end.

California stood on the threshold of a new era in the 1950s, fueled by defense spending and rapid population growth across the Southwest. The Colorado River became an increasingly valued resource, a factor that contributed to fears about the upper-basin project. Despite the 1922 and 1948 compacts—legal documents supposedly guaranteeing equal allocation of the Colorado River among both basins—southern Californians feared that too many upstream dams and power plants might adversely affect their own claims. Several public agencies in California had rights to Colorado River water—agencies such as the Department of Water and Power of the City of Los Angeles, a municipally owned electric system. This body had a contract with the federal government, by which it bought power from Hoover Dam and sold it to businesses and homes in Los Angeles. About 90 percent of the hydropower

Dams on the Lower Colorado River

from Hoover Dam went to southern California; another 10 percent to Las Vegas and southern Nevada. In addition, the Metropolitan Water District of Southern California, an entity representing several cities, brought water by aqueduct from the Colorado River to greater Los Angeles, San Diego, Long Beach, and Torrance. Southern Californians had taken pride in having funded the Colorado River Aqueduct, for it had enabled war industries to flourish. The aqueduct operated with pumping plants, which, in turn, were powered by Hoover Dam. The Imperial Irrigation District and the San Diego County Water Authority likewise had a large stake in water from the Colorado.[58]

Behind these powerful interests stood lawmakers such as Republican Senators William Knowland and Thomas Kuchel, and lawyers, the most prominent of whom was Northcutt Ely, who represented the Colorado River Board of California before House and Senate hearings on the CRSP.[59] Ely made clear that California had several concerns. First, the amount of storage planned for the upper basin was more than forty-eight million acre-feet, and California greatly feared being deprived of its rightful share of water and power during the time that it would take to fill the large reservoirs at Echo Park and Glen Canyon, something that would interrupt power generation at Hoover Dam.[60] Ely also pointed out that the CRSP did not guarantee an annual delivery of seven and a half million acre-feet of water to downstream states, but, instead, relied on an average amount, meaning that the lower basin might receive nine million acre-feet one year and six million the next. Sponsors of the bill vociferously maintained that this was not the case, but Ely and other attorneys expressed skepticism.[61] Other matters came to the fore as well. The California Division of Water Resources fretted that irrigation in the upper basin would introduce salts to the Colorado River that might threaten the state's agriculture. Congressman Craig Hosmer, a strident opponent of the CRSP, announced that the Colorado River already contained an overabundance of salinity, and he contended that the Bureau had ignored the problem entirely in formulating the CRSP.[62]

Californian's concerns also sprang from its conflict over the Colorado River with neighboring Arizona. That debate had grown more acrimonious over the years, and it had entered the judicial arena in 1952 with the case of *Arizona* v. *California,* eventually decided by the Supreme Court in 1963.[63] The two states wanted the case to be settled before agreeing to support the CRSP, and they took the position that judicial decisions on Colorado River apportionment should be made prior to any further legislative action. On July 15, 1954, California filed a motion with the Supreme Court to ask that upper-basin states be a party to the lawsuit, a move that inflamed the situation. The Court made no immediate decision on the motion, but upper-basin lawmakers and water officials feared that the CRSP might be delayed if they were made to await settlement of the lower-basin dispute.[64]

With the court case pending and other legal questions being raised, California lawmakers and newspaper editors regularly denounced the CRSP as a boondoggle. One article pointed out that Californians' share of the CRSP would be 93 million dollars—and San Francisco County alone stood to pay 5.5 million dollars. Fred Simpson, chairman of the California Colorado River Board, said that while the Board recognized the upper basin's right to use its water, that use should be "subject to the terms and conditions of [the Colorado River Compact and Boulder Canyon Project Act] as the Supreme Court construes them in the case of Arizona versus California now pending."[65] Meanwhile, congressmen such as Harry Sheppard missed no opportunities to denounce the upper-basin bill, saying that "if Congress thought the Central Arizona Project was unbelievable in its scope [with] its ridiculous financing, and its nefarious concepts, they have not seen anything. . . . [Like the Central Arizona Project, the CRSP is] designed to steal water from California [and] would seriously damage California's farms, homes, and industries. These projects would rock the foundations of billions of investments made by California citizens in good faith and with reliance on the integrity of the United States government."[66] Sheppard's colleague, Craig Hosmer, told fellow House members that "Congress might as well appropriate money to grow bananas on Pike's Peak as to approve the Colorado River Project."

Before long, states sharing the river became engaged in a shouting match, and the charges and countercharges escalated. Colorado Congressman Wayne Aspinall claimed that California had little to complain about, for she had benefited from federal water projects along the Colorado River for many years. Considering how much Hoover Dam and the Colorado River Aqueduct had fueled the state's tremendous growth it seemed intolerable that California, "awash in the ever-expanding benefits of basic water development, has the temerity and the gall to rise up in opposition to the legitimate development of the upper basin." Aspinall asserted that all of the complex legal questions raised by Californians about the CRSP were basically a smoke screen, and added that "the lower basin is not ready—in spite of legal agreement and sacred promises to the contrary—to help limit their own potential possibilities just as long as water must flow downhill. The hungry horde in the lower basin wants every possible drop." [67]

Some California lawmakers who denounced the CRSP included criticisms of Echo Park Dam, for it appeared obvious to them that conservationists offered excellent allies. And the alliance worked to the benefit of defenders of Dinosaur Monument as well, although the Sierra Club found itself constantly fending off charges that it was mainly a dupe of southern California water interests. [68] In the intense jockeying for position, upper-basin residents refused to see the alliance for what it was, and assumed that any and all denunciations of the Dinosaur dam merely disguised opposition to the CRSP from the "hungry horde" downstream. New Mexico's interstate streams commissioner, I. J. Coury, argued that California "very subtly, very ably, and very quietly picked out the Echo Park unit and began fighting [the CRSP]. . . . If it hadn't been the Echo Park [Dam,] it would have been another one." [69]

Convinced for years that a major purpose of Echo Park, Glen Canyon, and the other dams was to keep southern California from dominating the river, most upper-basin residents now became convinced that California's water and power interests constituted the major opposition to the project and to the dam in Dinosaur, and they all but discounted conservation interests completely. The Echo Park battle now became embroiled in the wrangling between the two basins. [70] Those who protested the Echo Park Dam were "mainly pawns of . . . California groups seeking to hog Colorado River water," declared the Casper, Wyoming, *Tribune-Herald*. The *Sheridan* (Wyoming) *Press* stated that, "we are positive that those who bludgeon the Echo Park dam project care nothing about Dinosaur National Monument, and care nothing about a national park. All they want is water belonging to others." "We might as well face it," proclaimed the Salt Lake City *Deseret News,* "California wants Colorado River water that belongs to Utah. As long as Echo Park and other dams go unbuilt, she will get that water." [71] From one standpoint, opponents of the dam were not hurt by these charges since their strength came from states outside the region, not counting the Sierra Club, which, in any event, only counted eight thousand members nationwide. Leaders of the Club worked hard to deny that their alliance with powerful forces in Los Angeles and in the Imperial Valley was deliberately meant to block the entire project.

Still, by associating conservationists with Californians, the upper basin became ever more determined to win approval of the CRSP and to keep Echo Park Dam in the package. Every objection that they heard against the upper-basin project from the "hungry horde" only hardened their conviction that Echo Park Dam was needed to capture their rightful share of the Colorado River; that California should object seemed to confirm the likelihood that Echo Park was truly their "last waterhole." As a result, it became that much easier for Senator Watkins and his colleagues to brush off conservationists like flies in the ointment and this, in turn, worked to discredit the latter's suggestions for "high" Glen Canyon and other alternative sites.[72] Fueled by resentment toward California, hostility toward opponents of Echo Park Dam emerged at the Senate hearings in March. Howard Zahniser, soft spoken and unfailingly polite, received a stiff welcome from New Mexico Senator Anderson, who said: "I see now that the program of the conservationists is not just to protect the Dinosaur National Monument. It is to prevent the entire construction of the upper Colorado Basin project. . . . [If] you take out Echo and Glen Canyon all we do is let the waters run to California, and California lets them run to the sea."[73]

From the upper basin's viewpoint, the problem with lower-basin opposition to the CRSP was that it played into the hands of opponents in other sections; only if western states were united, could the legislation have any chance to be enacted. After the war, there had been mounting criticism of reclamation projects in the West in Congress. Eastern and midwestern financial and agricultural interests had by now become so hostile toward reclamation for the West that no bill could be passed unless western states were united. Ironically, it was not until California began to understand this that progress on the CRSP was made. So, for the time being, the upper basin had to regroup, had to find ways to counter all of the criticisms about Echo Park Dam and the overall project. Throughout the fall of 1954, spearheaded by the Upper Colorado Grass Roots

Committee, proponents searched for ways to advertise the project more widely to the nation at large. Deeply frustrated with so much opposition to the CRSP bill, they realized they must do more to present the upper-basin plan as one that would benefit the whole nation, and try to gain approval of the bill in the Eighty-fourth Congress.[74]

Meanwhile, the conservationists realized that they must do some regrouping of their own. Much of the 1954 session had been spent in discussing issues far removed from the dam in Dinosaur, and while this had gained for them allies in the Congress, which delayed the project, it had also detracted from their purpose of upholding the sanctity of the national park system. In some respects, their campaign had been overtaken by the events of the 1954 session. In November 1954, Fred Packard and Sigurd Olson of the National Parks Association decided that too much energy had been devoted to challenging the Bureau's every move.[75] Above all, they must be united in their efforts and remain focused on Echo Park, partly because the great diversity of the organizations involved could be not be kept together around a campaign against the West's water development. Furthermore, as Zahniser, Packard, and other leaders recognized, their best strategy was to keep the public focused on Echo Park as a unique place. Only by capturing the public imagination for this magnificent wilderness and constantly reminding the public of the danger to the national park system could they continue to exert pressure on the Congress. Their campaign could not be sustained by raising questions about the entire CRSP or by wading too far off into its various features, but only by keeping their sights set on Steamboat Rock and the canyons of Dinosaur.

NINE

A Symbol of Wilderness

If the test of a national park must be its popularity with auto-
mobilists, then Dinosaur may not fare too well — and the San
Francisco Bay Bridge, Hollywood Freeway, and Holland Tun-
nel should become national parks (and the New York subway
system, which saves automobilists from themselves, a recre-
ation area.) If, on the other hand, the test of a park lies in
the quality of experience — unmechanized experience with the
magic of the natural world — then an unaltered park at Dino-
saur is, and can always be, as rewarding a source as any of the
great national parks.

—David Brower[1]

At the same time that the campaign to save Echo Park became
embroiled in congressional debate over the CRSP, conservationists,
buoyed by the triumph in the evaporation dispute, took the offensive to publicize Dinosaur National Monument and its wilderness
characteristics. Through a spate of articles in conservation publications as well as in the national press, Echo Park and the canyons
of Dinosaur became the focus of national attention and, indeed,
the primary example of the sort of "primeval" landscape that this
generation of preservationists cherished and sought to preserve. By
emphasizing its unique mixture of white-water rivers, looming
cliffs, archeological remains, and abundant wildlife, conservationists made the case that the monument was part of a rapidly disappearing wilderness, of "vanishing America," as a headline in the
New York Times proclaimed.[2] Here was the heart of their case—
that Dinosaur and other areas in the national park system protected some of the nation's great wilderness, and that these landscapes, at once unchanging, "primeval," and unique, were worthy
of preservation to foster the nation's physical and mental well being. Such claims came forth from a host of sources as the campaign
reached a peak, and before long, one of the least-known parts of

the national park system—a national monument little known to the public and with a vague public identity since its origins—became the primary symbol of wilderness in the United States. In Howard Zahniser's words, the dam "would destroy one of the unique, irreplaceable, scenic, wild wonders of the world."[3] That statement indicated that the Dinosaur controversy had firmly linked the fate of the national park system with the survival of wilderness.

At the beginning of the summer of 1954, just as the other forces aligned against the CRSP began to coalesce in Congress, the campaign to protect Dinosaur still appeared to confront difficult odds. Despite their clear triumph over the Bureau in the evaporation episode, the conservation leaders could not rest in the knowledge that the fight over the dam had been won. When a Senate subcommittee held hearings on the CRSP in June, Zahniser told a personal story that aptly described where matters stood. The Wilderness Society leader had visited Dinosaur Monument the previous summer with his family, and spent an exciting day strolling through Echo Park, "exhilarated and overawed." Standing across the Green River, directly opposite the sheer face of Steamboat Rock, he hollered at the top of his lungs: "Should we build a dam here?" The echo of his voice clipped off the first part of the sentence and the wind brought back the last two words—"dam here?" It was a ghostly feeling, and Zahniser told the subcommittee that the question still echoed in the halls of Congress, the Department of the Interior, and among the public.[4]

But the campaign took the offensive in the spring of 1954, with Zahniser, Brower, Penfold, and company determined to win converts to their side. Numerous articles appeared in conservation journals, including the *Living Wilderness*, the *National Parks Magazine*, the *Sierra Club Bulletin*, and *Planning and Civic Comment*. Some of the publications reached larger audiences, like Jack Breed's article on floating the Green and Yampa rivers in *National Geographic*. Wallace Stegner, then working on a biography of John Wesley Powell, published an essay in the *New Republic*, while Bernard DeVoto continued to harp at the Bureau from his "Easy Chair" column in *Harper's Magazine*.[5] Richard Leonard of the Sierra Club persuaded Eleanor Roosevelt to write against the dam in her syndicated column.[6]

Newspapers across the country joined in, their interest having been fueled by the Bureau's stumble in the evaporation debate. The National Parks Association, the Izaak Walton League, the Sierra Club, and other groups churned out regular press releases, and took satisfaction when major newspapers acknowledged the story. The *Christian Science Monitor* stood firmly against the dam in a number of editorials and news articles. David Perlman of the *San Francisco Chronicle* accompanied one of the Sierra Club float trips in the summer of 1954 before publishing a series of articles supporting the point of view that the monument ought to be preserved and alternate dam sites employed.[7] Meanwhile, the *Washington Post*, the *Milwaukee Journal*, the *Los Angeles Times*, the *New York Times*, and other papers editorialized against the dam.[8] Echo Park was fast becoming the primary conservation issue of the decade.

Defenders of Echo Park found strong support from John Oakes of the *New York Times*. Oakes had written a few articles in his monthly conservation column in the Sunday *Times* that caught the attention of David Brower as well as Otis Peterson of the Bureau of Reclamation, who invited Oakes to float the Yampa. When Brower learned of the trip, he could not resist inviting himself, in part out of fear that Oakes might succumb to the Bureau's propaganda. Brower met Oakes in Denver, and the two drove to Dinosaur and joined Peterson. Finding himself enchanted with the beauty, Oakes called Yampa canyon "some of the most strikingly beautiful canyon scenery in the world" and urged readers to make the trip down the river soon, before the dam erased the exciting white water.[9] At one point during their visit, Oakes and Brower strolled through Echo Park, beneath the shadow of Steamboat Rock. Noticing Peterson coming up behind they turned to hear him say quietly, "don't quote me, but if there is any other possible way to do this, the dam shouldn't be built here."[10] Peterson, though, proved less helpful to the campaign than Oakes, who published more than twenty articles about Dinosaur and gave the controversy prominent attention in the *Times*.[11]

Meanwhile, the Sierra Club continued to circulate Charles Eggert's film "Wilderness River Trail" at meetings of various groups around the country. The Club circulated the film among community organizations, garden clubs, and church and school groups in northern California, and, of course, among its own chapters. In Palo Alto, the Loma Prieta Sierra Club chapter purchased a copy of the film and showed it almost seventy times to more than forty-two hundred people in northern California during a three-month period, beginning in February 1954.[12] By the first part of 1955, the film was available in nearly a dozen cities, including Chicago, Boston, Portland, Denver, New York, and Los Angeles.[13] The Sierra Club planned a new round of float trips with Bus Hatch in 1954. Club members who had gone down the Green and Yampa through Echo Park the previous year had talked excitedly all winter about Dinosaur's massive rock walls, rushing rivers, and hidden wonders; and their enthusiasm sparked additional interest within the Club.

David Brower found time to produce another film linking the campaign to save Dinosaur with the failed effort to save Hetch Hetchy. Entitled "Two Yosemites," Brower aimed the film at members of Congress, to remind them that the earlier battle had ended in tragedy and had caused anguish to John Muir during the last crusade of his life.[14] Using still photographs of Hetch Hetchy Valley, taken early in the century, and contemporary pictures of its fluctuating reservoir, "Two Yosemites" sought to counter claims that Echo Park Dam would enhance the scenic backdrop by creating a beautiful lake. Hetch Hetchy Reservoir gave the lie to such claims, as the film revealed, with the raising and the lowering of the water level leaving ugly mud flats around the rim. Just eleven minutes long, Brower recalled later that "Two Yosemites" proved most effective in the halls of Congress, where Howard Zahniser showed it regularly with a home-movie projector.[15] Brower also published an article by Robert Cutter, entitled "Hetch Hetchy— Once Is Too Often." It reached the nearly nine thousand members of the Sierra Club and another thousand individuals who received copies printed for special distribution.[16] Speaking before the Senate subcommittee which held hearings late in June, Brower said that "if we heed the lesson learned from the tragedy of the misplaced dam in Hetch Hetchy, we can prevent a far more disastrous stumble in Dinosaur National Monument."[17]

By linking Echo Park with Hetch Hetchy, Brower tied the Dinosaur conflict to the most famous controversy in national park history and posed a similar question to a new generation of park supporters and wilderness lovers. Once again, a dam—itself a powerful symbol of human manipulation of the environment in the name of conservation—could intrude into a part of the national park system. This time, as Brower, Zahniser, Packard, and others repeatedly pointed out, the proposed dam presented a more direct threat to the sanctity of the national park system than the dam at Hetch Hetchy, which had come forth prior to the establishment of the National Park Service and its mandate to protect the parks "unimpaired" for future generations. Furthermore, the level of interest in national parks, forests, and other preserves was markedly higher than in Muir's time. The so-called wilderness cult at the turn of the century was confined to upper-class people with the means and leisure to make their way on first-class passenger trains to Yosemite, Yellowstone, and other national parks. Muir lacked the middle-class constituency of the postwar generation of wilderness lovers, who could point to a record number of visitors to parks to bolster their case for preserving Dinosaur.[18]

At the same time, the Echo Park controversy encouraged the view that national parks must be considered as more than playgrounds for newly mobile Americans. Certainly, the park system provided plentiful opportunities for hiking, boating, sightseeing, and a range of activities, and the postwar travel boom generated tremendous pressure from concessionaires and communities close to park boundaries for more roads, campgrounds, hotels, and other public facilities. During the 1950s, as these pressures grew, anxiety increased among leaders of the National Parks Association, the Wilderness Society, and others that the parks would be overrun with people and accommodations. This trend, coupled with mounting pressures from logging and mining interests and dam builders, sparked fears that the premier spectacles of nature contained in the national park system could be lost.[19] The threat to Dinosaur, then, provided a prime opportunity to emphasize the importance of parks as wilderness sanctuaries.

National parks and *wilderness* are often considered to be synonymous, but the two terms were not always interchangeable and it was the Echo Park controversy that did much to make them so. The Park Service had been founded with a dual mission of preservation and providing for public recreation, and the history of the NPS was one of constant tension between these divergent aims, with park administrators and the public constantly reshaping them over time.[20] The battle for Echo Park produced numerous pleas for keeping Dinosaur a "primitive" landscape, and for recognizing the role of national parks and monuments as wilderness sanctuaries. Herbert Levi, professor of botany at the University of Wisconsin, opposed the Echo Park Dam and the Bureau's case that it would attract visitors when he asked rhetorically, in the *New York Times,* whether "the number of visitors [should be] the criterion of national park standards? Is the Administration to place comic strips in the National Gallery of Art, or convert the Smithsonian Institution into a circus, to attract visitors?"[21]

This thread of thought ran through many of the published accounts aimed at gaining public support for an unimpaired monument, and Dinosaur now became a model of the type of primeval landscape that wilderness lovers sought to preserve. Olaus Murie referred to the monument as one of the "samples of original America," a phrase much like one that Illinois Senator Paul Douglas used in 1958, when he called the Indiana Dunes "a rare remnant of the original American wilderness."[22] Retaining the remnants of "primeval" America was not a new invention of park lovers of the 1950s; indeed, it linked them to earlier generations of park enthusiasts, but the kind of rhetoric employed during the Echo Park battle affirmed that this should be the primary purpose of the national park system for a new generation. Put simply, national parks were to be considered the true home of wilderness, of "primeval," untouched land. Such images of an unchanging landscape and the timeless qualities of nature surfaced repeatedly in articles and pictures devoted to Dinosaur. The Lodore and Yampa canyons offered "the most intimate possible contact with the timeless quality of earth," proclaimed one reporter.[23] Evoking such images, of hidden canyons unchanged for centuries, helped make the case that a dam constituted a shortsighted proposition. Seen from the point of view of "eons of time," the dam could only be considered temporary: Within a couple of hundred years its silt load would be enormous, greatly reducing its storage capacity, while its hydropower must soon become obsolete in the coming age of atomic power and other fuels like coal and oil shale.[24]

At the same time, Dinosaur possessed a rich human past, a point emphasized by David Perlman of the *San Francisco Chronicle,* who published a series of articles in June 1954, after a float trip down the Yampa with the Sierra Club. Throughout, he returned to phrases evoking a sense of antiquity in Lodore and Yampa canyons, linking their natural beauty with their history. Wanting to counter the claim that the canyons were an empty land never seen by humans—and therefore worthless if left in their natural state—

Perlman wrote that "the canyon country is the backdrop for a pageant of legend and history that stretches back to the dawn of man." There were caves with cultural artifacts, prehistoric food-storage bins, and walls and mortar that had stood for centuries. As for the recent past, Perlman referred to "tumbled log-and-sod cabins of cattle rustlers and bandits who hid out in this remote, impassable land during the last century."[25]

Taken as a whole, a portrait emerged of a landscape that stood to be lost beneath the dam, a landscape that was both a remnant of a bygone era and a place wholly unique in North America. A caption in the *Sierra Club Bulletin* stated that "the rainbow canyons of the Yampa and the Green, corridors through a primitive paradise unequalled anywhere, are a unique gem of the National Park System."[26] At bottom, the case for protecting the monument sprang from a passion for place, a particular place the likes of which existed nowhere else, and where the canyons, rivers, and wildlife provided "extraordinary opportunities for public enjoyment and refreshment of the most significant kind."[27] This appeal to protect unique places had long been a primary argument for setting aside national parks, ever since the first parks were established in the nineteenth century.[28] Now, the same passion to protect special places, the most unusual and awe-inspiring parcels of nature, underpinned the case for safeguarding Dinosaur and thereby ensuring the survival of the wilderness.

Such an argument should be viewed alongside swiftly changing cultural and economic patterns in the postwar years, and the growing sense of many social observers that American culture had suddenly become too homogenous, being inexorably shaped by suburbs, the automobile, television, and fast-food restaurants. With the nation's developed landscape expanding after the war, the Lodore and Yampa canyons and similar spectacles of nature could only become more valuable, much like a great work of art or a fine example of architecture. Like those cultural icons of civilization, no dollar sign could be placed on their value. To try to do so "would be like trying to arrive at the value of a great painting by adding the cost of the canvas, the paint and wages for the painter," wrote Harold Bradley. "Whether it be very great or very small will depend on your scale of values—whether you are perceptive or not to the intangibles that lie beyond a price tag."[29]

Put another way, the campaign to save Dinosaur did not arise from an interest in what Alfred Runte calls "total preservation" or in deep ecology. Those determined to preserve the monument had in mind the benefits of wilderness for humans rather than any concern for "the rights of nature."[30] As Wallace Stegner said in *This Is Dinosaur,* published by Alfred A. Knopf in 1955, "it is only for human use that [the monument] has any meaning, or is worth preserving." For the recent generation of wilderness lovers, the point may appear anthropocentric. At the time, appeals to protect national parks, monuments, and wilderness areas regularly drew on the needs of humans in an industrialized and urbanized society, where it was assumed that the opportunity to experience such places was vital to maintaining emotional and spiritual well being.[31] Sigurd Olson, president of the National Parks Association, told a Senate subcommittee in 1955 that the threat to Dinosaur must be turned back in order to protect one of the "sanctuaries of the spirit, places where men can find release from the tensions and pressures of a machine age." He continued:

> I sometimes wonder where our much-vaunted industrial civilization is leading us; if our country is going to become a sprawling industrial network that will engulf our quiet little villages; if all the land is going to be used up; if the population is going to go beyond the 200 million predicted for 1975; and eventually reach a point where there is standing room only and no longer any places of quiet and peace. . . . Dinosaur National Monument and the threat confronting it is a symptom of an era and a way of looking at the earth and its resources. It is also indicative of a way of life that is all speed and confusion and noise, where the so-called material values have become more important than the spiritual."[32]

Such appeals struck a chord with Americans of this generation, whose increasing numbers resided in cities and suburbs.

John Muir had made similar appeals in the effort to save Hetch Hetchy, of course; yet he lived at a time when the impulse to acquire a taste of the wilderness had not yet reached sufficient numbers, when the longing for an outing in the backcountry was still mainly confined to the leisured classes. By the 1950s, suburban life and culture had blossomed and taken root with growing numbers of the middle class. Suburbia, as Kenneth Jackson has pointed out, was in and of itself an expression of interest in escaping from cities and living closer to nature. Now the growing number of suburbanites responded to appeals for preserving parks and wilderness areas for recreational opportunities.[33]

In the face of this flurry of publicity about Dinosaur, supporters of Echo Park Dam found themselves on the defensive. They clearly disliked having the monument redefined as a great wilderness preserve, a unique place whose loss would obliterate a remnant of original America. It seemed remarkable to them that such an isolated and poorly funded national monument could achieve such an exalted reputation overnight, and so quickly capture the attention of the nation's press and prominent organizations. Consequently, they responded with their own version of events and their own definition of the national monument, and they sought to prove that its scenery, far from being unique, was replicated in countless other places across the Colorado Plateau. In their counterattack, they also revived words and phrases employed by the National Park Service and the Federal Power Commission officials during the detailed negotiations preceding the expansion of Dinosaur in the 1930s; and they argued that since the dams had been foreseen, Echo Park and surrounding canyons could hardly be considered to be such sacred ground. The ensuing debate centered on whether those "loopholes" for dams should still be respected and whether or not Dinosaur should be regarded with the same high esteem as more famous national parks.

Supporters of the dam attacked their adversaries in editorials and in other forums. Opinion pages in the *Salt Lake Tribune,* the *Deseret News* and the *Casper Tribune* charged conservationists with having undue influence in the national press and in the federal government. The *Salt Lake Tribune* charged that opponents of the project had "access to most of the national magazines and metropolitan newspapers. . . . Many editors . . . still regard the west as a kind of colony, sparsely settled by not-very-bright people."[34] Particularly irksome to the *Tribune* was all the attention lavished on Echo Park by John Oakes of the *New York Times:*

A newspaper of wide circulation and influence apparently bases its editorial policy on the conclusions of one man of limited experience and horizons. The Times staff would be surprised and amused if an editorial writer from the West concluded after a few days tour of New York—and we do go to New York—that the hundreds of millions of federal money spent on New York harbor and river work by the federal government was a waste of money and an imposition on the nation's taxpayers. The Times registers the altogether too-familiar contempt of the Eastern wiseacre for his own stereotype of the Westerner. Its theme closely follows the "West Against Itself" line of Bernard DeVoto—that the people of this region are too dumb or too selfish to know what is good for them.[35]

But more had to be done than merely taking swipes at the eastern press. Theron Liddle, editor of the *Deseret News,* sought to counter claims about the monument's beauty, which John Oakes had compared to other renowned preserves on the Colorado Plateau such as Capital Reef, Arches, Zion, or Bryce. Liddle refused to concede that Steamboat Rock and the environs along the Green and Yampa rivers were as impressive or in any way comparable. Seen from the road above Echo Park, the height of Steamboat Rock matched that of the surrounding canyon walls, leading him to point out that Oakes's objection to Echo Park Dam and "the altering of Steamboat Rock . . . would be similar to my objection to removal of one 20–story building from New York City on the grounds that it would ruin the famous skyline. You have more buildings than you can count—we have more stone mountains than we can count."[36]

G. E. Untermann, curator of the Utah Field Museum in Vernal, who had spent years exploring the canyons for fossils and rocks, claimed that no one should fear the loss of Echo Park because the great promontory in Jones Hole, a side canyon below Echo Park, offered "an exact duplication" of Steamboat Rock. In Untermann's mind "anything you do in the canyons from either a scenic or scientific or biotic point of view is not irreparably harming anything for the simple reason that you have the same thing duplicated in other areas of the monument, and outside the monument." Nor should the effects of Echo Park Dam be considered harmful because only the bottoms of the canyons stood to be covered with water—something that the dam's opponents did not understand. In Lodore Canyon, for instance, walls 3,000 feet high would hardly lose their scenic grandeur with the reservoir height slated to be 3 5 0 feet. Untermann conceded that depth at Echo Park itself would be great, but the pools would diminish in depth farther upstream, while "the lakes produced by Echo Park Dam will modify the *character* of the canyon country but will little effect their grandeur and scenic qualities."[37]

In an essay entitled "Realism and the Dinosaur National Monument Controversy," Untermann railed at the Sierra Club and "natural-born crusaders who are always ready to 'save' anything which they feel is worthy of their best efforts." [38] He refused to accept their claim that running rapids was safe, and he denounced the "insidious propaganda of the Sierra Club that anyone can blunder into the river and come through unscathed without the services of an experienced boatmen. . . . If the Sierra Clubbers want to commit suicide by going through the canyons without guides, that is their business. But if they encourage such foolishness for others they are guilty of homicide." Untermann predicted that river trips "will never be popular with the general public . . . and the monument's interior will remain little known." [39]

Untermann and his side took delight with the testimony of a young couple who ran the rapids in kayaks in the spring of 1954. After reading articles in *Sunset* and *National Geographic,* Barbara and Richard Schall entered the rivers feeling "perfectly safe," but their float trip proved less than enjoyable once they encountered rocks and turned over their boat. In a letter to the Sierra Club, they warned that "no one who is not trained nor has not the proper equipment should ever attempt this trip." On May 13, 1954, the Vernal *Express* publicized the incident with a front-page photo and story of the Schalls and their "narrow escape." [40]

Proponents of Echo Park Dam also argued that promises for the project had been given years before and that such pledges constituted a moral pledge that could not be rejected now. Much was made, for instance, about a series of hearings held in northeast Utah and northwest Colorado in the years prior to Roosevelt's proclamation, when a Park Service official, one David Madsen, a wildlife specialist at the monument, allegedly indicated that the National Park Service did not object to dams. After hearing the concerns of local residents from Vernal, Utah, and Craig, Colorado, Madsen went on record to say that expansion of the monument would not preclude grazing or reclamation projects in future years. Fifteen years later, Senator Watkins and other proponents learned of Madsen's remarks and persuaded him to sign an affidavit, which they then presented to Secretary Chapman in 1950. [41] David Madsen's name eventually became a symbol of promises made in the 1930s, and for John Will and the Upper Colorado River Commission, his affidavit carried indisputable proof that the National Park Service had foreseen and agreed to a dam in Dinosaur. Upper-basin editors made much of Madsen's promises, associating them with the official position of the National Park Service.[42]

Fred Packard of the National Parks Association believed that Madsen lacked the authority to make such promises. Packard told Congress that while Madsen may have given assurance that the Park Service had no objection to dams, he had not ever filed any such statement with the NPS director. Moreover, Madsen claimed in his 1950 affidavit to have been superintendent at Dinosaur in the 1930s, but this was impossible since the monument came under jurisdiction of the superintendent of Rocky Mountain National Park. Packard found Madsen's affidavit unconvincing, for it could not be reconciled with Madsen's own reports to the Park Service, compiled shortly after the local hearings; and Packard eventually concluded that it was "completely false, and . . . of no validity." According to Packard, Madsen did not refer to reclamation in his original report, only to grazing rights, which Packard maintained had been the fundamental issue debated in the 1930s. Packard also thought it was significant that Park Service files did not reveal any authorization to make promises for dams.[43] "Whatever he said," Packard concluded, "was probably in an informal manner, 'over a cup of coffee,' without any intention to commit the Park Service to more than it had already agreed to, and with no thought that his words would ever be used as an issue in an important controversy." Packard also knew that Madsen had a record of exceeding his own authority. Once dismissed by the Park Service, he held a grudge, and Packard contended that this was why he had written the 1950 affidavit in the manner that he employed.[44]

Still, proponents of the dam continued to find justification for the project among records buried in the archives. Along with Madsen's alleged pledge, Senator Watkins and others developed a similar argument, based upon President Roosevelt's proclamation of 1938. A key phrase of that document stated that Dinosaur Monument was made "subject to the Reclamation withdrawal of October 17, 1904, for the Brown's Park Reservoir Site in connection with the Green River Project." That provision offered additional evidence from the 1930s that some kind of water project had been envisioned. In 1955, Wilbur Dexheimer, commissioner of the Bureau, told a congressional hearing that Roosevelt's proclamation "specifically reserved this area for power and irrigation development sites." Dexheimer brought forth evidence of the negotiations between the Park Service, the Bureau of Reclamation, and the Federal Power Commission in the 1930s, when NPS personnel had accepted inclusion of the Browns Park phrase to indicate its recognition of the Bureau's interest in dams in the future. In other words, "Browns Park" had then been interpreted broadly to include Echo Park, Split Mountain, and other locations in the monument, and Dexheimer tried to revive the "broad" interpretation of the Browns Park provision.[45]

The argument carried little weight with U. S. Grant and Fred Packard. Grant preferred to think that writers of the proclamation "said what they meant and meant what they said." Browns Park dam site alone was granted, Echo Park was not.[46] Fred Packard offered a more detailed explanation to the Senate subcommittee. Bureau engineers, he argued, had already agreed before 1938 that the dam site in upper Lodore Canyon could never be utilized. When Park Service officials learned of this from the Bureau, they decided not to wait for the Bureau to nullify the Browns Park site through all of the cumbersome legal channels of the bureaucracy. So Roosevelt issued the proclamation with the Browns Park provision, but it was meaningless because both agencies understood that it was an "infeasible project. . . . The Park Service knew it could not be built, and so could never constitute an invasion of the national park system."[47] Nevertheless, this "broad interpretation" of the 1938 proclamation provided the key to the proponents' case.

Packard, U. S. Grant, and Joe Penfold discovered, much to their delight, that Secretary McKay disagreed with this broad interpretation, and during a meeting with them he said so. To explain his point, McKay laid his finger on the table in front of him and said, "If I authorize a man to use my desk, that does not mean he has the right to use the entire room."[48] Unfortunately for the opponents, McKay went no further to come to their aid, apparently only making the point about Browns Park in order to make some show of support. Frustrated with McKay's weak defense of the monument, Fred Packard told Rosalie Edge of the Emergency Conservation Committee:

> Secretary McKay's attitude towards national parks is decidedly alarming. He views them as glorified state parks, with recreation as their primary objective. He has on numerous occasions referred to

Lake Mead as an ideal national park, although most of us consider it a bastard area that should not properly be under the National Park Service at all, although they are doing a good job of administering it. He has no comprehension whatever of the fundamental concepts for which the national parks were established and for which they are being preserved. He is sincerely baffled that we have not leaped with joy because he is offering us twenty-one million dollars of Reclamation funds to develop Dinosaur National Monument. It is a very dangerous situation, and I do not know whether it will be possible to induce him to correct his thinking on this subject. Heaven knows we are trying.[49]

As McKay continued to waffle, supporters of Echo Park Dam pressed their case that the NPS had given its blessing to the project in the 1930s. Senator Watkins, the most eager and determined proponent of the dam among upper-basin political figures, developed yet another approach to make the point. In late March 1955, Watkins delivered a lengthy speech on the floor of the Senate, based on extensive research by his staff in records from the Department of the Interior and the Federal Power Commission. Calling his evidence indisputable, Watkins presented a great array of letters and documents to prove that the legal groundwork for dams had been established in the early 1920s, long before the enlargement of the monument around the Lodore and Yampa canyons. The heart of Watkins's speech concerned "power withdrawals," claims made by the Federal Power Commission or by the Interior secretary to lands within both canyons. Quoting the 1938 proclamation, the senator explained that administration of the monument had been made subject to the Browns Park reclamation withdrawal of 1904, and that the Park Service had recognized "the operation of the Federal Water Power Act of June 10, 1920, as amended." Roosevelt had also proclaimed the enlarged monument subject to "valid existing rights", a phrase that, Watkins said, "had in mind the water and power withdrawals."[50]

Watkins's determination to prove his case owed in part to the fact that the Bureau's earlier emphasis on minimizing evaporation as a rationale for the Dinosaur dam had been completely discredited, and he sought to bolster the new emphasis on the role of Echo Park Dam in generating hydroelectric power. Opponents of the dam, irritated that the Bureau had chosen not to emphasize its power capabilities during the first several years of the controversy,

had begun to remind the public that Congress had sought to prohibit power facilities inside the national park system under the
1935 amendment to the Federal Water Power Act, which restricted
the authority of the Federal Power Commission within the park
system.[51] Because the amendment had not been aimed at public
agencies like the Bureau, but at private firms, it did not necessarily
restrict the Bureau from building a power dam in Dinosaur. Still,
the amendment had been an indication that Congress intended to
thwart power facilities in national parks and monuments, and now
the amendment served to support those who defended Echo Park.
So Watkins, no longer able to justify the dam on evaporation, knew
he must find evidence that the Echo Park location had been coveted
as a power site from the beginning, and he wished to prove that
the necessary legal claims on the site had been made *before* the
1935 amendment.

Watkins brought forth documents, dating back to 1904, indicating that the secretary of the interior and the Federal Power Commission had withdrawn land for power development at Browns
Park, Echo Park, and Split Mountain. Over two decades, eleven
withdrawals were made; Watkins contended that all of them remained in effect, despite the 1938 proclamation giving the NPS
jurisdiction over the Green and Yampa rivers. He also revived correspondence among various agencies pertaining to the preparation
of Roosevelt's executive order of 1938 and argued that Frank
McNinch of the Federal Power Commission had then refused a
request from Secretary Ickes to surrender its power withdrawals.
McNinch's refusal seemed significant because he had mentioned
the Echo Park site's capacity for generating power. In Watkins's
mind, the 1938 proclamation had not nullified the "power withdrawals," and had protected the legal claims for an Echo Park
Dam. Thus, in his mind, the national monument "invaded" an area
properly reserved for power development.

Armed with this information, he pronounced that the National
Park Service had less of a claim over Dinosaur than power and
water interests and he concluded that "this puts the shoe on the
other foot. It is not a national monument that is being invaded. . . .
I am willing to go even further, and now state categorically, that
the areas now in controversy are not now and never have been
under the exclusive possession and jurisdiction of the National
Park [Service]." Dinosaur Monument had been subservient to the

power-site withdrawals all along, so invasion of the national park system by Echo Park Dam was an impossibility.[52]

It did not take long for a response to come. Fred Smith of New York City, chairman of the Council of Conservationists, told Watkins that the case he had tried to make was academic. Of course, Congress could sanction the dam, but there were thousands of Americans who thought that it should not.[53] Smith questioned Watkins's evidence as well. "It may be a neat legal question," he wrote the senator from Utah, "whether the Proclamation of 1938 actually creates a monument, or . . . only provides elbow room for mud flats." Still, the senator had made a weak case. The FPC refused to vacate its power withdrawals prior to 1938, and then confirmed this fact in a letter to Watkins nearly twenty years later. Watkins had presented the letter as proof. For Smith, this was tantamount to settling a difficult dispute by invoking the authority of a principal character in the controversy.[54] U. S. Grant III quickly responded to the Watkins history lesson too, insisting that "valid existing rights" were routinely included in such decrees so the federal government could protect itself from lawsuits that might arise if private property were obliterated without due process of law. The power withdrawals "in force in 1938 were not vested rights in private property, but merely the earmarking of certain areas by a bureau of the Federal Government."[55]

Congressman John Saylor supported Grant's point when he took his turn rebuking Watkins in the House of Representatives. Saylor relied on a legal brief requested by the Wilderness Society and prepared by William Norris, of a law firm in Washington, D. C. He agreed that the power-site withdrawals were not as legally sacred as the Utah senator suggested. This type of land withdrawal had been given to the executive branch simply as a way to reserve public lands by closing them off to private ownership. Withdrawals for dam sites had become routine; presidents and executive agencies filed for them often, and the number of withdrawals far outran the number of dams actually built. Presidents could file such withdrawals at a moment's notice and cancel them with other executive decrees. Roosevelt, Saylor concluded, saw no need to nullify withdrawals on land within the Yampa and Lodore canyons because his proclamation expanding the monument "revoked the withdrawals by implication."[56]

While Watkins did not convince any opponents of the dam, his efforts solidified public support for the project throughout the upper basin. The senator from Utah had obviously done some homework in trying to prove the prior reservation of the Echo Park site, and it followed that opponents of the project were simply bad historians, harping about a precedent against the national park system and completely unaware of the special circumstances behind the creation of Dinosaur. "The issue of an invasion of national parks is a complete phony," proclaimed an editorial in a Wyoming newspaper, "and was so established by Senator Watkins . . . The phony issue is being used, however, with some measure of success to deprive the upper Colorado Basin states of the waters which originate in them."[57] If there had ever been any doubt that Echo Park Dam should be built, Watkins removed it for good, and his proof thereby added to the notion that the conservation opposition masked California's interest in stealing the water. Furthermore, Watkins's argument appealed to many residents long frustrated with the federal management of public lands. According to the Utah senator, the upper Colorado basin contained as much acreage in national parks, monuments, and recreation areas as the total acreage in the New England states, a fact which caused the western states to be highly sensitive to any notion that conservationists regarded all of it as a huge playground for their own enjoyment.[58]

Debates such as these—detailed discussions over the meaning of the Browns Park proviso, the alleged remarks of David Madsen, and all the wrangling over Watkins's speech—offered a reminder that Dinosaur National Monument lacked a clear identity and had been ambiguously defined since its inception in 1915. Not surprisingly, as such debates continued, so did charges and countercharges about the prospect of flooded fossil bones and about the name *Echo Park* itself. In some cases, careless reporting or a lack of careful research probably contributed to confusion over flooded bones. In March 1954, William Blair incorrectly reported in the *New York Times* that Echo Park Dam would back water into the Green River gorge "at a point where skeletons of dinosaurs are imbedded in the cliffs."[59] In an article that outraged the monument's defenders, *Time* said that opponents of the dam had argued "for years" that it threatened fossil remains.[60] Fred Smith retorted that *Time* had employed McCarthy-like tactics by resorting to the "big lie." Smith accused *Time* of a highly biased article and of flagrantly engaging in propaganda. A furious David Brower sent a long letter to *Time* publisher Henry Luce, pointing out twenty errors and misleading statements in the short article and suggesting greater consultation with conservationists.[61]

While the name *Dinosaur* continued to be a source of confusion and debate, so, for that matter, did *Echo Park*. The word *park* was simply a descriptive term for open valleys in the high country of the West, but the word became embroiled in the wrangling. *Park* posed a problem for supporters of the dam because they had to stress that this was not a national park, a point that prompted opponents to reply that Dinosaur did belong to the national park system.[62]

The National Park Service was caught in the middle of this confusion and wrangling. Despite Director Wirth's interest in preserving Dinosaur, he could not offer any public opposition to the project, since Secretary McKay had given his approval, and as Wirth wrote to Harold Bradley, "you just cannot go out and criticize your boss."[63] Wirth, like Drury before him, found the constraints more than a little frustrating, and he could only encourage those fighting the dam from behind the scenes.

In 1950, the Park Service had renewed its presence at the quarry portion of Dinosaur Monument, and excavations on the fossil bones had been resumed. Paleontologist Theodore "Doc" White had uncovered additional bones, and plans had been made for building a public exhibit. The little station at the quarry became filled with curious sightseers. Yet, while interest in the dinosaur quarry had been revived, many travelers who took the road north from Jensen to the quarry often had more questions about Echo Park than about dinosaur fossils. And with the Park Service unable to take any public stand on the dam, visitors found themselves unable to obtain answers to their questions.

Ranger Harry Robinson found the situation more than a little trying. Every day people pressed him with questions about the dam, while stacks of letters demanding answers accumulated on his desk. Whose idea was it? How could a dam be built in the national park system? Robinson could only do his best to explain the situation. He could tell them the location of Echo Park and the location of the proposed reservoir but he could not mention the Bureau's evaporation errors or the state of Utah's keen interest in the CUP or other political aspects of the controversy. Obliged to answer the letters, Robinson could not reveal his own opinions or those of the Park Service.[64] Such were the hazards of belonging to the same department as the Bureau of Reclamation. At the same time the controversy provided much free publicity for the remote monument. In 1954, seventy-one thousand saw the quarry and the canyons, more than three times as many as in the previous year. The Echo Park controversy, Robinson informed a friend, "has given us a million dollars worth of free publicity."[65]

Robinson knew that residents of Vernal had not been idle during the controversy, and that the prospect of a huge construction project and reservoir had long been the talk of the town. Vernal depended on tourism greatly, and many residents had long believed that the Park Service had allowed the monument to languish. Echo Park Dam would change that situation dramatically, not only because a large reservoir would attract boaters and fishing enthusiasts, but because federal money would flow into Vernal and nearby communities for several years, during the construction period. Some estimates held that at least one thousand people would be hired for the construction project itself, and another two thousand jobs would arise in surrounding communities from the tourist dollars. Moreover, after completion of the dam, the Park Service stood to receive twenty-one million dollars from the Bureau of Reclamation for establishing roads, campgrounds, and other facilities alongside the reservoir.[66]

Local interest in the project ran so high that people in Vernal looked carefully upon the influx of newcomers to their town, ever suspicious that they might be connected to the opposition. Even local river runner Bus Hatch found himself ostracized. One local filling station had offered to provide Hatch with space for a booth, where he could arrange with tourists interested in joining him for a float trip. Viewing Hatch as a partner of the Sierra Club in its river trips, local residents boycotted the station and purchased their gasoline elsewhere, prompting its owner to scuttle the arrangement with Hatch.[67] Meanwhile, Vernal residents began a letter-writing campaign to members of Congress to urge their support for the dam. In a letter to Merrill J. Mattes of the National Park Service, Harry Robinson complained that it was "too bad we haven't had all this crusading back of the monument. If such energy had been unleashed a few years back in favor of development of the area as a park, we'd not have been in the midst of such a controversy today. And I have to keep my mouth shut."[68]

While Harry Robinson saw how the controversy looked from Vernal and the monument, members of Congress were engaged in a broader debate about the central principle considered to be at stake: was a national monument the same as a national park, and more to the point, should it be held in the same degree of esteem? This question had been bandied about within the Park Service for many years, particularly after the 1930s when many national monuments came under the wing of the NPS during the agency's reorganization and expansion. Yet the question had never been satisfactorily answered, and it certainly had not been debated in public in quite so dramatic a fashion as now. Of course, the legal difference was clear enough, with parks being created by an act of Congress and monuments by presidential proclamation under the Antiquities Act. Supporters of the parks maintained that the names meant little, that any distinctions made between monuments and parks detracted from the fact that both belonged to the park system, and that both types of preserves deserved the same level of protection. In hearings before a Senate subcommittee in March 1955, Fred Packard said that "anything happening to damage a national monument would serve as a precedent for damage to a national park."[69]

Yet drawing a distinction between parks and monuments offered a weapon to backers of the dam. In April 1955, Wyoming Senator Joseph O'Mahoney seized on the point in a debate with Oregon Senator Richard Neuberger, who had just then introduced an amendment deleting the dam from the CRSP. O'Mahoney asserted that national monuments had a different history than parks, and that because they were established by executive decree rather than by Congress they did not deserve the same degree of public support. O'Mahoney, along with his colleague in the Senate, Frank Barrett, had special reason to be sensitive to this question about presidential power and the Antiquities Act, for they were fresh from a long struggle in their state over the creation of Grand Teton National Park, a battle that left them scarred by presidential use of the Antiquities Act. In 1929, Congress had established this park to protect the Teton range itself; and in the following years, conservationists, led by National Park Service Director Horace Albright and John D. Rockefeller, Jr., worked to expand the boundaries to include the lakes and flatlands of Jackson Hole. Rockefeller purchased thousands of acres in the valley, intending to donate them as an addition to the park. Congress could not pass such a bill, largely due to western opposition to such federal "land grabs." President Roosevelt, though, felt compelled to find a way to accept Rockefeller's gift, and he did so by circumventing Congress and proclaiming the creation of Jackson Hole National Monument in 1943, a move that enraged the state of Wyoming. O'Mahoney and Barrett tried to nullify the action in Congress, but failed to do so; and after a bitter battle, they conceded to a bill in Congress that made the monument part of an enlarged Grand Teton National Park.[70]

What made Roosevelt's move so galling to the state of Wyoming was that Jackson Hole National Monument had been designed to protect the sage-covered flatlands of Jackson Hole, not the rugged and picturesque Teton range, which had been protected as a national park since 1929. O'Mahoney now seized upon that fact to make a case that national monuments traditionally lacked the scenic qualities of parks. He argued that none of the famous national parks, including Yellowstone, Glacier (or Grand Teton!), had ever been monuments and that only national parks—established by Congress—protected great scenic preserves. O'Mahoney held that the original Dinosaur Monument had been in compliance with the Antiquities Act to protect scientific resources, but its expansion in 1938 to protect the Yampa and Lodore canyons had violated the spirit of that law. Furthermore, O'Mahoney argued that Dinosaur lacked the scenic qualities ever to be designated as a national park, so there should be no reason to fear that approval of Echo Park Dam would injure the integrity of the national park system. "Whatever may be done with respect to a monument," O'Mahoney said, "is by no means a precedent with respect to what may be done with a national park."[71]

By making this distinction between parks and monuments, O'Mahoney sought to bolster Senator Watkins and those who pointed to loopholes that allowed for dams in the presidential proclamation enlarging Dinosaur. "I venture to say," the Wyoming senator proclaimed, "that the national parks have not had a more persistent defender than I . . . but when the senator from Oregon suggests that a national monument created by an Executive Order which specifically preserved existing rights would set a precedent for invading national parks, I say the facts are all against him."[72] O'Mahoney made a neat case meant to calm fears of fellow lawmakers about the national park system being at stake in the Echo Park project.

Yet his argument had flaws, most obviously his point that Grand Teton National Park had not been preceded by a national monument when only a few years earlier Jackson Hole National Monument had caused him to rage. Olaus Murie, angry that O'Mahoney raised the false issue of flooded dinosaur bones, now had further reason to scold him for misleading lawmakers about the recent history of Jackson Hole. And to suggest that Dinosaur could not become a national park due to inferior scenery was simply a personal opinion, one that Murie totally rejected.[73] Furthermore, as historian Robert Righter has recently shown, a number of national monuments did become national parks, because supporters of these preserves relied on presidential use of the Antiquities Act as a delaying tactic to protect a place until Congress could act.[74]

This discussion about the distinction between parks and monuments did not change opinions on either side of the Echo Park debate, and in that sense, resembled the other disagreements involving David Madsen and "power withdrawals." But the open discussion about parks and monuments and about Dinosaur itself indicated to conservationists that they must do more to demonstrate that the monument contained more than sagebrush or other pedestrian kind of scenery. By 1955, several hundred Sierra Club members and many other people had discovered the stunning beauty in the Lodore and Yampa canyons, and few considered its quality anything less than that of other western national parks. Articles in the *Sierra Club Bulletin, Living Wilderness,* and other publications had testified to its grandeur, but now it seemed clear that such publicity had reached too limited an audience.

David Brower decided that the time had come to reach a larger public with a book, and he began to search for a publisher and editor. For Brower, the book should serve the same purpose as "Wilderness River Trail," "Two Yosemites," and the illustrated articles that he had published in the *Sierra Club Bulletin:* to reveal the monument's landscape and its varied resources. In his words, the "primary object of the book is to let the people know what's there [and] whether it is to be a milestone or a headstone—this question is for the people and Congress to decide."[75]

Brower found a publisher who had a deep love of the national parks and of conservation. Alfred A. Knopf had long supported the parks and decided that a volume of essays about Dinosaur would be a fitting way to demonstrate his long love affair with them. Knopf thought profits were a secondary consideration in this case, with the main task being to persuade Americans that Dinosaur deserved more than a giant dam. Encouraged by Knopf's willingness to publish a book on short notice, Brower approached Wallace Stegner to edit the work. Having just published a biography of John Wesley Powell, *Beyond the Hundredth Meridian,* Stegner seemed a natural choice. He had immersed himself in western history for nearly a decade while writing the book on Powell. He had written about the canyons through which Powell had gone and, having lived part of his youth in Utah, he had a feel for the Colorado River and Plateau. For Stegner, Powell embodied the devoted conservationist of the nineteenth century, advocating careful land and water management, and near the end of the biography Stegner speculated that Powell would not have thought well of a dam at Echo Park. Powell might have been displeased, Stegner mused, with a plan of such giant proportions.[76] Stegner also agreed with Brower that the book should above all reveal the monument's vast size and enormous beauty, and demonstrate its national significance in the park system. "The mere weight of a book does some good; anything worth making a book about should be worth saving," he wrote to Harry Robinson. "And so I think without entering the controversy too directly, we can do something to counteract the yawp of publicity and misrepresentation that the Vernal boys are putting up."[77]

This Is Dinosaur became the first book-length publication in conservation history that sought to publicize a park or wilderness preserve and to aid in a preservation campaign. It became the foundation for one of David Brower's central contributions to the environmental movement in the decade to come: the Sierra Club's series of Exhibit Format books.[78] Stegner's editing of *This Is Dinosaur* proved to be just what the Sierra Club and other groups in the campaign wanted. Filled with colorful pictures of Echo Park and "its magic rivers," the book was equally impressive for its excellent essays. Knopf himself wrote one, insisting that "the very special purposes of recreation, education, refreshment, and inspiration for which Parks and Monuments have been set aside prohibit many economic uses which are thoroughly legitimate elsewhere."[79] Eliot Blackwelder contributed an essay on geology, and David Bradley entitled his entry "A Short Look at Eden." Joe Penfold and Olaus Murie focused on "The Natural World of Dinosaur," and archeologist Robert Lister supplied a readable essay on prehistoric life in the canyons. Otis Marston, an experienced river-runner and historian of the Colorado River, recounted the history of those who had descended the Green and Yampa. Stegner supplied an introduction to the volume, as well as an essay that placed the region's past into the context of western American history.

Stegner insisted in his preface that the authors deliberately chose not to "make this book into a fighting document" because too much "bad feeling and bad prose" had already been generated by Echo Park Dam. *This Is Dinosaur* had a different aim, to provide a survey of "the national resource [and] its possibilities for human rest and recreation and inspiration."[80] Stegner hoped that the book might show Americans what they stood to lose from Echo Park Dam. To help ensure its effectiveness, Knopf agreed to donate a copy to each member of Congress. Knopf told Otis Marston he was delighted with the book and hoped "that it will do the work in the House."[81] Because Brower wanted to leave no doubt about the book's message, *This Is Dinosaur* included a brochure inside its back cover, displaying a picture of mud flats at Lake Mead with a quote from Secretary McKay as its caption: "What We Have Done At Lake Mead is What We Have In Mind For Dinosaur."[82]

This Is Dinosaur firmly linked the sanctity of the national park system with preserving wilderness. Behind the pictures and eloquent words of the book came forth the philosophy of wilderness activists in the postwar years. The Echo Park campaign had begun in order to reaffirm the nation's commitment to the national park system—to strengthen the National Park Service Act of 1916 with its mandate to conserve the parks "unimpaired for future generations." But over the course of this battle the fate of the monument had become inextricably tied to wilderness preservation, thanks to all the intensive publicity of Echo Park and the canyons of Dinosaur, where the spectacles of rock and white water captured the essence of "wild" land to this generation of preservationists. Stegner offered the clearest exposition of the relationship between parks and wilderness in the preface of *This Is Dinosaur*. While an actual wilderness preservation act lay years in the future, a decision to preserve Echo Park in the present would represent a vitally important step in that direction. In an eloquent passage that foreshadowed the language of his "Wilderness Letter" of 1960 as well as the Wilderness Act of 1964, Stegner caught the larger moral dimension of the decision at Echo Park:

> It is legitimate to hope that there may [be] left in Dinosaur the special kind of human mark, the special record of human passage, that distinguished man from all other species. It is rare enough among men, impossible to any other form of life. It is simply the deliberate and chosen refusal to make any marks at all. . . . We are the most dangerous species of life on the planet, and every other species, even the earth itself, has cause to fear our power to exterminate. But we are also the only species which, when it chooses to do so, will go to great effort to save what it might destroy . . . in the decades to come, it will not be only the buffalo and the trumpeter swan who need sanctuaries. Our own species is going to need them too. It needs them now.[83]

This Is Dinosaur appeared in the spring of 1955 when the battle had reached a peak in Congress and conservation groups were exerting every ounce of pressure that they could muster. By linking the fate of the national parks with the survival of the American wilderness, *This Is Dinosaur* set the stage for the final thrust of the campaign.

That campaign had undergone some important changes late in 1954, changes mainly in organization and funding. In 1954, the Supreme Court issued a decision upholding the Federal Lobbying Act of 1946, which made clear to conservation groups that they could no longer lobby against the CRSP without threatening their tax-deductible status. For groups like the Sierra Club and the Wilderness Society, as Michael Cohen has pointed out, a choice had to be made either to register as lobbyists or to renounce any kind of activity (such as direct-mail campaigns) that could be construed as lobbying. The Sierra Club, for its part, debated the issue carefully. David Brower thought the Club should become a registered lobbyist and proceed with its vigorous battle against the dam. He argued that the growth of the Club was generating sufficient monies to fund the lobbying efforts and that the Club could afford to dispose of its tax-deductible status. Richard Leonard, a tax lawyer and former president of the Club disagreed. He maintained that a vital source of the Club's strength resided in its image as a nonpartisan and educational organization. Leonard worried about reaction from members if the Club decided to join the ranks of lobbyists in Washington. His view prevailed.

Consequently, the Sierra Club helped to organize a separate entity, Trustees for Conservation, one of three organizations with lobbying status that prosecuted the Dinosaur campaign until its end. The Trustees, like the Sierra Club, was a nonprofit group, but donations to it could not be deducted. Donations sent to the Sierra Club or to the Wilderness Society for the Dinosaur campaign now had to be returned to the sender with a request to mail the check to one of the lobbying groups. Together with the other two lobbying groups, the Trustees for Conservation coordinated the lobbying efforts against Echo Park Dam for the remainder of the campaign in 1955 and in early 1956.[84]

While the Trustees came together in San Francisco, the Council of Conservationists arose in New York City, following a meeting of twenty-eight organizations at the Plaza Hotel in November 1954. Those present included the General Federation of Women's Clubs, the American Nature Association, the American Planning and Civic Association, the American Museum of Natural History, and the Conservation Foundation, as well as the Wilderness Society and the National Parks Association. The Council had first been established several months before, during a campaign in New York state to block a dam at Panther Mountain in the Adirondacks. At that time, funding for the organization came from a wealthy St. Louis businessman and Sierra Club member, Edward Mallinckrodt, Jr., who owned a summer home in the Adirondacks and opposed the dam. On November 8, 1954, New York voters turned down the controversial dam, giving satisfaction to Mallinckrodt that his money had been well spent and encouraging him to support the Council in its campaign to protect Dinosaur.[85] Supplied with his funding, the Council of Conservationists became the best supported of the three umbrella lobbying groups. Public relations expert Fred Smith ran the Council, while Brower, Zahniser, Ira Gabrielson, Joseph Penfold, and Anthony Smith sat on the Executive Board.[86]

The Council's Board members, it should be understood, did not formally represent their organizations. Instead, they served as individuals and claimed to represent the "conservation sentiment of the country."[87] In a very practical way, however, formation of the Council, Trustees, and Citizens Committee on Natural Resources welded the NPA, the Wilderness Society, the Izaak Walton League, the Sierra Club, and other groups into a powerful force and imposed unity on the Echo Park campaign. Despite their differences, the multiple threats to national parks, monuments, and wilderness areas, which erupted after the war, brought conservationists together in a show of force and revealed the strength they possessed in numbers. On the other hand, staying focused on Dinosaur had become increasingly difficult, especially once the battle had reached Congress and questions had arisen about the economics of the whole upper-basin project. At that point, as we have seen, strategy and tactics had become a source of disagreement. Some, such as Brower and DeVoto, had been inclined to challenge the entire project, while others had constantly worried that the campaign had lost its focus and had wandered too far off into questions about the Colorado River Compact and the cost accounting of the CRSP. Concerns began to emerge that the campaign had lost some of its bearings and had dispersed its energies in joining with other forces.

Although gaining allies had been vitally important, it also seemed clear to various leaders of the campaign that conservationists must remain distinct from other forces in battling the bill, and that they must keep their sights firmly set on Dinosaur. Sigurd Olson, a major figure in the Wilderness Society and the National Parks Association, indicated this to Fred Packard in early November 1954. Olson believed that although much had been learned about the shortcomings of the overall project, there was little to be gained in trying to force an overhaul of the Bureau's whole plan. He thought that there might be danger in continuing to question the economics of the CRSP, and he insisted that "we should be in the position of fighting for a clear cut principle, rather than get involved in the multitudinous ramifications of the entire Upper Colorado Project."

Olson's point concerned tactics, not philosophy. He thought that the campaign against the dam had gained a substantial amount of momentum, but he also felt sure that unless the focus remained on guarding the national park system, the coalition might weaken and would risk losing the battle. Signs had begun to emerge from the upper-basin states that they might be willing to abandon Echo Park Dam if conservationists agreed to support the rest of the project. As early as September 1954, the Colorado Water Conservation

Board conceded that opposition to the dam was mounting and might be insurmountable.[88]

Olson had helped to organize the strategy session at the Plaza Hotel, where he urged unity on the goals and tactics of the campaign during the upcoming (Eighty-fourth) session of Congress, when the upper basin would renew the fight. He reminded those present that the campaign had begun over protecting the "primeval" park system and that it must continue in that vein. "The mere existence of [wilderness and parks] serves as a reminder of our past, gives us respect for the courage, hardship and vision of our forefathers, and serves as balance wheels to the speed and pressures of a high-powered civilization. It is good for the moderns to experience the wilderness," he added. "It is part of the cultural background of America. . . . Any development in any national park or monument which destroys [this]," he declared, "is breaking faith with the original intent of Congress to pass these areas on unimpaired."[89]

The meeting in New York led to agreement on several principles, all meant to refocus the campaign. First, because the national parks had become ever more popular they deserved full protection; second, the coalition must oppose legislation to authorize Echo Park Dam or any dam inside the park system; third, it should recognize the importance of water in western states and agree to support the CRSP once threats to the park system were eliminated; and finally, the groups agreed that the Bureau had yet to respond adequately to proposals for alternative sites and to prove the need for Echo Park Dam.[90] All of these goals helped to redirect the campaign and provide a foundation for the next session of Congress. Ironically, the lobbying groups would soon find themselves publicly opposing the entire CRSP, though this was because of the way that the climax of the battle unfolded in the following session of Congress. In any event, the upper basin was coming back to Congress to fight a second round over the dam. The battle entered its final phase.

Triumph for the Park System

I have a big future ahead of me. And I want to be able to meet it here in my own home state. But if we run out of water what future do I have to look forward to here? I truly believe the great Echo Park Dam Project, on the mighty Colorado River, will be our salvation.
—thirteen-year-old Melvin Besser of Springville, Utah[1]

Senator [O'Mahoney] in his appeal that the rest of the country be merciful to those arid or semiarid States has forced me to say something I had hoped I would not be compelled to say. It is that already the rest of the country is paying through the nose for the 16 votes which the great Mountain States have in the United States Senate.
—Senator Paul Douglas[2]

When the Eighty-fourth Congress convened in January 1955, the Colorado River Storage Project was a major item on the agenda. The upper-basin states had girded themselves to renew the fight, and they came prepared to argue their case for Echo Park Dam and for the rest of the project. Battered in the previous session of Congress by eastern and midwestern farmers and by organizations such as the Engineers Joint Council and the Hoover Commission, they realized they must overcome the notion that the CRSP had been designed primarily to serve regional interests and therefore constituted "pork barrel" legislation. The bill had failed in the previous session for several reasons, but as upper-basin lawmakers, members of the Upper Colorado River Commission, and other strategists saw it, not enough members had been adequately informed about its benefits for the nation. Too many doubts had been raised about the 1.5 billion dollar bill, and too many members of Congress believed that it would be of benefit only to the region. That impression had to be changed, so supporters of the bill set out to persuade members of Congress from the East, South, and

Midwest, and to advertise the project as widely as possible with civic leaders, chambers of commerce, and newspaper editors around the country.[3]

The bill had also been hammered by the state of California, which upper-basin strategists felt was by far the most critical opponent they needed to appease. The problem with California's intransigence on the bill was that it played into the hands of easterners, midwesterners, and bodies like the Hoover Commission that sought to reform such expensive water projects and reclamation policy generally. California did not present the strongest force against the CRSP, but so long as western states remained divided on the legislation, the upper basin faced difficult odds in gaining approval of the bill. In the face of so much hostility toward reclamation from other sections, the West had to present a united front. As a result, upper-basin lawmakers worked tirelessly to court California Senators William Knowland and Thomas Kuchel, as well as Northcutt Ely and other legal representatives of lower-basin interests, by repeatedly assuring them that the CRSP did not seek to overturn the Compact or to threaten California's rights to Colorado River water. But this proved to be difficult, and as southern California lawmakers continued their verbal barrage against the bill, suspicions continued to grow in upstream states that California had designs on seizing as much Colorado River as it could.

Meanwhile, the impression remained widespread throughout the upper basin that opponents of Echo Park Dam were simply a tool of water and power interests downstream and that all talk about wilderness and the sanctity of the national parks was hyperbole—a disguise for the real opponents of the project. Given that perception, upper-basin lawmakers basically ignored the conservation opposition and stoutly refused to surrender the controversial dam in Dinosaur National Monument, insisting that this would undermine the entire upper-basin project. When Congress convened in January, Senators Watkins, Anderson, and O'Mahoney, along with their counterparts on the House Interior Committee, sought to push ahead with the bill and try to mollify other opponents of the project, whom they considered far more important in terms of their political strength than the preservationists. Such a strategy made sense, particularly for upper-basin senators who expected a favorable vote on the bill by the full Senate, where the conservation vote could much more easily be ignored. Senators did not fully recognize the mounting pressure against the dam in the House, and in any event, they chose to ignore it. When Senator Watkins delivered his speech on power withdrawals in March 1955, he emphasized that "the only water left to us, for development and growth in [Utah], is the water of the upper Colorado. The Echo Park Dam is a key dam which is necessary for the successful operation of all the other dams we have, in order to make this entire project feasible."[4]

Watkins and his upper-basin colleagues thereby renewed their efforts to publicize the CRSP and to extol its merits to the nation at large, and they had reason to believe that the legislation might finally be approved during the upcoming session of Congress. Their optimism grew in January, when President Eisenhower mentioned the CRSP bill in his State of the Union message, urging Congress to approve it. Eisenhower's support of the bill ensured a certain number of votes from Republicans in Congress.[5] Meanwhile, newspapers from Colorado, Utah, Wyoming, and New Mexico trumpeted the CRSP with their best rhetoric. The *Salt Lake Tribune,* in its annual "Empire Edition," focused on the project with a series of articles, asserting that the Colorado River was "the last big water hole" in the intermountain region. Secretary McKay contributed an article to the edition, calling the CRSP "the key to the future agricultural and industrial growth of the Rocky Mountain West." The *Tribune* published thousands of extra copies and delivered them to members of Congress, business people, governors, and editors across the nation, making sure that they understood the benefits of an Echo Park Dam to the region and to recreationists everywhere:

> The reservoir would become both a means of transport and the center of recreation within the monument. Boats would ply its calm waters in utmost safety. Resorts would spring up along its shores and a variety of game fish would find the waters habitable, just as they have in Lake Mead behind Hoover Dam. . . . No scenic features of the monument will be destroyed. Instead, they will be enhanced with the addition of a vast lake, and made available to all."[6]

Perhaps the most effective source of propaganda came from the Upper Colorado River Grass Roots, Inc., an organization with more than 100,000 members throughout the region. The "Aqualantes" found able leadership under Tom Bolack of Farmington, New Mexico, who told a Senate subcommittee that the group was "dedicated to the proposition of defending their right to develop the water resources of their States."[7] Bolack used contributions from members to print and distribute handsome flyers and hand-bills as well as to make a film entitled "Birth of a Basin"—all supporting the case that the project offered benefits to the federal treasury from an increased tax base in the West, and to the national defense. The Aqualantes sent letters to daily newspapers, solicited funds from local civic groups, and urged upper-basin business leaders to write their parent companies in the east to bring pressure to bear on their congressional representatives. This kind of advertising of the project to the nation at large was central to the strategy in the early going of 1955.[8]

To strengthen that argument, supporters stressed how the project would aid in developing the vast reserves of coal, uranium, and oil shale from the upper Colorado basin, and that these resources ensured industrial growth in the West, a region whose strategic importance in military affairs had become markedly more critical since the war. With such a spate of reserves waiting to be tapped and with tensions with the Soviet Union growing apace, water and power from the Colorado River might make the difference in the event of this generation's worst fears. "Dispersion of industry could mean the difference between winning or losing a war," Wyoming Governor Milward Simpson proclaimed to a House subcommittee. "The great distances available only in this area are the best means of escaping from the deadly atomic fallout."[9]

New Mexico Senator Anderson led the charge on the bill in the Eighty-fourth Congress. He introduced the legislation (S. 500) on January 18, 1955, with a bipartisan group of western senators acting as cosponsors. Anderson, who succeeded Colorado Senator Millikin as chair of the Subcommittee on Irrigation and Reclamation, announced that hearings before the subcommittee would begin on February 28. They took place in the Senate Caucus Room where the Army-McCarthy hearings had been held in 1954.[10]

Anderson wished to thwart opposition from the East and the Midwest, and, in Richard Baker's words, to avoid "a battle of the sections."[11] He knew full well that dozens of members of Congress who were reluctant to vote for the bill shrank from the thought of adding agricultural land at a time of surplus crops. Les Miller had written an article for *Reader's Digest,* in which he declared that "no new farm land is needed to feed America today. To force arid lands to produce food in the teeth of an already surfeited market may bring disaster."[12] Anderson, former secretary of agriculture as well as chairman of the World Food Board, defended the CRSP from Miller and other such critics, arguing that the nation's present surpluses of food would disappear within a decade due to population growth in the United States, which he projected to be two hundred million by 1975. He stressed that the country must remain self-sufficient in its food supply, and pointed to the steady reduction of farm land due to the construction of highways and cities.[13] Anderson also pointed out that world-population growth would increase the demand for food from industrialized nations, an argument that carried weight before the "green revolution" of the 1970s. Finally, he emphasized how the reclamation projects would aid farmers and ranchers across the upper basin, burdened by recent droughts in 1953 and 1954.[14] Lower than average snowfall over two winters made for exceedingly dry rangelands, and in southwest Wyoming and western Colorado the situation had become desperate. Ranchers feared that their stock would starve because of poor feed crops, and many had to sell stock at low prices for lack of feed. "The past two years have been made of the stuff that tries men's souls," cried Clifford Hansen, president of the Wyoming Stockgrowers Association, and urged passage of the CRSP.[15]

Even before Senator Anderson's subcommittee completed the hearings in March, there was no doubt that the Senate would approve of an upper-basin bill. The major question was whether the Senate could then persuade the House to accept its version of the legislation, and that appeared to be highly unlikely. In the first place, S. 500 called for a project costing nearly 1.5 billion dollars, making it substantially larger and more expensive than the version being developed in the House Interior Committee. The Senate's bill sought authorization of Echo Park, Glen Canyon, and Navajo dams in the first phase of the project, and provided funding to complete engineering studies of the Curecanti Dam on the Gunnison River in western Colorado. The bill also included thirty participating projects, many designed to irrigate rangelands in each of the four states. Senators did not expect to win approval of all that they wanted, but they did want to keep as many dams and irrigation projects in the package as they could, so as to have room to maneuver with House members when the bill reached a conference committee. Generally, they hoped to demonstrate the value of the CRSP to the nation, keep the Dinosaur dam in the bill, and force recalcitrant House members to support the whole package. Such a strategy, though logical in one way, did not take into account the tremendous opposition to Echo Park Dam building in the House, and that pressure eventually proved fatal to the bill during the session.

Meanwhile, the Senate Subcommittee on Irrigation and Reclamation finished its hearings on March 5, then eagerly waited for the Interior committee to act. When that time came in late April, Oregon Senator Richard Neuberger, fresh from his debate with Senator O'Mahoney about the distinction between parks and monuments, offered an amendment to remove Echo Park Dam from the Senate bill. It failed by a vote of 52 to 30. On the same day, April 20, the full Senate approved of the CRSP with Echo Park Dam by 58 to 23, with a bipartisan block of thirty-one Democrats and twenty-seven Republicans in support.[16] Only two Republicans from the West disliked the bill, California Senators Knowland and Kuchel. Since attempts to satisfy their concerns had obviously failed, western states remained divided on the legislation, a clear disappointment to upper-basin lawmakers. Suspicions toward the "hungry horde" persisted throughout the year, and tensions between the basins continued to grow.

Failing to win support from California was one problem, but ignoring the aroused conservationists soon created another. After the Senate passed its bill, the Council of Conservationists, the Trustees for Conservation, and the Citizens Committee on Natural Resources immediately shifted their lobbying strategy into high gear and announced their opposition to the entire upper-basin project. Because the Senate had refused to detach Echo Park Dam from the rest of the CRSP, they felt compelled to reestablish their alliance with farm states and anti-reclamation forces and revive the coalition that had blocked the legislation in 1954. During the previous five years, conservationists had used moral arguments and suggested alternate sites to make their case for an unimpaired park system, but while that strategy had generated a national outcry against the dam, it had not been adequate to force its removal from the project. Now the advantage of the three lobbying groups became significant. They intended to show the upper basin that no bill could pass so long as Echo Park Dam remained in it. "Our reason for opposing the dam," one of the Council's news releases put it, "is still the same; they must not build the Dam in Echo Park, and if the project has to be killed to prevent it, then we must do everything possible to help kill it."[17]

In taking a stand against the whole bill, the umbrella lobbying groups repeated the familiar criticisms of Raymond Moley, the Hoover Commission, and the Engineers Joint Council, and made the case that Echo Park Dam only added to what promised to be "a great waste of precious water and a cataclysmic economic mistake as a source of power."[18] The Wildlife Management Institute took note of the "$1.5 billions price tab and the $3 billions of hidden subsidies that are involved in this proposal," while the Sierra Club called for a delay in the legislation to wait for the Hoover Commission's final report on the nation's water policy, scheduled to appear late in May.[19]

Opposing the entire project was a tactic, not a goal, of their campaign, although some of the more aggressive conservationists like Martin Litton, David Brower, and Bernard DeVoto had been raising questions about the economics of the Bureau's plan for some time. Brower had been attacking the CRSP ever since the evaporation dispute, and now he felt vindicated. To him, the Bureau of Reclamation had been careless with its evaporative rates, dishonest in its public statements about the purposes of the project, and politically motivated in its preferred combination of dams. Brower had to be careful in how he used the *Sierra Club Bulletin,* so as not to court trouble with the Internal Revenue Service; therefore, he brought the fruits of his research efforts to the Council, the Trustees, and the Citizens Committee, and he served on all three boards.[20] Privately, Brower continued to oppose the entire project, repeatedly saying that the numerous dams would lose upward of one million acre-feet per year through evaporation—a waste of precious water—all to supply power to a region where other power sources "are abundant."[21]

But the campaign's purpose had not fundamentally changed. Echo Park Dam remained the target, and the lobbying groups maintained their position, first announced in November of the previous year, of supporting the CRSP as soon as lawmakers agreed to remove it from the bill. Opposing the total project may have appealed to Brower and Litton and a few others, but most regarded it as only a delaying tactic and one that could not fail to bring results. With their single demand left unsatisfied, they could, by dint of their alliance, block the entire package; at the same time, their position constituted a tempting offer to the upper basin to take out one dam and break the logjam on the whole bill. Having positioned themselves in this way, conservationists knew that they were the one element of the opposition that could most easily be mollified.

It was remarkable how the threat to Dinosaur brought together such a wide variety of organizations—from the small, single issue (and in the case of the Sierra Club, regional) groups like the Wilderness Society and the National Parks Association, to the larger sporting groups like the Izaak Walton League and the National Wildlife Federation. Never before had so many sporting, wildlife, and wilderness organizations come together into such a powerful coalition. As the climax of the controversy approached in the spring of 1955, their strength became clear. Wallace Stegner later remembered that the coalition of groups had amassed "astonishing political muscle" and allied with other opponents of the CRSP, had the power to stop the dam.[22]

After the Senate's approval of the bill in April, the campaign to save Dinosaur Monument began to crest. Generously funded by Edward Mallinckrodt, Jr., the Council of Conservationists published and distributed pamphlets that reached hundreds of newspaper editors, state and community officials, and members of Congress.[23] The propaganda against the dam in the spring and summer of 1955 made John Muir's final efforts to save Hetch Hetchy pale by comparison. No single conservation issue had ever been given such national publicity. One of the Council's flyers reprinted editorials from major newspapers including the *Christian Science Monitor*, the St. Louis *Post-Dispatch*, the *Knoxville Journal*, and the *Milwaukee Journal*, all opposing the dam. By assembling them in a single leaflet, the Council effectively displayed the rising outcry to wavering members of Congress.[24] Another of the Council's flyers, "Why Echo Park Dam Must Be Stopped," quoted Elmer Peterson, author of *Big Dam Foolishness*, who warned that dams offered a target to enemy nations, and said that the Soviets had "already tested out the technique and efficiency of dam bombing." Accordingly, any unnecessary dams ought not to be built in the first place.[25]

The Council and the two other lobbying groups accompanied their leaflet barrage with one of the largest mail campaigns in conservation history. Addresses of thousands of Americans were obtained from the gamut of groups enlisted in the campaign, and the political clout of some of the larger organizations now became decisive. Groups like the Garden Club of America and the General Federation of Women's Clubs exerted substantially more influence among eastern and midwestern members of Congress than the Sierra Club, the Wilderness Society, or the National Parks Association. The General Federation of Women's Clubs counted more than five million members in its forty-eight state organizations, and forty-seven of them supported the Federation's resolution against Echo Park Dam.[26] Larger conservation groups weighed in, too. In March 1955, the National Wildlife Federation, after a spirited discussion at its annual meeting in Montreal, voted decisively to oppose the dam. The Wildlife Federation, a conglomeration of sporting groups and state and local wildlife organizations, represented upward of three million people. At the convention it resolved, by a vote of 30 to 12, to oppose the dam "and do everything possible to see that our national park system is not needlessly invaded or despoiled."[27]

With numbers like these under their command, the lobbying groups generated mail that several House members later remembered as remarkable, with some claiming that the mail running against the dam was 80 to 1.[28] Of course, the effect of such letters depended on the political sensitivity of each recipient to conservation interests, and that obviously varied among House members. Certainly the deluge of mail did not affect each of them equally; but dozens of representatives clearly felt the pressure of the fierce opposition to the Echo Park Project, and they made clear their intent to wait on supporting the CRSP until the matter had been resolved.

To make certain that the canyons of Dinosaur remained in the spotlight, Howard Zahniser camped out in the halls of the House with a movie projector and David Brower's film "Two Yosemites." Showing the film to members of Congress on their way to lunch or to their offices probably did not change the minds of those in favor of the dam, but it did help to solidify support of those unfamiliar with Echo Park who had been badgered by their constituents to safeguard it. Zahniser later reported that few passersby could resist stopping to watch. For his part, Alfred A. Knopf donated a copy of *This Is Dinosaur* to each House member, with the copies arriving when the climax of the battle peaked in the spring of 1955.

By May, the scene had shifted to the House where the legislation underwent scrutiny by several committees. Senator O'Mahoney, delighted with the Senate's action on the bill, had claimed to be "rather optimistic" about its chances in the House Interior Committee, scheduled to meet during early June.[29] But like Watkins and Anderson, O'Mahoney was being coy with his constituents, for he realized the growing pressure against the dam in the House. O'Mahoney's Wyoming colleague in the House, Congressman Keith Thomson, who was as eager as Barrett and O'Mahoney to gain passage of the bill, estimated that any version of the bill that included Echo Park Dam would fail by several votes in the House Interior Committee, but would pass with a strong majority once the dam was removed.[30]

In early June, the Subcommittee on Irrigation and Reclamation took the first of several steps on the bill in the House. As members prepared to vote on the legislation, Pennsylvania Congressman John Saylor offered an amendment to eliminate Echo Park Dam, only to see the subcommittee turn it down just as it had done the previous May.[31] Still, subcommittee members favoring the dam realized that the extensive publicity about Dinosaur Monument did not bode well for the project on the House floor. They knew that in this instance many House members would vote according to their constituents' wishes rather than accept the recommendations from a committee. And they knew the bill had a greater chance of passage on the House floor with Echo Park Dam removed. Colorado Congressman Wayne Aspinall and Utah's William Dawson now found themselves in anguish, wanting to do whatever they could to salvage the dam, but acutely aware of the strength of its opponents on the floor. Members jockeyed behind the scenes.

Finally, a compromise emerged. On June 8, two days after rejecting Saylor's amendment, the subcommittee reconsidered and voted 15 to 9 to remove Echo Park Dam from the bill. Then, in a move that infuriated conservationists, it approved of another amendment put forth by Aspinall to appoint a board of engineers to reexamine all the alternative dam sites to compare their quality with that of Echo Park, and to make a final recommendation to the president by the end of 1958. Aspinall's amendment kept the proposed dam alive, or, as Congressman Dawson explained, put it "in a category where it isn't lost by any means." [32] With the dam placed in this ambiguous category, the subcommittee passed the bill on to the Interior Committee. [33]

It so happened that lawmakers on the subcommittee also sat on the Interior committee, and that body had an obligation to vote on the legislation separately. On June 14, the committee approved a 760 million dollar project, with storage facilities at Glen Canyon and Flaming Gorge and on the San Juan River in New Mexico— the Navajo Dam so dear to Senator Anderson and to the Navajo Tribal Council. Funding was approved for engineering studies to begin on Curecanti Dam and for eleven participating projects, less than half the number in the Senate's version. [34] Aspinall's amendment remained, but the price of Echo Park Dam, approximately 176 million dollars (and by now a significant factor at this stage in the controversy) did not appear in the bill. The House bill amounted to 760 million dollars with the dam removed, and this improved the bill's prospects on the House floor. As time on the session continued to run, representatives in favor of the CRSP pushed for a vote by the full House sometime in July.

Members of the House Interior Committee thought that their job was done. They soon realized that they might have to reconsider the legislation still again in light of continued opposition from conservationists, who regarded Aspinall's amendment with great skepticism. To them the move was simply a ploy designed to push the legislation through committee, and it did not prohibit revival of the dam later in the process. In light of his amendment, the Council, the Trustees, and the Citizens Committee refused to withdraw their opposition, and with time on the session running out and public pressure continuing to mount, members of the Interior committee realized that the bill must be changed. On June 28, by a vote of 20 to 6, the committee deleted Aspinall's amendment and took Echo Park Dam out for good. [35]

Conservationists remained dubious that the committee had taken it out for good and leaders of the campaign felt that this was no time to relax. In their minds the dam had only been temporarily removed. After the full House voted on the CRSP without the controversial dam the bill would move to a House-Senate conference committee for final approval, and the likelihood of the dam being reinserted at that point seemed strong. After all, the Senate had retained the dam in its version of the bill and the Bureau of Reclamation had given no indication that it stood ready to scrap it. In addition, the House-Senate conference would contain members of both Interior committees supportive of the dam, and they might be able to restore it and slip the bill through the full House just prior to adjournment. Despite deletion of Aspinall's amendment by the House committee, then, the three lobbying groups maintained their opposition to the entire project, assuming that the committee had removed the dam only as a tactic to move the bill along and with designs of restoring it in conference. Utah's *Deseret News* claimed that this was indeed the strategy.[36] Tensions had become so great at this point that upper-basin residents became convinced of an opposite scenario. Echo Park Dam had been removed, so the reasoning went, in order that Congressman Saylor "and his allies from southern California" would deliberately restore it "with the aim of getting the whole plan killed by the house."[37]

Finally, in a move foreshadowing the end of the dam, two key members of the House Interior Committee made an announcement that shocked everyone. Chairman Clair Engle of California and Wayne Aspinall reversed their position and agreed to oppose any bill with Echo Park Dam included and they pledged to thwart any attempt to reinsert it in the conference committee.[38] Their change of mind did not cause the lobbying groups to withdraw their opposition because neither the Senate or the Bureau made any moves to take the dam out of the bill or indicate they would consider doing so. As it developed, those promises became the final part of the conservationists' campaign. But the announcement from Engle and Aspinall had a powerful effect on upper-basin senators, who now understood, perhaps for the first time, that they could not hope to get a bill with the Dinosaur dam.

Clair Engle did not change his mind because of a sudden devotion to the sanctity of national parks and monuments. Rather, he did so because he understood and feared the high tensions in Congress in the middle of the 1950s over federal water projects like the CRSP. Engle had recently sponsored legislation for a water project on the Trinity River in California, only to see the House approve of the bill by a narrow margin. Some negative votes on the Trinity Project had come from western lawmakers, making Engle realize that divisions within the West posed great danger to such projects. He knew that the West must be united in the face of so much anti-reclamation sentiment from other regions.[39] For this reason, Engle understood the problems that southern California had been making by opposing the CRSP and the Central Arizona Project, and he concluded that he must support the CRSP. If he did not, it might become a victim of the powerful coalition of states in the East, the Midwest, and the South, and such a move could hardly bode well for western reclamation in the future. Engle thus agreed to surrender Echo Park Dam and promise that it could not be restored. His announcement ended California's monolithic opposition to the bill, although it did not greatly alter upper-basin suspicions toward the lower basin.

For a time, it looked as if the CRSP would be voted on by the full House. After the Rules Committee approved the bill in mid-July, House Majority Leader John McCormack of Massachusetts announced a floor vote the following week. Supporters realized that it faced strong opposition, partly because of doubts about its cost and the much-discussed problem with agricultural surpluses, but mostly because of steady pressure from the Council, the Trustees, and the Citizens Committee who maintained their stand. They refused to withdraw their opposition until both the Senate and the Bureau agreed to abandon the dam and guarantee the protection of Dinosaur Monument. Until such assurances came forward, the lobbying groups worked to keep the bill from reaching the floor. Congressman Saylor, their major spokesman in the House, kept the pressure on by reminding members of the exorbitant costs of reclamation projects which invariably outran initial estimates. Although the present House version approved of 760 million dollars in dams and participating projects, Saylor warned that the ultimate price of building them could reach 3.5 billion dollars.[40] Congressman Craig Hosmer of Long Beach also weighed in and said that "Congress might as well appropriate money to grow bananas on Pikes Peak as to approve the upper Colorado River project."[41] Of course, the House version of the bill remained considerably less than the Senate's, meaning that when and if the Echo Park matter could be settled the bill had a good chance of being passed.

The matter had reached a stalemate. Too many House members had been urged to turn down the bill until greater assurance had been given that Echo Park Dam could not be reinserted in conference. Supporters of the legislation in the House now feared that they had insufficient votes and decided to wait until the following year to try again. Once again, a combination of forces had blocked the bill, and the full House did not vote on the project in 1955. On July 26, Speaker Sam Rayburn, Democrat from Texas, and Minority Leader Joe Martin, Republican from Massachusetts, decided to wait until the 1956 session to reconsider the bill.[42]

Upper-basin residents reacted by blaming California, and justifiably so to some extent because Congressman Engle's remarks that the western states should unite had not persuaded lawmakers from Los Angeles and the Imperial Valley to back the bill. Senator Watkins spoke to the Senate in late July and pointed out that lobbyists from California had spent nearly one million dollars to fight the upper-basin project. The *Salt Lake Tribune* agreed that "the opposition to the Upper Colorado project is based fundamentally on California selfishness and refusal to live up to compact pledges for division of Colorado water."[43] The disappointed upper basin blamed Secretary McKay and the Eisenhower administration for being lackadaisical in lobbying Congress for the project. Eisenhower had experienced difficulty with his position on federal water projects, never satisfactorily explaining why he scrapped some and encouraged others, and such inconsistency had also helped to block the CRSP bill.[44] Then, too, upper-basin states themselves remained divided, remarkably enough. Obviously, the four states needed to concur on which dams and participating projects they wished to include, but the split between western and eastern Colorado had caused considerable debate within that state and produced disagreement between Colorado, New Mexico and Utah. Colorado Governor Ed Johnson had been critical of the CRSP out of his belief that his state did not receive benefits proportional to the amount of water that it contributed to the Colorado River system. A meeting among the four governors, in Cheyenne in January, had endeavored to settle those differences, but Johnson remained unhappy with the bill.[45]

While all of these factors had delayed the legislation, Watkins, Anderson, and other senators realized that pressure against the Dinosaur dam had been relentless and would not subside, and they now concluded that the time had come to scrap the project. While they did not accept the argument that the dam would set a precedent that would weaken the national park system, they understood political reality and the poor prospects of the bill in the full House.

Engle and Aspinall had made it clear that if senators did not remove the dam, they could not hope to pass the bill in 1956. Not
long afterwards, Watkins and his colleagues began to say publicly
that the Dinosaur dam had to be jettisoned. Senator Barrett, in
a talk to the Wyoming Development Association convention on
October 4, said that it should be removed from the bill in order
to win the support of more House members.[46] Senator Anderson
agreed, and he suggested a special meeting of upper-basin senators
in Denver on November 1.[47] When the meeting convened, Anderson told his colleagues that Echo Park Dam must be taken out to
break the logjam on the bill.[48] Each of them agreed, for they knew
that no bill could be approved otherwise.

Conservationists sensed that victory on Dinosaur might be at
hand, but they still worried about last-ditch efforts to salvage the
dam. A few days before the meeting in Denver, Secretary McKay
stated his support for the Senate's version of the bill, a clear sign
that he had little interest in surrendering Echo Park Dam or a number of much-coveted participating projects that western senators
yearned to have approved.[49] With McKay's announcement casting
doubt on the Bureau's willingness to concede the dam, the Council
of Conservationists took a bold step to make clear its determination to oppose the legislation until the controversial dam was removed. On October 31, one day before the Denver meeting, the
Council placed a full-page advertisement in the *Denver Post*. Paid
for by Mallinckrodt and signed by the Council's executive committee of David Brower, Joe Penfold, Fred Smith, Howard Zahniser,
and Ira Gabrielson, the ad recounted all of the familiar charges
against the CRSP and announced their continued opposition so
long as any version of the bill "needs, encompasses, anticipates, or
secretly hopes for a dam or reservoir in a National Park or Monument."[50]

The Council knew that it had the Senators trapped. If they refused to give in and tried to renew the fight in the following session,
they could expect to face the same coalition of opponents. Anderson and his colleagues understood that conservationists were the
only element in that coalition that they had the power to satisfy.[51]
So long as the dam remained in the bill, they could hold out little
hope for Glen Canyon or Navajo dams, or for the CUP or other
participating projects.[52]

Having said for years that Echo Park Dam was the "piston in

the engine" of the CRSP, the linchpin of the entire plan, upper-basin senators gave in. They removed the dam from the bill and adopted a resolution that "in the hope of getting action on an Upper Colorado River Storage Project bill in the present Congress, the Senators and Representatives present agree that they will not try to reinsert the Echo Park dam." Another resolution called for storage dams at Glen Canyon and Flaming Gorge and for the Curecanti and Navajo projects. Senator Anderson assured Fred Smith two days later that Echo Park Dam could not be reinserted in the bill, and on November 29, Secretary McKay officially announced its deletion from the CRSP.[53] McKay indicated that the Bureau would reopen its previous studies of alternate sites. He also promised that the Park Service, then making plans for its Mission 66 program of improving park facilities, would build a road into Dinosaur Monument to allow visitors to view Echo Park from Harper's Corner, a stunning vantage point high above the river canyons.[54] After six years of fighting, the battle appeared to be won.

But not quite. Wyoming Governor Milward Simpson spoke before the chamber of commerce in Rock Springs in late November, and said that Echo Park Dam was "only temporarily forsaken," reviving conservationists' fears that it might be restored in the bill in 1956. This, coupled with their desire to obtain a stronger guarantee that the CRSP would not harm the national park system, led to several more weeks of delicate negotiations over the wording of the final bill. They proved to be an important legacy of the battle for the wilderness movement.

After spending more than five years on the campaign, Brower, Penfold, Gabrielson, and others felt unable merely to accept a pledge from senators not to reinsert Echo Park Dam into the legislation in 1956. They wanted more. First, the bill must state explicitly that the legislation would not permit any dam or reservoir from intruding into any part of the national park system. Second, provision must be made for protecting Rainbow Bridge from the reservoir behind Glen Canyon Dam.[55] Inclusion of such provisions would reaffirm the sanctity of the national park system and underscore the mandate given to the National Park Service in its enabling act to conserve the parks "unimpaired." Furthermore, the provisos would serve as a reminder that "primeval parks" should offer wilderness recreation rather than the kind of sporting activities associated with dams and reservoirs. Howard Zahniser thought that the first proviso was especially critical, for it would reaffirm "the sanctity of dedicated areas" and thereby bolster the effort to protect a broader range of wilderness lands.

For Senators Watkins, Anderson, and upper-basin members of the House, the additional demands from conservationists proved to be trying. With Echo Park Dam out of the bill, lawmakers wanted to consider their differences with conservationists resolved, to take control of the legislative process, and to generate momentum for the next session of Congress, set to begin in January 1956. Above all, they wanted the lobbying groups to make good on their pledge to support the CRSP, now that Echo Park Dam had been jettisoned. Further demands only held up matters and encouraged California and other opponents to continue their broadsides, and that weakened the bill going into the next session. Both sides remained anxious and worked feverishly in the final weeks of 1955—the proponents of the CRSP did so because they wanted quick approval of the bill from conservationists, while the lobbying groups pressed to gain approval of their much-desired provisos.[56]

As a result of these different agendas, negotiations over the two provisos proved exceedingly delicate, and no one understood that quite so well as Howard Zahniser. From the beginning of the battle in 1949, Zahniser understood the importance of guarding the park system to provide a foundation for wilderness preservation, and he believed that inclusion of the provisos in the final bill was the last vital step. From the moment that the Denver meeting adjourned in early November until late January 1956, Zahniser worked unceasingly to include them in the final bill and he found himself constantly pressed to assure lawmakers that the provisos did not mask any hidden agendas. At a time when conservation organizations relied on letter writing, face-to-face diplomacy with members of Congress, and an informal network of alliances with other groups, Zahniser was a model of tact, persistence, and vision. He had a methodical, patient style that, from the vantage of point of contemporary environmental affairs, appears slow, cautious, and conservative. In the context of the 1950s, however, Zahniser provided unexcelled leadership for the fledgling wilderness movement.

On December 20, Zahniser went to Capitol Hill to see Utah Congressman Dawson and ask his support for the desired provisos.[57] The encounter was friendly, but Zahniser needed all of his grace and tact to persuade Dawson to do so. He began by telling

Dawson that he "felt friendly" toward the CRSP and that he and members of the coalition looked forward to supporting the legislation as soon as the desired provisos could be secured. Presenting a letter on behalf of the Council of Conservationists, Zahniser asked Dawson and his Interior committee colleagues to include "protective wording" so that "no project authorized by this legislation should alter or impair any unit of the National Park System." [58]

Dawson hesitated and seemed wary of the request. Since the Council, the Trustees, and the Citizens Committee had been opposed to the CRSP throughout the recent session of Congress, how could he be sure of the sincerity of their pledge to support the project now that Echo Park Dam had been taken out? He also remained convinced that conservationists acted as a front for southern California water and power interests, an impression that made him especially wary of Brower and the Sierra Club. Dawson recalled to Zahniser Brower's *Pacific Spectator* article, where he denounced the Colorado River Compact and took swipes at the upper-basin project, as well as his detailed critique of Glen Canyon Dam. Dawson knew that the effort by river runners and wilderness activists in Utah to save Glen Canyon had become more prominent in recent months, and he now feared that with Brower's help, they might launch a campaign to do so. From Dawson's point of view, conservationists had been enough trouble already, and any additional demands merely delayed matters further.

Zahniser tried to be reassuring. He said that Brower had been committed all along to the preservation of Echo Park and that his raising questions about Glen Canyon Dam had merely been a tactic to delay the overall project. Zahniser admitted that he too regretted that stunning beauty would soon be lost under Glen Canyon Dam—that he and many others had become aware of this in recent months—but he indicated that there had been no decision to alter course and mount an effort against that dam too. He said the Sierra Club fully agreed with the Wilderness Society and other opponents and that Brower "is right along with the rest of us in our attitude toward the Glen Canyon project in the circumstances at present." [59]

In most respects, Zahniser spoke accurately although Brower had indeed come very close to leading a last-ditch effort to stop Glen Canyon Dam. Over the previous year, Brower had completely altered his position on the huge dam slated for construction near the Utah-Arizona border just north of Lee's Ferry. In 1954, he had taken the lead advocating a higher dam at Glen Canyon to replace Echo Park. Then, over the following months, as he listened to the Bureau denounce that suggestion and express fears of the rock structure at the Glen Canyon site and as he became aware of the vast scope of this upper-basin project, his doubts grew. By the summer of 1955, Brower was pouring an enormous amount of energy into investigating every aspect of Glen Canyon Dam. At one point he prepared a study of the dam that ran to many pages and circulated it among other conservationists and engineers for comment. Brower had by then decided that its costly hydropower was an unwarranted expense "in the dawn of the era of atomic power."[60] At the House subcommittee hearings in March, Joe Penfold proposed a resolution for the Echo Park controversy and other contentious aspects of the legislation by urging the Bureau to make Glen Canyon Dam the sole provider of storage and power in the project's first stage. Brower disliked that idea, not because he had been converted to the proposition that Glen Canyon ought to be designated a national park, but because he thought the need for its power and storage had not been demonstrated. He came very close to a conclusion later advanced by Luna Leopold and Walter Langbein well after the controversy, that Glen Canyon Dam was unneeded for river regulation because it duplicated the function of Hoover Dam. Finally, by October 1955, Brower realized that Glen Canyon Dam also threatened a beautiful stretch of the Colorado River, and while the area did not belong to the park system, its inundation beneath the reservoir would be tragic. Since the Sierra Club had a policy of opposing power projects in scenic areas, Brower moved to oppose Glen Canyon Dam on those grounds.[61] At the Club's Board of Directors executive meeting in late 1955, he urged the Club to remain in league with other opponents of the CRSP and to hold up the project until the upper basin surrendered Glen Canyon Dam.[62]

The Board of Directors refused on the grounds that the Sierra Club and its allies had fulfilled their original objective in the Dinosaur controversy and should not alter course now. Conservationists had pledged themselves to back the CRSP once Echo Park Dam was removed, and to change that pledge might backfire seriously with lawmakers from western states. The Board's decision not to fight Glen Canyon Dam reflected the consensus among conservationists that the guarantee of national park sanctity would represent a major accomplishment of the campaign—it was after all what they had started out to do—and would demonstrate that wilderness and park protection could be made compatible with the economic growth of the American West. Brower was disappointed, and he has maintained ever since that the coalition of forces had the power to stop Glen Canyon Dam. Perhaps. But Brower's own disappointment with the loss of Glen Canyon is more attributable to his discovery of the scenery in the years after the CRSP bill passed. Brower's hindsight—not to speak of that of almost every other environmentalist in the West—has been perfect on Glen Canyon Dam. The truth is that too few of those involved in the campaign had an opportunity to see Glen Canyon before agreeing to support the legislation, and the political pressures of the battle compelled them to support the dam in the end.[63]

In any event, Zahniser found that Dawson's suspicions could not easily be overcome. As he continued to press Dawson to agree to the provisos, Dawson revealed yet another concern: Could such provisos be interpreted broadly, in a way that would recognize wilderness areas? Dawson knew that the Wilderness Society had been interested in doing this, and he had heard that the secretary of the interior's advisory committee had begun talking about wilderness protection on various public lands. Zahniser had to assure him that the first proviso was in no way intended to designate wilderness, but merely to restate the commitment to protecting the national park system.

Zahniser then asked Dawson to provide a public statement that the provisos would be included in the final bill, and he added that once senators on the conference committee stated their support for the language, "that is all that we'll need," indicating he would take them at their word. Dawson then asked Zahniser if members of the Council would provide written assurance of their own that they had withdrawn their opposition to the CRSP. Zahniser said they would do so, once the necessary provisions had been adopted. Dawson then indicated that the CRSP bill was not going back to committee, but directly to the House floor. More negotiations followed after Christmas and in the first part of January, with Senator Anderson expressing concern that the desired provisos "could only be an effort to assist the California lobbyists in their desire to kill the legislation." Zahniser had to assure him that this was not the case.[64]

Under Zahniser's direction, and with prodding from leaders of all the major organizations, the final agreement fell into place. Senators and congressmen agreed to the two provisos to reaffirm the protection of the park system. The first proviso stated that "it is the intention of Congress that no dam or reservoir constructed under the authorization of this Act shall be within any national park or monument." The second proviso declared that "as part of the construction, operation, and maintenance of the Glen Canyon unit the Secretary of the Interior shall take adequate protective measures to preclude impairment of the Rainbow Bridge National Monument."[65] The provisos in the final bill bore the personal stamp of Howard Zahniser, and capped the long campaign to save Echo Park. They reaffirmed the "sanctity of dedicated areas," thereby linking the successful campaign with the effort to establish a national wilderness system, which Zahniser revived immediately. With the agreements in place, the three lobbying groups withdrew their opposition, and leaders of the National Wildlife Federation, the Izaak Walton League, the Sierra Club, and a number of other groups followed their lead and did the same, effectively ending the conservation opposition.[66]

Some final skirmishes arose early in 1956, largely because southern California lawmakers realized that they had lost their major ally in blocking the bill. For several months, beginning in 1955, Californians had held out hope that the CRSP might be delayed pending the outcome of the lawsuit between Arizona and California over the apportioning of lower Colorado River water. Lawyers representing California's water interests tried to make the upperbasin states a party to that lawsuit, but in 1955, the Supreme Court announced that those states would not be made a party to the case. California lawyers continued to press the matter but in January 1956, the Court refused to grant them an opportunity to argue their case, and this ruling finally settled the matter.[67] The Supreme Court's ruling meant that Congress had no obligation to await outcome of the lawsuit before voting on the CRSP, and it cleared the way for enactment of the bill later that session.

When conservationists removed their opposition at the same time in late January 1956, southern Californians realized that pas-

sage of the CRSP was imminent, prompting Long Beach Congressman Craig Hosmer to make a last-ditch effort to discredit the
upper-basin project. He did so in dramatic fashion. In late January,
a few days after the lobbying groups announced their support of
the bill, Hosmer's picture appeared in the *Los Angeles Times,* pouring water on a piece of chinle shale, the rock formation at the Glen
Canyon dam site. Within five minutes the rock had completely dissolved, leaving Hosmer to warn that taxpayers who approved of
the CRSP "would be left with a billion dollar mud puddle."[68]
Hosmer sent samples of the rock to all 435 members of the House,
urging them to try the experiment themselves and calling for additional surveys of the Glen Canyon site by independent geologists.

Hosmer's effort was obviously desperate. Upper-basin editorials
pointed out that Hosmer conveniently ignored studies of the dam
site by Bureau engineers, adding that "no trick is too foul, no lie is
too big, no technique too devious for this well-heeled lobby in its
desperate campaign to block the development of the Upper Basin
and to grab more Colorado River water."[69] Senator Watkins likened Hosmer's move to tactics employed by Joe McCarthy, using
the "big lie" in an effort to discredit the CRSP.[70] Wilbur Dexheimer
and E. O. Larson of the Bureau of Reclamation appeared in public
to provide assurance of the safety of the Glen Canyon site. They
said that seven hundred feet of sandstone covered the chinle shale
and provided a sound geological base for the dam. Besides, questions about the chinle shale had been taken up thoroughly and put
to rest in the House subcommittee hearing in 1955.[71]

With Hosmer's effort discredited, the final steps in Congress
took place. On March 1, 1956, the House voted for the legislation
by a vote of 256 to 136, with 136 Democrats and 120 Republicans
favoring the bill. A week later, members of the House-Senate conference committee, including Senators Anderson, Watkins, Millikin, and O'Mahoney, accepted the removal of Echo Park Dam,
with Senator Anderson saying afterwards that "not a soul raised
an objection." The conference committee members also agreed to
the amendments to protect Rainbow Bridge and the national park
system from any dams contained within the CRSP.[72] The final bill,
approved by Congress on March 28, also included an amendment
introduced by Congressman William Avery of Kansas, which prohibited newly irrigated lands from being used to produce any of
six major crops then in surplus (including wheat, cotton, and corn)

for the first ten years of the project.[73] On April 11, 1956, a few days after attending the Master's golf tournament in Augusta, Georgia, President Eisenhower signed the 760–million-dollar bill into law. The Echo Park controversy was over, and conservationists claimed a great victory on behalf of the park system and the American wilderness.

Epilogue

Looking back on the Echo Park controversy from a present-day vantage point risks becoming an exercise in nostalgia. The battle had little to do with concerns typically labeled "environmental" in our own time. Americans in the 1950s had barely begun to consider the ramifications of their industrial society on public health or the environment, while adverse effects of pesticides and atomic fallout had only just emerged in the public discourse. The Echo Park battle obviously had little to do with such matters. Nor did the campaign focus on biological or ecological concerns such as wildlife habitat, endangered species, or what is now commonly called "the rights of nature." In addition, the controversy did not play out in the legal arena, as environmental issues so often do in our own times. Had Brower, Zahniser, and their friends possessed a crystal ball, they might well have envied the next generation of environmentalists who blocked Tellico Dam because it threatened the snail darter, an endangered species, or delayed the New Melones Dam, on the Stanislaus River in California because of inadequate assessments of its possible "environmental impacts."[1]

Nevertheless, the successful effort to preserve Echo Park and Dinosaur National Monument revealed the dawning of a new era in the nation's environmental history, by signifying growing public support for parks and wilderness that few could have imagined in John Muir's day. The outcome of the Echo Park struggle gratified those conservationists who remembered and lamented the loss of Hetch Hetchy, partly because the essential question in that conflict had again been at the center of debate: Should national monuments and parks be left unimpaired in order to provide for those eager to escape to the wilderness? This time, fueled by rapid changes in the economy and society in the postwar years, the public provided a decidedly different answer. The show of strength appeared in the large number of conservation-minded Americans who dutifully took pen in hand to write to their congressional representatives, pleading to protect what they regarded as a cherished American institution. We have seen that they owed a considerable debt to powerful allies from other regions, who blocked the CRSP long enough to convince the upper basin that it must abandon the Dinosaur dam in order to gain passage of the rest of the project. Nevertheless, the solidarity of millions of Americans and dozens of organizations which stood united against Echo Park Dam had been decisive in its removal from the CRSP bill.

The outcome revealed an important shift in public attitudes toward the national park system, demarcating the postwar generation with its interest in protecting national parks and wilderness areas. Park Service leaders in the early years of the agency, notably Stephen Mather and Horace Albright, had felt compelled to stress the role of national parks as playgrounds rather than as wilderness sanctuaries in order to generate political support. In so doing they had deliberately emphasized that portion of the National Park Service Act of 1916 calling on the agency to provide for public enjoyment of the parks. In their generation, public interest in the parks sprang more from their recreational opportunities than from a strong national commitment to protecting their resources. From the 1930s to the 1950s, as a growing number of threats came to the fore, the public commitment to the preservationist mandate of the NPS Act increasingly came into question, and met its sternest test in the Echo Park affair. The end of Echo Park Dam thereby signaled a new era in national park affairs marked by public support for unimpaired preserves.

It was a milestone in the history of the parks and the American conservation movement. By blocking the dam, the coalition of organizations resoundingly reaffirmed the sanctity of the parks and strengthened the mandate for protecting them contained in the National Park Service Act of 1916. While it can hardly be said that defeat of Echo Park Dam ended all threats to parks, it became substantially more difficult to offer serious proposals for dams, mines, or other intrusive structures inside park boundaries. Even in the more famous conflict over the Grand Canyon in the 1960s, the Bureau of Reclamation did not propose to locate its dams inside the national monument or national park boundaries, but just outside them. With the Bureau claiming that the two dams posed no direct intrusion, the controversy revolved around their potential to alter the "living Colorado."[2] From the 1960s onward, threats to parks increasingly came from outside sources, often from activities beyond their boundaries such as oil and gas drilling and operations of power plants. While such threats could be as potentially damaging to park resources as dams, they were often more difficult to identify and to arrest. The end of Echo Park Dam strengthened the barriers against outside threats.

Ironically, just as those barriers had been strengthened, some observers of the parks began to fear that their integrity might be compromised by too many hotels, curio shops, and other facilities designed to make a visitor's stay more comfortable.[3] With the postwar travel boom continuing to flourish, some wondered if the greatest threat to the parks might be coming from the throngs of people flocking to them. After Echo Park, it was clear that conservationists often found themselves with a difficult dilemma, for they had now won the political support for preserving parks that Mather and Albright had achieved in developing them into national playgrounds. Both mandates of the National Park Service Act of 1916 now had popular support, setting the stage for the next era of national park history when the NPS would find its balancing act of preservation and public enjoyment ever more challenging.

In another way, the outcome of the controversy strengthened the hand of the various and sundry conservation groups, who discovered a remarkable degree of strength in uniting in a common cause. As Roderick Nash has pointed out, some seventy-eight organizations joined the Echo Park campaign, far outnumbering the handful which aided John Muir with the Hetch Hetchy battle.[4] Numbers alone did not guarantee victory, of course, because the single-interest groups had never before worked together on this scale, and much time had to be spent in formulating strategy and in coordinating efforts. Yet no one could deny the impressiveness of their coalition. Irving Brant told President Truman early in the controversy that it was first time he could recall when so many groups "have been stirred up to a joint campaign," and later, he told Adlai Stevenson that "there is no issue on which conservationists are more deeply stirred, or which has done more to galvanize them into a nationwide organization."[5] Harold Bradley recalled that the controversy unified "the conservation forces into a single, very formidable fighting force," while *New York Times* columnist John Oakes, who had done so much to publicize the threatened monument in his columns, wrote that it demonstrated "what strength the conservation movement has at last achieved in the United States."[6] Almost exactly thirty years after the end of the controversy, Oakes recalled in an interview that Echo Park had been "of vital importance in demonstrating for the future that these kinds of battles could be won."[7]

The controversy had special significance for those organizations most interested in parks and wilderness, namely the Sierra Club, the National Parks Association, and the Wilderness Society. Small by today's standards, with only a few thousand members each, these groups could not exert much pressure on Congress by themselves, and they had counted heavily on larger conservation groups like the National Wildlife Federation and the Izaak Walton League. But if the larger groups provided essential political clout, the preservationists had the most to gain, for the threat to Echo Park provided them with an excellent opportunity to laud "primeval parks" and to incorporate their agenda into the broader conservation movement. The Wilderness Society gained national attention from the battle and took advantage of the triumph to resume its campaign for a national wilderness system. Echo Park offered a major opportunity to the society to publicize the kind of commercial pressures on public lands that its leaders contended would soon eliminate remaining parcels of wilderness unless they were permanently protected by Congress. Howard Zahniser recognized, at the outset, that upholding the "sanctity of dedicated areas" must be the basis of any wilderness system, and his effort to do so had culminated in the proviso in the final CRSP bill.

It was Zahniser who best knew how to connect the nascent campaign for the wilderness with the effort to preserve Echo Park, and he felt that the society had a major stake in the outcome. He was immeasurably encouraged by the campaign's success, for the outpouring of public support demonstrated something that he already deeply believed: that the last remaining wilderness offered a type of outdoor experience unavailable elsewhere, and that an important segment of the public was willing to help protect such lands. Zahniser, whose role in the Echo Park controversy has been for too long overshadowed by Brower's, took satisfaction in the knowledge that the Wilderness Society no longer had to rely solely on a small group of scientists and academics, as it had done since its founding in 1935. Because he knew that the public had spoken decisively, he lost no time in reviving the wilderness campaign. On April 11, the same day that President Eisenhower signed the CRSP legislation, Zahniser sent a letter to key members of Congress to ask them to sponsor a bill for a wilderness system. That campaign proved to be long and exceedingly difficult, but Zahniser had the support of organizations that had become committed to wilderness during the Echo Park affair.[8]

While Zahniser found encouragement, David Brower and the Sierra Club gained a new public image and an important new place in the pantheon of conservation organizations. Echo Park proved to be a key episode in the Club's history, taking it away from its traditional focus on the Sierra Nevada. In 1953, the Sierra Club Board of Directors had given its blessing to a new chapter in the Pacific Northwest, and agreed to add more chapters elsewhere as increased membership warranted. These formal steps were necessary to enlarge the Club's sphere, but it was the battle over Echo Park that launched the Club from its California base into the national conservation arena, and gave it a reputation for a hard-driving and uncompromising approach.

Brower, who symbolized the Club's stridency, developed his own national reputation from the Echo Park battle. In his first major campaign as executive director of the Club, he became renowned as a bold champion of the wilderness and a fierce combatant against those who threatened it—in this case, the Bureau of Reclamation. Brower also showed signs of a trait that dogged him throughout his career—a certain disregard for working within manageable limits, even a kind of recklessness at times, which caused anxiety among some of his counterparts. In June 1954, just prior to the Senate hearings, Fred Packard wrote Brower a long letter, warning him against antagonizing lawmakers with sharp critiques of the CRSP or with tirades against the Bureau.[9] Yet in spite of Brower's tendency to wander off course, everyone considered him invaluable to the campaign and understood him to be a new voice for conservation that was bound to be heard from again. More than any of his counterparts, Brower knew instinctively how to grab public attention for threatened preserves. He knew the power of pictures and films; "Wilderness River Trail" and *This Is Dinosaur,* two highly successful tools of the campaign, had been his ideas. He knew how to dramatize a sense of place, and Echo Park and the canyons of Dinosaur lent themselves to rhetorical flourish about the merits of wilderness. Here was refreshment of the most exciting kind, a tonic of white water and silence and the grandeur of the inner gorge.

Brower's passion for protecting places lay at the core of the embryonic wilderness movement in the 1950s, and beyond. Such an appeal to safeguard places represented a kind of scenic "monumentalism" (to borrow Alfred Runte's phrase) for a new generation.[10] Taken together with the full page advertisements pioneered by the Council of Conservationists, Brower spearheaded new tactics of wilderness propaganda. He developed filmmaking and book publishing skills that became primary tools in the 1960s to establish Point Reyes National Seashore and North Cascades and Redwoods National Parks, and in the battle over dams in the Grand Canyon which the Club led.

Perhaps the greatest impact of the controversy was in the American West. In regional terms, the elimination of Echo Park Dam from the CRSP brought to an end the first great clash between water developers and preservationists, and it established a major theme of environmental conflict in the West for the next two generations. For westerners most closely involved in the battle, the memories of the great conflict remained for years to come. As Roy Webb has pointed out, residents of small towns near Dinosaur Monument, like Vernal, Utah, remained bitter for years after the loss of the dam, and such memories remain even today.[11] One reason they have persisted is that the passions of the battle have hardly disappeared in their region and in many parts of the West, and their resentment against "environmentalists" remains strong. The loss of Echo Park Dam caused rural westerners to be aware of a new force in their lives, a force that seemed to many of them to be ignorant of water and power needs in their region, but which nevertheless seemed able to determine their fate. Local people sensed it might be a force not easily turned back, and in their frustration, some of them found a new dimension to the old colonialism that held sway in the region's past. Conservationists from the urban East and West Coast had now become a force to be reckoned with. With Echo Park it became clear that guardians of the public lands had suddenly become more prominent.

From the vantage point of those who won, Echo Park not only reversed the verdict of Hetch Hetchy but proved to be an important victory at the dawn of a new era in the West. World War II had brought the region into the national spotlight, had spurred its economic growth, and made it a valuable part of the nation for military training, atomic testing, and manufacturing. The war and its aftermath brought sweeping changes to the region, including a strong increase in pressures on natural resources on public lands. Those pressures were by no means new, for western extractive industries had long been central to the region's economy. To the opponents of a dam in Dinosaur, the Bureau of Reclamation, with its managed view of rivers and natural resources, had become associated with loggers, miners, and ranchers in viewing nature's bounty as a resource to be exploited. By defeating Echo Park Dam, conservation groups turned back the development-minded interests in a decisive way. They made their own presence known and indicated that the West would not always be dominated by miners, ranchers, loggers, or the builders of dams. As for the Bureau, the battle taught David Brower and his counterparts a great deal about the economic, political, and technical aspects of water projects in the American West that carried over into subsequent wilderness and water controversies. As a result of discovering the evaporation errors and learning how the Bureau shifted its rationale for the dam, conservationists became ever suspicious of the agency's publicly stated positions, and sought to honestly display the political and economic motives behind the agency's ambitious plans. The Bureau earned a reputation among its opponents for a kind of "credibility gap" in the Echo Park affair, and found many of its projects in the years ahead carefully scrutinized.

The Echo Park affair represents the classic tale of environmental controversy in the American West. Indeed, the battle instigated a pattern of conflict between the kind of managed use of natural resources that became woven into the fabric of the West in the first part of the twentieth century—and the postwar interest in wilderness and environmentalism. This conflict has played itself out across the West in the decades since the Echo Park battle. From the energy boom of the 1970s to the Sagebrush Rebellion to the spotted owl, westerners have faced the same difficult and sometimes agonizing conflict between the traditional brand of conservation, first espoused by Gifford Pinchot and Theodore Roosevelt to ensure a constant supply of water, timber, and rangelands, and the postwar interest in preserving the aesthetic qualities of public lands. In that sense, Echo Park is a story that lies at the birth of the modern American West. It is a touchstone for a pattern of conflict that has shaped the region's history in the last half century. In a letter to the author in 1984, Wallace Stegner wrote of Echo Park that "in many ways it is a key squabble, the interests involved are the classical and unavoidable interests, the warring points of view, the endlessly repetitive exploitation v. preservation points of view." [12]

Given this pattern, the battle was also a sign of what was to come. In fact, if we look more closely at events that followed the end of the controversy, it is striking to note how little had actually been resolved. Defeat of the dam by no means ended the upper basin's interest in water or power, and from those states' point of view, passage of the CRSP proved more important over time than the loss of one dam. The Bureau continued to be a major force in the upper-Colorado basin, and as the 1950s came to an end, residents of the region took satisfaction with the large construction projects of Flaming Gorge and Glen Canyon dams. In the years to come, millions of federal dollars were poured into the Central Utah Project and other participating projects authorized by the CRSP.

In fact, some upper-basin residents had not yet abandoned hope for Echo Park Dam, which dismayed conservationists and kept them at the barricades in the months after President Eisenhower signed the CRSP into law. Interest in the dam remained sufficiently high that some lawmakers tried to revive the project, and their efforts effectively thwarted a move to create Dinosaur National Park. This proved to be the greatest irony of the battle: despite all of the attention paid to the sanctity of the national park system, conservationists could not capitalize on their success to help establish a new national park. Failure to do so was particularly disappointing because, for a time after the passage of the bill, everyone seemed to favor the idea. Conservationists were happy to consider adding the Lodore and Yampa canyons to the much-lauded "national park" category, for it would thwart efforts to resurrect Echo Park

Dam and it seemed a fitting way to cap their long campaign on behalf of the national park system. Upper-basin states also seized on the idea, knowing that the "national park" label meant increased funding for roads, campgrounds, visitor centers, and other public facilities. In late April 1956, with everyone seemingly in agreement, Congressmen Aspinall and Saylor, at loggerheads for many months over the dam, each introduced legislation in the House of Representatives to begin the conversion.[13]

Skirmishes on the issue began almost at once, and the congenial atmosphere quickly disappeared. Senator Watkins resented the sudden effort to attain national park status for Dinosaur. He thought that the existence of a park could only hamper the secretary of the interior, who, now that the Echo Park project had been scrapped, had an obligation to make a recommendation on an alternate dam site outside the monument. With a certain amount of twisted logic, Watkins demanded that no changes be made in the status of Dinosaur until the secretary had made his recommendation on dam sites outside of it. In truth, he and other lawmakers in the upper basin realized that any hope of gaining approval of Echo Park Dam in the future was lost as soon as the monument became a national park and joined the more exalted park category with Yellowstone, Grand Canyon, and Yosemite. Watkins had not abandoned hope for the dam, a fact that the Citizens Committee on Natural Resources and its fellow lobbying groups pointed out.[14]

But Watkins did not have the center of attention as much as Colorado Senator Gordon Allott, who introduced the legislation for the monument's conversion. Allott envisioned a substantial increase in funding for the preserve once it acquired national park status, promising to generate more tourist dollars to the northwest corner of his state. Like Watkins, Allott had not given up on Echo Park Dam; so he inserted a provision in his bill to permit the Bureau of Reclamation to continue to investigate the site.[15] Such a provision obviously challenged the proviso in the CRSP legislation specifically prohibiting dams in any part of the park system, but Allott tried to make the case that his provision was acceptable since it belonged to an entirely separate piece of legislation.[16] In any event, debate over the Allott bill reopened the passions of the Echo Park controversy and drove everyone into a flurry. Charges and countercharges flew once again. Supporters of the bill argued that opponents merely wanted to keep Dinosaur undeveloped so as to

retain the preserve as their own playground, and they revived familiar charges about the elitist nature of the wilderness crowd.[17] Conservationists insisted that Allott and Watkins had still not accepted
the public's interest in maintaining the sanctity of the park system.

For their part, the Allott bill put conservationists on the horns
of a difficult dilemma. They wanted very much to convert Dinosaur
into a park and add it to the "crown jewels" of the system, but
they clearly did not want to risk reviving the possibility of the dam
itself, and most felt certain that this was Allott's intent. However,
they did not agree on the merits of the bill among themselves, and
they soon fell to wrangling. Fred Smith, chairman of the Council
of Conservationists, insisted that Allott's provision permitting the
Interior secretary to investigate dam sites posed no problem, for
the secretary already had such authority, and Smith considered the
provision harmless. He urged quick action on the park bill while
Dinosaur remained fresh in the public mind. He believed that once
Dinosaur became a park, it would become far more difficult to
reintroduce the proposal for the dam, and that by supporting the
bill, conservationists could demonstrate their sincerity in making
the Lodore and Yampa canyons accessible—and put to rest that
question which had been at the heart of the controversy.[18]

But Fred Smith's view did not prevail, and his own actions
proved damaging to relations among the conservationists. Brower,
Zahniser, Packard, Penfold, and Charles Callison strongly disagreed that Allott's provision posed no danger, and they refused to
support the legislation because of it. When Smith suggested that
Allott's provision permitting the secretary of the interior to investigate all potential dam sites was merely routine, Olaus Murie replied that the Bureau had "routinely" inserted the Echo Park Dam
in the Colorado River Storage Project in the first place. Most organizations, including the Sierra Club, the National Parks Association, and the Wilderness Society were exceedingly wary of Allott's
bill and refused to support it so long as that provision remained.[19]
Congressman Saylor introduced a bill in the House with Allott's
provision removed, prompting Smith and Edward Mallinckrodt to
accuse him of being overly cautious and delaying the park. Fred
Smith then sent a stinging letter to Fred Packard, assailing him for
opposing the Allott bill, a move which raised tempers all around
and stirred Zahniser, Murie, and others to mount a vigorous de-

fense of Packard and to denounce Smith's tone. In January 1957, when the time came to reconsider the makeup of the Council of Conservationists, Smith dropped Penfold, Zahniser, and Brower from the organization, and Ira Gabrielson resigned shortly thereafter.[20]

The wrangling among conservationists did not help the bill move along, and Allott's refusal to submit the bill with the offensive provision removed ensured that it would not pass the House of Representatives. With a majority of conservationists fearful of reviving Echo Park Dam, and with Allott refusing to concede, the effort to convert Dinosaur National Monument into a national park failed. While most of the conservationists preferred such an outcome because it prevented a revival of the dam, they could not help feeling sharp disappointment. Having just achieved a great symbolic victory for the sanctity of the national park system, they had not been able to cap their successful effort with an addition of a new park to the system.[21] Their disappointment also served as a reminder of the still powerful forces in the West eager for water and power.

Further reminders were forthcoming. As the Bureau began to construct Glen Canyon Dam, conservationists began to suspect a deliberate effort to violate the pledge to protect Rainbow Bridge from the reservoir behind the dam, an important provision of the CRSP. Brower and Zahniser blamed Secretary of the Interior Stewart Udall for the broken promise, partly because he had been involved in agreeing to the provision when he served on the House Interior committee. In 1960, while still a member of Congress from Arizona, Udall joined Floyd Dominy and officials of the Bureau of Reclamation on a foray through the remote desert country to Rainbow Bridge, and undertook a firsthand study of the options for protecting the arch from Glen Canyon Dam. Udall and the Bureau later concluded that among the various options, such as building a small abutment to protect the bridge, the best course appeared to be to do nothing, for any kind of added construction would only scar the landscape and detract from the scenery around Rainbow Bridge. Udall made this recommendation to his fellow committee members, who, in turn, rejected the funding for the necessary abutment that the Bureau had originally made a part of the expense of Glen Canyon Dam. In practical terms, this meant that the reservoir behind the dam would one day snake into the narrow canyon below Rainbow Bridge and enter the national monument just beneath the arch. That prospect raised concerns about Lake Powell undermining its base and weakening the foundation of the arch, but at the time, such a prospect seemed less troubling than did added abutments.[22] When Udall became secretary of the interior in the John F. Kennedy administration, he did recommend that protective measures be taken as part of his department's proposed budget to Congress, yet the Bureau worked to persuade key members of Congress that the 25-million-dollar estimated cost of an abutment dam was too high and the threat to the Bridge too slight.[23]

In any event, the decision effectively violated one of the key provisos in the CRSP legislation, and conservationists lost no time in firing protest letters to Congress, the Bureau, and Udall. Howard Zahniser expressed dismay to Udall, contending that the Bureau had altered course on its original pledge by citing the high cost of an abutment, but Zahniser maintained that such cost had been calculated in the expense of constructing Glen Canyon Dam; therefore, scrapping the abutment was to admit that the Bureau could not afford to build Glen Canyon Dam in the manner that it had originally anticipated.[24] Brower badgered Udall over the issue for the next two years, but without success, and he expressed his own disappointment in the *Sierra Club Bulletin,* in an article he called "Wilderness River Betrayal."[25]

While the decision to overlook the original CRSP proviso on Rainbow Bridge proved dismaying to conservationists, their disappointment with the inundation of Glen Canyon itself became the most bitter part of the controversy's aftermath. River runners and a handful of outdoor lovers from Utah had tried to focus more attention on the scenery of Glen Canyon, urging that it be designated as a national park; but their voices had been drowned out beneath the chorus of protests against the Dinosaur dam.[26] Conservationists had committed themselves to backing Glen Canyon Dam and the rest of the project so long as Echo Park was saved. Later in the 1950s, with Bureau of Reclamation engineers and building contractors erecting Glen Canyon Dam, the Sierra Club and other groups took river trips through the canyon, and discovered the stunning scenery that few had known about and now regretted must be lost. So many eloquent accounts of their journeys appeared that it is illuminating to consider the impressions of Harold Bradley, who floated through Glen Canyon in 1955, while the fate of Echo Park remained uncertain:

> It lacks that intimate quality that one has in the Dinosaur—it's all too big, often too far away. The color scheme tends to grow monotonous after a few days—for the most part dull red, with a good deal of black desert varnish making the impression still duller. . . . Glen Canyon could well be a National Park for its magnificence. But I think it can be well used if it is dammed. . . . I feel content to see Glen go under water, provided the Dinosaur does not, and of course provided the plans for Glen are sound.[27]

Few of those who followed Bradley during the next few years shared his assessment. Instead, they were struck by Glen Canyon's wondrous beauty, its hanging gardens in quiet and sublime side canyons, its lofty cliffs and pink and red monuments of stone—the whole a highly unique and remarkable slice of the Colorado River and Plateau. And much of what they found would soon be submerged by hundreds of feet of water. "Glen Canyon reservoir is going to flood some of the most outstanding God-created places on Earth," Charles Eggert lamented, "Hidden Passage, Music Temple, Twilight Amphitheater, Guardian Pool to name only a very few. Glen—from the San Juan down has the most extraordinary scenery I have ever seen."[28] Sierra Club float trips through Glen Canyon soon gave birth to another publication, one of the first Exhibit Format books, and the successor to *This Is Dinosaur*. With photographs by Eliot Porter and a gamut of poets providing the captions, *The Place No One Knew* became a memorial to Glen Canyon and a bitter reminder of the place that had been largely overlooked during the Dinosaur battle.[29] "Glen Canyon is gone," declared Richard Bradley in 1964, "and with it some of the most remarkable canyon country on the continent. It died its needless death just as people were beginning to learn of its wonders."[30]

The loss of Glen Canyon proved to be more than just a highly regretful occurrence: it became a touchstone for environmental action along the Colorado River and Plateau for the next generation. It steeled wilderness activists against any further dams along the Colorado River.[31] *The Place No One Knew* signaled to the Bureau of Reclamation, the Interior Department, and the Federal Power Commission that the Sierra Club and its friends were prepared to

oppose other dams and power projects that might be planned.
Within the following year, the Bureau had proposed dams at Marble Canyon and Bridge Canyon, setting off a battle that overshadowed the Dinosaur conflict by the time it was over.[32]

Bitterness at losing Glen Canyon sprang not only from the discovery of its scenery, but from conclusions reached by Luna Leopold and Walter Langbein, in two USGS publications appearing in
1959. These technical reports did not reach a wide public audience, but they hit home with Richard Bradley, Brower, and others
among the conservationists who had spent so much time in studying the hydrologic aspects of the CRSP. In detailed analyses replete
with graphs and mathematical analysis, Leopold and Langbein
concluded that the amount of storage from large dams on the Colorado and in some other locales did not sufficiently account for
water loss from evaporation. In brief, they argued that the number
of dams the Bureau was erecting on the Colorado constituted a
form of overkill. Hoover Dam already stored twenty-seven million
acre-feet; and once Glen Canyon Reservoir filled up, the total storage would exceed fifty million acre-feet. According to Leopold and
Langbein, forty million acre-feet was the upper limit of storage,
beyond which evaporation losses would be of such magnitude as
to nullify the benefits of additional storage. Taken together, Leopold and Langbein implied that Glen Canyon Dam was unnecessary in order to regulate the Colorado River, and was wasteful of
precious water; moreover, Hoover Dam provided the necessary
storage west of the Grand Canyon. Of course, their conclusions
did not take into account the Colorado River Compact, which
marked the basins at Lee's Ferry; but they did add to the bitterness
among conservationists at the loss of Glen Canyon, which, it now
appeared to them, had been unnecessary. Having assumed
throughout the Dinosaur battle that Glen Canyon Dam played a
vital role in river regulation, Brower and his friends were shaken.[33]

In many ways, the aftermath of Echo Park—the foiled effort to
convert Dinosaur into a park, the Bureau's change of mind on
Rainbow Bridge, and the discovery of Glen Canyon—provided
conservationists with ample reminders of the limits of their victory.
The controversy had ended with a strengthened national park system and with greater recognition of wilderness lands on the Colorado Plateau, but the events of subsequent years made starkly clear
that western states and the Bureau of Reclamation represented

forces that would continue to challenge wilderness and parks in the future. The Colorado River had become a key center of conflict among two divergent forces in the postwar West, forces of development and preservation that appeared dramatically in the clash at Echo Park. From the vantage point of western history, the Dinosaur controversy provided the first sign of that conflict, as well as the first indication that the West would emerge as a primary region of environmental conflict in the United States.

Notes

Chapter 1

1. Olaus Murie to Frank Barrett, Jan. 12, 1954, Box 15, Frank Barrett Papers, American Heritage Center, University of Wyoming (hereafter cited as Barrett Papers).

2. Murie to O'Mahoney, May 10, 1955, Box 356, Joseph C. O'Mahoney Papers, American Heritage Center, University of Wyoming (hereafter cited as O'Mahoney Papers). See also Fred Packard to John Will, Nov. 2, 1951, and Will to Packard, Dec. 3, 1951, Box 3847, Department of the Interior Central Classified Files, 1937–1953, Record Group 48, National Archives (hereafter cited as CCF/Interior, RG 48).

3. *Time*, 23 August 1954, 16.

4. See Joe Penfold, "Reclamation's Plan for Invasion," *Sierra Club Bulletin* 37 (May 1952): 10–14.

5. See Mark Harvey, "Utah, the National Park Service, and Dinosaur National Monument, 1909–56," *Utah Historical Quarterly* 59 (Summer 1991): 243–63.

6. Earl Douglass to W. J. Holland, Dec. 2, 1909, and June 4, 1910, in "Douglass Letters to the Carnegie Museum," Dinosaur National Monument.

7. Harvey, "Utah," 251.

8. Roger W. Toll, "Report to the Director, National Park Service, on Dinosaur National Monument," Nov. 29, 1929, Box 580, National Park Service Central Classified Files, Record Group 79, National Archives (hereafter cited as NPS/CCF, RG 79).

9. Harvey, "Utah," passim.

10. For studies of the Fremont, see Jesse D. Jennings, *Prehistory of Utah and the Eastern Great Basin* (Salt Lake City: University of Utah Press, 1978); James H. Gunnerson, *The Fremont Culture: A Study in Culture Dynamics on the Northern Anasazi Frontier* (Cambridge, Mass.: Peabody Museum, 1969); Robert H. Lister and Florence C. Lister, *Those Who Came Before* (Tucson: University of Arizona Press, 1983).

11. J. Donald Hughes, *American Indians in Colorado* (Boulder, Colo.: Pruett Publishing Co., 1977), 26–29; Duane Vandenbusche and Duane A. Smith, *A Land Alone: Colorado's Western Slope* (Boulder, Colo.: Pruett Publishing Co.,

1981), 12–14; James Jefferson, Robert W. Delaney, and Gregory C. Thompson, *The Southern Utes: A Tribal History* (Ignacio, Colo.: Southern Ute Tribe, 1972), vii–x.

12. Wallace Stegner, ed., *This Is Dinosaur: Echo Park Country and Its Magic Rivers,* 2d ed. (Boulder, Colo.: Roberts Rinehart, 1985), 7–8.

13. William Culp Darrah, *Powell of the Colorado* (Princeton, N.J.: Princeton University Press, 1951), 108–13; Wallace Stegner, *Beyond the Hundredth Meridian* (Boston: Houghton Mifflin Co., 1954), 45.

14. Stegner, *This Is Dinosaur,* 10–11; Stegner, *Beyond the Hundredth Meridian,* 63–64.

15. Roy Webb, *If We Had A Boat: Green River Explorers, Adventurers, and Runners* (Salt Lake City: University of Utah Press, 1986), 58–64, 77–78; Stegner, *Beyond the Hundredth Meridian,* 67–69; Powell, quoted in Eliot Porter, *Down the Colorado* (New York: E. P. Dutton, 1969), 52–53.

16. The best source for the history of this remote border region is Frederic J. Athearn, *An Isolated Empire: A History of Northwest Colorado* (Denver: Colo. State Office, Bureau of Land Management, 1977), 77. Information can also be gleaned from Vandenbusche and Smith, *A Land Alone.* See also Ernest Staples Osgood, *The Day of the Cattleman* (Chicago: University of Chicago Press, 1929), 22; John Rolfe Burroughs, *Where the Old West Stayed Young* (New York: William Morrow and Co., 1962), 113.

17. An editor in Craig, Colorado who visited the mountains to the west claimed to find gold, iron, copper, and other minerals, and looked forward to completion of a railroad so that the underground wealth could be shipped to market. See *Moffat County Courier,* Jan. 22, 1914. Frederick Bonfils, publisher of the *Denver Post,* and a promoter named Volney Hoggatt established the colony near Craig in 1915, but it languished after a just few years because the occupants could not make the arid land produce and because grain prices slumped after World War I. Bonfils encouraged settlers with articles in *The Great Divide,* a weekly newspaper that spoke glowingly of Colorado's soils and climate. See Robert Athearn, *The Coloradoans* (Albuquerque: University of New Mexico Press, 1976), 252–53.

18. Robert Athearn, *Rebel of the Rockies: A History of the Denver and Rio Grande Western Railroad* (New Haven: Yale University Press, 1962), 288; Steven F. Mehls, "David H. Moffat, Jr.: Early Colorado Business Leader," (Ph.D. diss., University of Colorado, 1982).

19. *Denver Post,* Aug. 17 and 19, 1928.

20. Ibid, Sept. 5, 1928; other accounts of Birch's journey appeared in the *Post* on Aug. 29 and Sept. 4, 1928.

21. Roger W. Toll, "Report on the Yampa Canyon Proposed National Monument," submitted to Arno B. Cammerer, Dec. 18, 1933, 7. The Rocky Mountain Regional Office of the National Park Service in Denver generously provided a copy of this report to the author.

22. Ibid., 7.

23. Ibid., 22.

24. Ibid., 1 and 27.

25. C. E. Seymour to C. F. Moore, Oct. 28, 1937; Directors of National Park Service and Division of Grazing to Secretary of the Interior, Mar. 1, 1938; "History of the Relationships between the National Park Service and the Division of Grazing in Connection with the Escalante and Green River National Monument

Projects in Utah and Colorado," all in Box 2163, NPS/CCF, 1907–1949, RG 79. On the matter of grazing rights, see also A. E. Demaray to William King, May 20, 1938, Box 2159, NPS/CCF, RG 79.

26. *Vernal Express,* Dec. 16, 1937, and Aug. 18, 1938.

27. Ibid., Oct. 14, 1937; J. A. Cheney to A. E. Demaray, Nov. 18, 1937, Box 2164, NPS/CCF, 1933–1949, RG 79.

28. Frederick H. Reid to Oscar Chapman, Jan. 18, 1934; Craig, Colo. Lions Club to Edward Taylor, Feb. 12, 1935; Colorado Mountain Club to A. E. Demaray, Mar. 20, 1935, Box 2164, NPS/CCF, 1933–1949, RG 79.

29. Barry Mackintosh, *The National Parks: Shaping the System* (Washington, D.C.: U.S. Department of the Interior, 1985), 24–35.

30. Donald C. Swain, "The National Park Service and the New Deal, 1933–1940," *Pacific Historical Review* 41 (August 1972): 327–29.

31. Barry Mackintosh, "Harold L. Ickes and the National Park Service," *Journal of Forest History* 29 (April 1985): 82–83; Susan Schrepfer, *The Fight to Save the Redwoods: A History of Environmental Reform, 1917–1978* (Madison: University of Wisconsin Press, 1983), 60–61.

32. Ickes to Roosevelt, Apr. 13, 1938, Box 2159, NPS/CCF, 1933–1949, RG 79.

33. A. E. Demaray to NPS Regional Director (Omaha), May 17, 1938; Demaray to Commissioner of General Land Office, Aug. 19, 1938, both in Box 2159, NPS/CCF, 1907–1949, RG 79; Executive Proclamation, "Enlarging Dinosaur National Monument," July 14, 1938, *Federal Register,* Vol. 3, Order no. 140.

34. *Vernal Express/ Roosevelt Standard,* "Scenic Edition," Aug. 18, 1938; *Vernal Express,* July 28, 1938.

35. Hal Rothman, *Preserving Different Pasts: The American National Monuments* (Urbana: University of Illinois Press, 1989).

36. Quoted in Toll, "Report on the Yampa Canyon," 23.

37. Ibid.

38. Ibid.; "Echo Park Project Holds Key to Development of Utah," *Salt Lake Tribune,* Feb. 20, 1950; Donald B. Alexander to Conrad Wirth, July 2, 1934, Box 2164, NPS/CCF, RG 79.

39. Samuel P. Hays, *Conservation and the Gospel of Efficiency: The Progressive Conservation Movement* (New York: Atheneum, 1975), 239.

40. Frank McNinch to Arno Cammerer, Dec. 13, 1934, Box 2164, NPS/CCF, RG 79.

41. Ibid.

42. Ickes to McNinch, Nov. 6, 1935, Box 2164, NPS/CCF, RG 79; Arno Cammerer to Ickes, July 30, 1935, "National Monuments, Dinosaur" folder, Box 1978, CCF/Interior, RG 48.

43. McNinch to Cammerer, Dec. 13, 1934; McNinch to Ickes, Jan. 9, 1936, Box 2164, NPS/CCF, RG 79.

44. Cammerer to Ickes, July 30, 1935, CCF/Interior, RG 48; Susan Rhoades Neel, "Irreconcilable Differences: Reclamation, Preservation, and the Origins of the Echo Park Dam Controversy," (Ph.D. diss., University of California, Los Angeles, 1990), 121–211.

45. Fred Fischler, interview with author by telephone, June 7, 1991.

46. Memo from "W.E.L." (NPS Planning Coordinator) to Ben Thompson, Feb. 2, 1938, Box 2159, NPS/CCF, RG 79.

47. Demaray Memorandum to NPS Director, Sept. 9, 1943, Box 2163, NPS/CCF, RG 79. This is a very important two-page memorandum, which details some of the discussions between the Park Service and the Bureau in the 1930s; see also Neel, "Irreconcilable Differences," 177–88, 192–95.

48. Executive Proclamation "Enlarging Dinosaur National Monument," July 14, 1938, *Federal Register,* Vol. 3, Order no. 2290.

49. Demaray to NPS Director, Sept. 9, 1943, Box 2163, NPS/CCF, RG 79.

50. Ibid.; Neel, "Irreconcilable Differences," 195–203; J. Lee Brown to Conrad Wirth, Jan. 17, 1936, Box 2164; Demaray to Senator William King, May 20, 1938, Box 2159, NPS/CCF, RG 79.

Chapter 2

1. Senate Committee on Interior and Insular Affairs, *Hearings on Colorado River Storage Project on S. 1555,* 83d Cong., 2d sess., June 28–July 3, 1954, 170–72.

2. Hal Rothman, *Preserving Different Pasts: The American National Monuments* (Urbana: University of Illinois Press, 1989), 89–116.

3. Carlos Schwantes, *The Pacific Northwest, An Interpretive History* (Lincoln: University of Nebraska Press, 1989), 306–9; Michael P. Malone, Richard B. Roeder, William L. Lang, *Montana: A History of Two Centuries,* 2d ed. (Seattle: University of Washington Press, 1991), 300–2; Richard Lowitt, *The New Deal and the West* (Bloomington: Indiana University Press, 1984), 82–87, 193–202; Norris Hundley, Jr., *The Great Thirst: Californians and Water, 1770s–1990s* (Berkeley and Los Angeles: University of California Press, 1992), 248–57.

4. Marc Reisner, *Cadillac Desert: The American West and Its Disappearing Water* (New York: Viking Penguin, 1986), 165. Another important work on this subject is Donald Worster, *Rivers of Empire: Water, Aridity, and the Growth of the American West* (New York: Pantheon Books, 1985, repr., Oxford University Press, 1992). See also Donald C. Swain, "The Bureau of Reclamation and the New Deal, 1933–1940," *Pacific Northwest Quarterly* 61 (July 1970): 137–46.

5. Robert G. Athearn, *The Mythic West in Twentieth Century America* (Lawrence: University Press of Kansas, 1986), 92–96.

6. Michael P. Malone and Richard W. Etulain, *The American West, A Twentieth Century History* (Lincoln: University of Nebraska Press, 1989), 94–107; Lowitt, *New Deal,* 28–30; John F. Bluth and Wayne K. Hinton, "The Great Depression," in Richard Poll, Thomas G. Alexander, Eugene E. Campbell, and David E. Miller, eds., *Utah's History* (Provo: Brigham Young University Press, 1978), 481–96, and T. A. Larson, *History of Wyoming,* 2d ed. (Lincoln: University of Nebraska Press, 1978), 411–46.

7. Joseph E. Stevens, *Hoover Dam: An American Adventure* (Norman: University of Oklahoma Press, 1988); Hundley, *Great Thirst,* 224–25.

8. Eugene P. Moehring, *Resort City in the Sunbelt: Las Vegas, 1930–1970* (Reno and Las Vegas: University of Nevada Press, 1989), 14–22.

9. Hundley, *Great Thirst,* 220–30 (quote on 230); Lowitt, *New Deal,* 82–86.

10. Norris Hundley, Jr., *Water and the West: The Colorado River Compact and the Politics of Water in the American West* (Berkeley: University of California

Press, 1975), 182–85; Hundley, *Great Thirst,* 209–13; Robert Dunbar, *Forging New Rights in Western Waters* (Lincoln: University of Nebraska Press, 1983), 137–39.

11. Utah Governor Henry Blood spoke at length about Utah's interest in the Colorado River, and the speech appears in the *Vernal Express,* Sept. 8, 1938. E. O. Larson of the Bureau of Reclamation fully explained the purposes of the Colorado River Storage Project at congressional hearings during the 1950s. See House Committee on Interior and Insular Affairs, *Hearings on H.R. 4449, 4443, and 4463 on Colorado River Storage Project,* 83d Cong., 2d sess. (Jan 18–28, 1954), 49–89.

12. "Review of Some Phases of the Colorado River Storage Project and Participating Projects," in *Project Planning Report No. 4–8a.* 1–1, 7–9, Bureau of Reclamation Records, Bureau of Reclamation Regional Office, Salt Lake City (hereafter cited as BRO, SLC).

13. Reprinted in *Vernal Express,* Mar. 2, 1939.

14. A. E. Demaray to Acting Superintendent Rocky Mountain National Park, Feb. 18, 1939, Box 2160, National Park Service Central Classified Files, 1933–1949, Record Group 79, National Archives (hereafter cited as NPS/CCF, RG 79).

15. Ibid.

16. Quoted in "Echo Park Project Holds Key to Development of Utah" *Salt Lake Tribune,* Feb. 20, 1950.

17. David Canfield to Edwin Nielsen, June 24, 1940, Box 2160, NPS/CCF, RG 79.

18. Neel, "Irreconcilable Differences," 227–326.

19. Ibid, 122–30; Mackintosh, *National Parks,* 52.

20. Harry B. Robinson, interview with author, July 4, 1985, Columbia, Mo.

21. "Memorandum of Understanding Re Proposed Survey of Recreational Resources of the Colorado River Basin," Nov. 4, 1941, Box 22, Irving Brant Papers, Manuscript Division, Library of Congress (hereafter cited as Brant Papers).

22. Ibid.; John Page to Drury, Sept. 22, 1942, Box 22, Brant Papers; Neel, "Irreconcilable Differences," 281–302, 304–7; Elmo Richardson, *Dams, Parks and Politics: Resource Development and Preservation in the Truman-Eisenhower Era* (Lexington: University Press of Kentucky, 1973), 50–53.

23. Neel, "Irreconcilable Differences," 275–464.

24. Tom Turner, "The Dinosaur Story," *Sierra* 76 (Nov.–Dec. 1991): 28. Some writers have claimed that Drury only learned about the Dinosaur dams when he read a copy of the *Federal Register* in 1950. See Dyan Zaslowsky and the Wilderness Society, *These American Lands* (New York: Henry Holt and Co., 1986), 33.

25. Paul Brown to NPS Director, Aug. 6, 1938, Box 2159, NPS/CCF, RG 79.

26. John C. Page to Newton Drury, Sept. 22, 1942, Box 22, Brant Papers; Hillory Tolson to Frank Setzler, May 5, 1942, Box 2160, NPS/CCF, RG 79. See also Elmo Richardson, "Federal Park Policy in Utah: The Escalante National Monument Controversy of 1935–1940," *Utah Historical Quarterly* 33 (Spring 1965): 122.

27. Neel, "Irreconcilable Differences," 280 and 227–326, passim; Neel has also explained Drury's accommodating strategy in two papers: "Recreation, Reclamation, and Preservation: National Park Service Policy in the Colorado River Basin, 1933–1940," presented at the annual meeting of the Pacific Coast Branch of the

American Historical Association, Salt Lake City, Utah, August 1990, and "Newton Drury and the Echo Park Dam Controversy," presented at a meeting of the American Society for Environmental History, Houston, Tex., Mar. 1991.

28. Michael Straus to the Secretary of the Interior, Dec. 20, 1949, "Water Power, Echo Park Dam Site" folder, Official Records of the National Park Service, 1945–1950, Dinosaur National Monument Library (hereafter cited as NPS Records, 1945–1950, DNM).

29. Hillory Tolson to A. E. Demaray, July 21, 1943; Demaray to NPS Director, Sept. 9, 1943; Newton Drury to Bureau Commissioner, Dec. 1, 1943, Box 2163, NPS/CCF, RG 79. See also Page to Drury, Sept. 22, 1942, Box 22, Brant Papers.

30. Michael Straus to Oscar Chapman, Dec. 20, 1949, NPS Records, 1945–1950, DNM.

31. Hillory Tolson to A. E. Demaray, July 21, 1943, Box 2163, NPS/CCF, RG 79.

32. Ibid.

33. Lawrence Merriam to NPS Director, Dec. 4, 1943; Conrad Wirth to NPS Director, Aug. 21, 1944; and Tolson memorandum, Aug. 2, 1944, Box 2163, NPS/CCF, RG 79; Neel, "Irreconcilable Differences," 327–32.

34. Beard to NPS Regional Director, Region Two, Oct. 5, 1943, Box 2163, NPS/CCF, RG 79.

35. A. E. Demaray to Regional Director, Region Two, Jan. 5, 1948, Lawrence Merriam to Superintendent of Rocky Mountain National Park, Jan. 13, 1948, NPS Records, 1945–1950, DNM.

36. *The American West Transformed: The Impact of the Second World War* (Bloomington: Indiana University Press, 1985), 17; Malone and Etulain, *American West,* 107–19.

37. Nash, *American West Transformed,* 156–57; Malone and Etulain, *American West,* 110; Marshall Sprague, *Colorado: A Bicentennial History* (New York: W. W. Norton and Co., 1976), 170–75, and Larson, *History of Wyoming,* 487–93.

38. Nash, *American West Transformed,* 23–24, 29–30; see also John E. Christensen, "The Impact of World War II," in Poll et al., *Utah's History,* 497–514, and James L. Clayton, "Contemporary Economic Development," in Poll et al., *Utah's History,* 531–44.

39. *Salt Lake Tribune* "Empire Edition," Jan. 14, 1951; see also "Water for Utah: Report of Utah Water and Power Board," 51–56, July 1, 1948, Utah Water and Power Board Records, Utah State Archives, Salt Lake City (hereafter cited as Water and Power Board Records, USA).

40. Herbert Maw, "More Power to Utah from the Multi-Million Dollar Central Utah Project," *Reclamation Era* 33 (Feb. 1947): 30–32.

41. "Upper Basin Lags in Development," *Salt Lake Tribune,* Feb. 21, 1950; Larson, *History of Wyoming,* 508–9, and J. T. Banner, "Report on Development and Utilization of Wyoming's Colorado River Basin Water Resources" (Laramie, Wyo.: J. T. Banner and Associates, 1955), 30–34.

42. Maw, "More Power to Utah," 30; "Water for Utah" Report, 104, Water and Power Board Records, USA.

43. Nine major streams tumbled down the southwest slope of the High Uintas and passed through the Uinta Basin before running into the Green River, the big-

gest tributary to the Colorado. See the testimony of George Clyde, Senate Committee on Interior and Insular Affairs, *Hearings on S. 1555 on CRSP*, (1954), 147.

44. *Deseret News*, Dec. 11, 1946. The newspaper published a series of pieces on the CUP in 1946, including the issues of Dec. 3, 10, and 14. See also *Vernal Express*, May 23, 1946.

45. "Water for Utah" Report, 102–12, Water and Power Board Records, USA; *Salt Lake Tribune*, Apr. 9, 1946.

46. "Water for Utah" Report, Water and Power Board Records, USA.

47. Dean Mann, Gary Weatherford, and Phillip Nichols, "Legal-Political History of Water Resource Development in the Upper Colorado River Basin" (National Science Foundation, Lake Powell Research Project, 1974), 29.

48. Ibid.; see also the Bureau of Reclamation Project Planning Report, Colorado River Storage Project, 8–9, June 15, 1951, BRO, SLC.

49. E. O. Larson to Lawrence Merriam, Dec. 3, 1948, NPS Records, 1945–1950, DNM.

50. Neeley, from the Bureau of Reclamation's office in Vernal, pointed out the role of Echo Park Dam's power in funding the CUP in a speech before several groups in Vernal in Nov. 1947. G. Lee Sneddon, on the staff at Dinosaur Monument, attended the meeting. See Sneddon's "Memorandum for the Files," Dec. 23, 1947, NPS Records, 1945–1950, DNM.

51. "Water for Utah" Report, 91–93, Water and Power Board Records, USA.

52. See chapter 1.

53. *Vernal Express*, Oct. 18, 1945; G. Lee Sneddon "Memorandum," Dec. 23, 1947, and Ben H. Thompson to Newton Drury, Sept. 13, 1949, NPS Records, 1945–1950, DNM; W. D. Romig to N. B. Bennett, Sept. 19, 1949, Central Classified Files of the Bureau of Reclamation, 1945–1950. The Bureau's office in Washington, D.C. provided microfiche copies of these records to the author, and remain in my possession (hereafter references will be cited as CCF/BR).

54. E. O. Larson to Michael Straus, Sept. 30, 1949, CCF/BR.

55. The Bureau's regional office in Salt Lake City provided detailed figures of the advantage of the Echo Park site for power output, in a "Brief Report," Dec. 5, 1949, CCF/BR.

56. The Bureau also stressed how Echo Park Dam's power-plant capacity was substantially greater than that planned at Flaming Gorge Dam upstream on the Green River. Echo Park would have an installed capacity of 200,000 kilowatts, Split Mountain 100,000 kilowatts, and Flaming Gorge 72,000 kilowatts. See E. O. Larson's testimony to the Upper Colorado River Commission, Mar. 20, 1950, Official Record of Upper Colorado River Commission, 90, Box 44, Harry Bashore Papers, American Heritage Center, University of Wyoming (hereafter cited as Bashore Papers).

57. "Federal Bureau Backs Central Utah Project," *Salt Lake Tribune*, Apr. 9, 1946; "U.S. Works Out Nine-Project Upper Colorado River Plan," *Salt Lake Tribune*, Oct. 16, 1949; Craig (Colo.) *Empire-Courier*, Oct. 5, 1949.

58. Bureau of Reclamation, *The Colorado River*, "A Natural Menace Becomes a Natural Resource," *A Comprehensive Report on the Development of the Water Resources of the Colorado River Basin for Irrigation, Power Production, and other Beneficial Uses in Arizona, California, Colorado, Nevada, New Mexico, Utah, and Wyoming* (Washington, D.C.: Department of the Interior, 1946), 25.

59. Ibid., 11–19.

60. Ibid., 244.

61. Mann, et al., " Legal-Political History," 7.

62. Bureau of Reclamation, *Colorado River,* 7; William Warne to Harry Bashore, Mar. 24, 1948, Box 37, Bashore Papers.

63. Floyd Bishop, interview with author, Feb. 9, 1982, Laramie, Wyo.

64. The negotiations leading to the Upper Colorado River Compact can be found in three bound volumes entitled "Upper Colorado River Compact Commission," available in the William Robertson Coe Library, University of Wyoming. See also Hundley, *Water and the West,* 300–301.

65. J. A. Howell to Bryant Stringham, Aug. 3, 1949, Box 14, Bryant Stringham Papers, Department of Special Collections, Utah State University (hereafter cited as Stringham Papers).

66. The text of the Compact is available in U.S. Congress, House of Representatives, *Hearings before a Subcommittee on Irrigation and Reclamation,* Committee on Public Lands, 81st Cong., 1st sess., Mar. 1949, 1–10; see also Jean S. Breitenstein, "The Upper Colorado River Basin Compact," *State Government* 22 (1949): 214–16, 225; Clifford H. Stone, "Looking 50 Years Ahead on the Colorado," in Box 13, Stringham Papers.

67. See chapter 3.

68. *Deseret News,* Dec. 12, 1946; *Vernal Express,* Dec. 20, 1945, and May 23, 1946.

69. Hundley, *Great Thirst,* 220–32.

70. Quoted in Butcher, "This Is Dinosaur," *National Parks Magazine* 24 (Oct.–Dec. 1950), repr., *Congressional Record Appendix,* 81st Cong., 2d sess., Dec. 20, 1950, 96, pt. 18: A7891.

71. *Rivers of Empire,* 259–60.

72. Ibid, 260.

Chapter 3

1. Butcher, "This Is Dinosaur," 122–36.

2. Olaus Murie to Oscar Chapman, Mar. 21, 1950, CCF/BR.

3. Roderick Nash, *Wilderness and the American Mind,* 3d ed. (New Haven: Yale University Press, 1982), 209–19.

4. Ibid., 210, 219.

5. Besides Nash's book, see Victor B. Scheffer, *The Shaping of Environmentalism in America* (Seattle: University of Washington Press, 1991), 118.

6. "A Statement on Wilderness Preservation in Reply to A Questionnaire," (Memorandum for the Legislative Reference Service), 6–7, Folder no. 22, Box 41, Wilderness Society Papers, Western History Department, Denver Public Library (hereafter cited as WS Papers).

7. Nash, *Wilderness,* 214–17.

8. John Diggins, *The Proud Decades: America in War and Peace, 1941–1960* (New York: W. W. Norton and Co., 1988), 177–219; William O'Neill, *American High: The Years of Confidence* (New York: Free Press, 1986), 9–50.

9. Sigurd Olson, "Why We Need Wilderness," *National Parks Magazine* 20 (Jan.-Mar. 1946): 19; Olson, "The Preservation of Wilderness," *Living Wilder-*

ness 13 (Autumn 1948): 1–8; Howard Zahniser and F. S. Baker, "'We Certainly Need A Sound Philosophy,'" *Living Wilderness* 12 (Winter 1947–1948): 1–5.

10. Murie quoted in Nash, *Wilderness,* 213–14; Samuel P. Hays, *Beauty, Health, and Permanence: Environmental Politics in the United States, 1955–1985* (New York: Cambridge University Press, 1987), 2–5, 22–24.

11. The best comprehensive history of American conservation organizations and their leaders is Stephen Fox, *John Muir and His Legacy: The American Conservation Movement* (Boston: Little, Brown, and Co., 1981; repr., Madison: University of Wisconsin Press, 1985).

12. Zahniser to John Saylor, July 29, 1955, Folder no. 9, Box 34, WS Papers.

13. Howard Zahniser to Editor of the *Washington Post,* Feb. 20, 1951, Box 102, Waldo Leland Papers, Manuscript Division, Library of Congress (hereafter cited as Leland Papers).

14. Zahniser made this remark at the 1956 annual meeting of the Wilderness Society Council held in Ranier, Minn., the minutes of which are in Box 18, WS Papers; Legislative Reference Report, 18–19, Folder no. 22, Box 41, WS Papers. Bestor Robinson of the Sierra Club also pointed out the connection between the Dinosaur campaign and the wilderness preservation effort. See Michael P. Cohen, *The History of the Sierra Club, 1892–1970* (San Francisco: Sierra Club Books, 1988), 155.

15. Bradley's analysis was not published. It appears on handwritten sheets under the title "Lost Valley," Box 115, Harold Bradley Papers, Sierra Club Records, Bancroft Library (hereafter cited as HB Papers). For published accounts of the Hetch Hetchy battle, see Nash, *Wilderness,* 161–81; Kendrick Clements, "Politics and the Park: San Francisco's Fight for Hetch Hetchy, 1908–1913," *Pacific Historical Review* 48 (May 1979): 185–215; Holway Jones, *John Muir and the Sierra Club, The Battle for Yosemite* (San Francisco: Sierra Club, 1965).

16. Alfred Runte, *National Parks: The American Experience,* 2d ed. (Lincoln: University of Nebraska Press, 1987), 155–59.

17. Quoted in Barry Mackintosh, *The National Parks: Shaping the System,* 2d ed. (Washington D.C.: Department of the Interior, 1991), 19.

18. John Ise, *Our National Park Policy: A Critical History* (Baltimore: Johns Hopkins Press, 1961), 447–53; editorial, "The Rocky Mountain Tunnel Threat," *National Parks Bulletin* 13 (June 1937): 9–10; C. W. Buchholtz, *Rocky Mountain National Park: A History* (Boulder: Colorado Associated University Press, 1983), 189–90; Zaslowsky and the Wilderness Society, *These American Lands,* 32. See also Annual Report of the National Park Service, in *Annual Report of the Secretary of the Interior for Fiscal Year Ended 30 June 1943* (Washington, D.C.: Department of the Interior), 198–207.

19. See Annual Report of the National Park Service, in *Annual Report of the Secretary of the Interior* (Washington, D.C.: Department of the Interior, 1946), 307. Drury's outlook following the war can be seen in his remarks before a National Conservation Policy Conference in 1947. A copy of the proceedings is in Box 7, Dale E. Doty Papers, Harry S. Truman Library, Independence, Mo. (hereafter cited as Doty Papers); Olson, "Preservation of Wilderness," 1–8.

20. Editorial, "Krug Stimulates Cooperation," *National Parks Magazine* 22 (Jan.–Mar. 1948): 3; Olaus Murie to Newton Drury, Nov. 22, 1948, Box 265, Olaus Murie Papers, Western History Department, Denver Public Library (hereafter cited as Murie Papers); see also Fred Packard's testimony in House Committee

on Interior and Insular Affairs, *Hearings on H.R. 4449 et al. on CRSP* (1954), 811–12.

21. Annual Report of the National Park Service, in *Annual Report of the Secretary of the Interior for Fiscal Year Ended 30 June 1947* (Washington, D.C.: Department of the Interior), 344.

22. Hays, *Beauty, Health, and Permanence,* 2–5, 22–24 (quote on 23); Runte, *National Parks,* 171.

23. Runte, *National Parks,* 132; William P. Wharton, "The National Primeval Parks," *National Parks Bulletin* 13 (Feb. 1937): 3–5; news item, "National Primeval Park Standards: A Declaration of Policy," *National Parks Magazine* 19 (Oct.–Dec. 1945): 6–11.

24. Runte, *National Parks,* 117–18.

25. Quoted in Legislative Reference Report, 26, Folder no. 22, Box 41, WS Papers. See also a piece probably written by Howard Zahniser, "Why We Cherish the San Gorgonio Primitive," *Living Wilderness* 12 (Mar. 1947): 1–7; Olson, "Preservation of Wilderness," 1–8; Aldo Leopold, "Wilderness," in *A Sand County Almanac* (New York: Oxford University Press, 1968), 188–93; Runte, *National Parks,* 109–11.

26. Murie to Oscar Chapman, Mar. 21, 1950, CCF/BR.

27. Runte, *National Parks,* 171, and Bernard DeVoto, "Let's Close the National Parks," *Harper's Magazine* 207 (Oct. 1953): 49–52.

28. Drury to DeVoto, June 10, 1948, Box 6, Records of Newton B. Drury, Central Classified Files of the National Park Service, Record Group 79, National Archives (hereafter cited as Drury Records, CCF/NPS, RG 79). See also Lester Hunt to Harry S. Truman, Nov. 8, 1952, Box 55, Official File of Harry S. Truman, Harry S. Truman Library, and DeVoto, "Let's Close the National Parks," 49–52.

29. Irving Clark to Wilderness Society, Mar. 24, 1947, Box 361, Murie Papers; Irving Brant to Harry S. Truman, Dec. 28, 1947, Box 15, Brant Papers. Threats to the national parks and conservationists' fears about them can be found in a number of primary and secondary sources. See Weldon F. Heald, "The Squeeze Is On the National Parks," *National Parks Magazine* 24 (Jan.–Mar. 1950): 3–4; Devereux Butcher to Dale Despain, Dec. 20, 1950, Box 14, Stringham Papers.

30. DeVoto to a Mr. Phillips, Mar. 9, 1949, Box 5, Bernard DeVoto Papers, Department of Special Collections, Stanford University (hereafter cited as DeVoto Papers); DeVoto, "The Western Land Grab," *Harper's Magazine* 194 (June 1947): 543–46; Wallace Stegner, *The Uneasy Chair: A Biography of Bernard DeVoto* (Garden City, New York: Doubleday & Co. Inc., 1974), 301–3. On the "landgrab" see also Kenneth A. Reid, "Your Heritage: A Statement Clarifying the Threatened Federal Landgrab," *National Parks Magazine* 21 (July–Sept. 1947): 13; Robert Righter, *Crucible for Conservation: The Creation of Grand Teton National Park* (Boulder: Colorado Associated University Press, 1982), 117–25.

31. Harold Bradley, "The Dinosaur Case," *Garden Club Bulletin* (Mar. 1953): 89.

32. Editorial, "Glacier View Dam," *National Parks Magazine* 22 (Oct.–Dec. 1948): 3–4; Olaus Murie, "Primitive Area on Trial," *Living Wilderness* 13 (Spring 1948): 16–25.

33. Fred Packard, "Grand Canyon Park and Dinosaur National Monument in Danger," *National Parks Magazine* 23 (Oct.–Dec. 1949): 11–13.

34. Waldo Leland and Frank Setzler to Morris L. Cooke, Mar. 31, 1950, Box

18; Olaus Murie to Cooke, Apr. 8, 1950 and Newton Drury to Leland Olds, Sept. 13, 1950, Box 37, Records of the President's Water Resources Policy Commission, Harry S. Truman Library (hereafter cited as PWRPC Records).

35. Schaefer quoted in Legislative Reference Report, 20, Folder no. 22, Box 41, WS Papers.

36. Fred Packard to Phillip Sirotkin, Aug. 26, 1955, Box 34, WS Papers.

37. Fred Packard, "Grand Canyon and Dinosaur," 12; Devereux Butcher, "Stop the Dinosaur Power Grab," *National Parks Magazine* 24 (Apr.–June 1950): 61–65.

38. Butcher, "This Is Dinosaur," 122–36; John B. Oakes, telephone conservation with author, Feb. 27, 1986; David R. Brower, interview with author, San Francisco, July 24, 1985; Margaret Murie, interview with author, Moose, Wyo., Aug. 5, 1988.

39. Drury, interviewed some years later, referred to this discussion within the Park Service about the merits of National Recreation Areas. See Newton Bishop Drury, "Parks and Redwoods, 1919–1971," an oral history conducted 1960–1970 by Amelia Fry and Susan R. Schrepfer, Vol. 2, 516, Regional Oral History Office, The Bancroft Library, University of California, Berkeley, 1972 (Courtesy of the Bancroft Library); see also Herbert Evison and Newton Bishop Drury, "The National Park Service and Civil Conservation Corps," an oral history conducted 1961–1962 by Amelia Fry, 117–118, Regional Oral History Office, The Bancroft Library, University of California, Berkeley, 1963 (Courtesy of the Bancroft Library).

40. Untermann to Doerr, Oct. 1, 1949, NPS Records, 1945–1950, DNM.

41. Righter, *Crucible for Conservation*, 89–92, 127. See the editorial "Krug Stimulates Cooperation," and Isabelle F. Story, "Water Recreation in the Desert," *National Parks Magazine* 22 (January–March 1948): 3, 25–26.

42. "Krug Stimulates Cooperation," 3; Butcher, "Stop the Dinosaur Power Grab," 61–64; Olaus Murie to Newton Drury, Nov. 22, 1948, Box 265, Murie Papers.

43. Martin quoted in Legislative Reference Report, 22, Folder no. 22, Box 41, WS Papers.

44. Rosalie Edge to Newton Drury, June 28, 1947, Box 21, Brant Papers; Neel, "Irreconcilable Differences," 398–410.

45. Drury made this comment to a group in Miami, Fl., Mar. 18, 1952, in a speech entitled "Preserving Nature in a National Resources Policy"; a copy is contained in Box 14, WS Papers; see also Fry and Schrepfer, interview of Newton B. Drury, Vol. 2, 516, Bancroft Library.

46. Fry and Schrepfer, interview of Newton B. Drury, Vol. 2, 516, Bancroft Library; Fry, interview of Evison and Drury, 117–18, Bancroft Library; Neel, "Irreconcilable Differences," 398–410.

47. There were several exchanges between the Bureau and the Park Service about the Bureau gaining access to the monument. See Hillory Tolson to Bureau Commissioner, June 6, 1946; Lawrence Merriam to E. O. Larson, Oct. 4, 1946, and A. E. Demaray to Bureau Commissioner, July 19, 1949, CCF/BR; see also Neel, "Irreconcilable Differences," 415–23.

48. Drury to Oscar Chapman, Jan. 24, 1950, and Mar. 3, 1950, "Dinosaur National Monument, 1949–1951" folder, Box 6, Drury Records, CCF/ NPS, RG 79.

49. Discussion of this point appears in the first history of the Echo Park contro-

versy written by two political scientists: Owen Stratton and Phillip Sirotkin, *The Echo Park Controversy* (University of Alabama, [Interuniversity Case Program, ICP Case Series no. 46], 1959), 23.

50. Lombard to David Canfield, Apr. 26, 1946, NPS Records, 1945–1950, DNM.

51. Clinton Woods to Bureau of Reclamation Commissioner, Sept. 2, 1949, CCF/BR; "Reconnaissance Report on Echo Park, Split Mountain, Flaming Gorge, and Cross Mountain Units," NPS Records, 1945–1950, DNM.

52. Bureau of Reclamation, *Interim Report: Colorado River Storage Project and Participating Projects* (Salt Lake City, 1949), BRO, SLC.

53. A. E. Demaray to Michael Straus, Sept. 13, 1949, NPS Records, 1945–1950, DNM.

54. Drury to Chapman, Dec. 30, 1949, NPS Records, 1945–1950, DNM. Drury noted in his annual report of 1946 that the two agencies agreed in a memorandum that the Bureau would fully consult with the Park Service "in advance of the detailed planning of any reclamation projects which would in any way affect the areas within the National Park System." Annual Report of the National Park Service in *Annual Report of the Secretary of the Interior* (Washington, D.C.: Department of the Interior, 1946), 323; Fry and Schrepfer, interview of Newton Drury, Vol. 2, 512, Bancroft Library.

55. Drury to Chapman, Dec. 30, 1949, NPS Records, 1945–1950, DNM.

56. Lombard to Canfield, Dec. 3, 1948, NPS Records, 1945–1950, DNM.

57. *A Survey of the Recreational Resources of the Colorado River Basin* (Washington, D.C.: U. S. Department of the Interior, 1950), 197–98; Olaus Murie to Drury, Nov. 22, 1948, "National Park Service" folder, Box 265, Murie Papers.

58. Untermann to Devereux Butcher, Nov. 1, 1950, Box 10, G. E. Untermann Papers, Department of Special Collections, University of Utah (hereafter cited as Untermann Papers); see also the author's article, "Utah, the National Park Service, and Dinosaur National Monument, 1909–1956," *Utah Historical Quarterly* (Summer 1991): 243–63.

59. William Bruce Wheeler and Michael J. McDonald, *TVA and the Tellico Dam, 1936–1979: A Bureaucratic Crisis in Post-Industrial America* (Knoxville: University of Tennessee Press, 1986).

60. David A. Kathka, "The Bureau of Reclamation in the Truman Administration: Personnel, Politics, and Policy" (Ph.D. diss., University of Missouri, 1976), 27. In 1946 alone, Congress appropriated a record ninety million dollars for the agency; "Annual Report of the Bureau of Reclamation," in *Annual Report of the Secretary of the Interior for 1946* (Washington, D.C.: Department of the Interior, 1946), 57.

61. Ise, *Our National Park Policy,* 309.

62. See Hays's superb essay "Three Decades of Environmental Politics: The Historical Context," in Michael J. Lacey, ed., *Government and Environmental Politics: Essays on Historical Development Since World War Two* (Washington, D.C.: Woodrow Wilson Center Press; Baltimore: Johns Hopkins University Press, 1991), 19–79 (quote on 22).

63. In addition to the essay noted above Hays's ideas are also found throughout his book *Beauty, Health, and Permanence.*

Chapter 4

1. June 29, 1951, Box 121, "Reading File 1951," Oscar Chapman Papers, Harry S. Truman Library (hereafter cited as Chapman Papers).

2. Oct. 30, 1951, "Water Power, Echo Park Dam Site," Official Records of the National Park Service, NPS Records, 1951–1953, DNM.

3. Straus to Chapman, Dec. 20, 1949, "Water Power, Echo Park Dam Site" folder, NPS Records, 1945–1950, DNM.

4. Reisner, *Cadillac Desert,* 143.

5. Straus to Chapman, Dec. 20, 1949, NPS Records, 1945–1950, DNM; Clinton Woods to Straus, Sept. 2, 1949; Straus to Chapman, Jan. 13, 1950, CCF/BR.

6. Straus to Chapman, December 1949, "Brief Report on the Importance of the Echo Park and Split Mountain Units, Colorado River Storage Project and Their Relation to the Existing Dinosaur National Monument," CCF/BR.

7. On the relationship of politics and federal water projects in this period, see Elmo Richardson, *Dams, Parks and Politics,* passim, and "GOP's Policy for West: Reclamation," *Salt Lake Tribune,* June 6, 1948.

8. Quoted in Clayton R. Koppes, "Environmental Policy and American Liberalism: The Department of the Interior, 1933–1953," *Environmental Review* 7 (Spring 1983): 27. See also Chapman's introductory remarks in the *Annual Report of the Secretary of the Interior* (Washington, D.C.: Department of the Interior, 1950), vi.

9. Clayton R. Koppes, "Oscar L. Chapman: A Liberal at the Interior Department, 1933–1953" (Ph.D. diss., University of Kansas, 1974), 103; Fry and Schrepfer, interview of Newton B. Drury, Vol. 2, 523, Bancroft Library.

10. Bureau of Reclamation, *Interim Report,* 1949, BRO, SLC.

11. Arthur Watkins to Chapman, Feb. 4, 1950, Box 14, Stringham Papers. This box of Stringham's papers also contains letters and resolutions sent to Chapman from business groups in Colorado and Utah; "Five Western States Ask Okeh of Dam," *Salt Lake Tribune,* Feb. 10, 1950; "Minutes of Mar. 20, 1950," Official Record of the Upper Colorado River Commission, Box 44, Bashore Papers.

12. Fry and Schrepfer, interview of Newton Drury, Vol. 2, 523, Bancroft Library; Richardson, *Dams, Parks and Politics,* 55.

13. A typescript copy of the proceedings, "Shall Dams Be Built in Dinosaur National Monument?" is in Box 114, HB Papers. A list of those whom Chapman invited to the hearing is in "Miscellaneous Records," Reading File, 1949–1950, Box 114, Chapman Papers.

14. Olaus Murie to Oscar Chapman, Mar. 21, 1950; Harold Ickes to Chapman, Apr. 10, 1950, CCF/BR.

15. Donald McCoy, *The Presidency of Harry S. Truman* (Lawrence: University Press of Kansas, 1984), 173–90.

16. *Saturday Evening Post* 221 (May 14, 1949), 30–31, 160–62.

17. Brant to Chapman, Apr. 3, 1950, Box 20, Brant Papers.

18. Zahniser testimony, 73; Horace Albright testimony, 40–41, Chapman Transcript, HB Papers.

19. Setzler testimony, 23, Chapman Transcript, HB Papers.

20. Olmsted testimony, 17–19, Chapman Transcript, HB Papers.

21. Gabrielson testimony, 10–14; Charles Sauers testimony, 36–39, Chapman Transcript, HB Papers.

22. Oscar Chapman to William Voigt, Jr., Jan. 25, 1951, CCF/BR.

23. Drury testimony, 7–9; Fred Packard testimony, 78, Chapman Transcript, HB Papers.

24. Sauers testimony, 38, Chapman Transcript, HB Papers. A few days after the hearing, the *New York Times* editorialized that alternate dam sites outside Dinosaur should be carefully examined; otherwise, "the unique beauty and interest of this area will be irretrievably wrecked." See editorial, "The Dinosaur Dams," *New York Times,* Apr. 7, 1950.

25. Brant to Chapman, Apr. 3, 1950, Box 20, Brant Papers; Olaus Murie to Chapman, Mar. 21, 1950, CCF/BR; Drury testimony, 7–9; Horace Albright testimony, 41–42, Chapman Transcript, HB Papers. The Wilderness Society opposed dams designed to control floods or produce power that threatened parks or wilderness areas, especially when alternate sites could be used and when the dams in question mainly benefited local economies.

26. Arthur Carhart to Bernard DeVoto, Mar. 11, 1954, Box 120, Arthur Carhart Papers, Western History Department, Denver Public Library (hereafter cited as Carhart Papers). See also "Minutes of Mar. 20, 1950," Official Record of the Upper Colorado River Commission, 117, Box 44, Bashore Papers.

27. A copy of John Geoffrey Will's statement is found in Box 2, Lester Hunt Papers, American Heritage Center, University of Wyoming (hereafter cited as Hunt Papers).

28. "Minutes of Mar. 20, 1950," Official Record of the Upper Colorado River Commission, 137, Box 44, Bashore Papers; Will statement, Apr. 3, 1950, Box 2, Hunt Papers; Clifford Stone testimony, 88, Chapman Transcript, HB Papers.

29. N. B. Bennett testimony, 4–6, Chapman Transcript, HB Papers.

30. Ibid., 4. The Bureau also pointed out the erratic flow of the Colorado River in a "Brief Report on the Importance of the Echo Park and Split Mountain Units," Dec. 1949, CCF/BR.

31. Bennett testimony, 6; Stone testimony, 91, Chapman Transcript, HB Papers. E. O. Larson of the Bureau also used the figure of 350,000 acre feet to indicate added evaporation when he spoke to the Upper Colorado River Commission two weeks before Chapman's hearing. See "Minutes of Mar. 20, 1950," Official Record of the Upper Colorado River Commission, 90, Box 44, Bashore Papers; "Chapman Takes Echo under Advisement," *Salt Lake Tribune,* Apr. 4, 1950. See also the text of the Upper Colorado River Basin Compact, in U.S. Congress, House of Representatives, *Hearings before a Subcommittee on Irrigation and Reclamation,* Committee on Public Lands, 81st Cong., 1st sess., 1949, 1–10.

32. Quoted in "Chapman Takes Echo under Advisement," *Salt Lake Tribune,* Apr. 4, 1950.

33. Press Release, Secretary of the Interior, June 27, 1950, Box 14, Stringham Papers; Chapman to Drury and Straus, June 27, 1950, CCF/BR; editorial, "Echo Park Dam Gets Approval," *Salt Lake Tribune,* June 28, 1950.

34. Stratton and Sirotkin make this point in their work, *Echo Park Controversy,* 42; Richardson, *Dams, Parks and Politics,* 27–30, 66; see also a memorandum by Irving Brant, Mar. 16, 1957, Box 21, Brant Papers.

35. Fred Packard to Phillip Sirotkin, Aug. 26, 1955, Box 34, WS Papers; Oscar Chapman to William Voigt, Jr., Feb. 5, 1951, CCF/BR.

36. Provo (Utah) *Daily Herald,* June 27, 1950; Arthur Carhart to Bernard DeVoto, Mar. 11, 1954, Box 120, Carhart Papers; Richardson, *Dams, Parks and*

Politics, 58–59; Robert J. Donovan, *Tumultuous Years: The Presidency of Harry S. Truman, 1949–1953* (New York: W. W. Norton and Co., 1982), 154–57; Mc-Coy, *Presidency of Harry S. Truman,* 224.

37. Telegrams from Vernal, Utah, and Craig, Colorado, Chambers of Commerce are found in CCF/BR; see also John G. Will to Chapman, July 6, 1950, CCF/BR; Thomas quote is in *Salt Lake Tribune,* June 28, 1950, and the editorial "Echo Park Dam Approval Gives Green Light to Western Water, Scenic Developments," Ibid., June 29, 1950.

38. "Competition for Water from Colorado River Looms as Western Battle of the Century," *Salt Lake Tribune,* July 31, 1950; Floyd Bishop, interview with author, Feb. 9, 1982, Laramie, Wyo.

39. Carhart to Chapman, June 30, 1950, CCF/BR; other letters from conservationists denouncing the decision appear in NPS Records, 1950–1951, DNM.

40. A recent biography is T. H. Watkins, *Righteous Pilgrim: The Life and Times of Harold Ickes* (New York: Henry Holt, 1990).

41. Fry and Schrepfer, interview of Newton Drury, Vol. 2, 523, Bancroft Library.

42. Carhart to Chapman, June 30, 1950, CCF/BR; Irving Brant to Harry Truman, Feb. 16, 1951, Box 21, Brant Papers; Devereux Butcher to Harold Ickes, Oct. 25, 1950, Box 73, Harold Ickes Papers, Manuscript Division, Library of Congress (hereafter cited as Ickes Papers).

43. DeVoto published a number of articles on the landgrab. For a complete bibliography of DeVoto's writings, see Julius P. Barclay's list in Catherine Drinker Bowen et al., eds., *Four Portraits and One Subject: Bernard DeVoto* (Boston: Houghton Mifflin Company, 1963), 117–206; Stegner, *Uneasy Chair,* 322.

44. DeVoto to Raymond Moley, Feb. 8, 1955, Box 6, DeVoto Papers.

45. DeVoto, "One Hundred Year Plan," *Harper's Magazine* 201 (Aug. 1950): 60–64; DeVoto to General S. D. Sturgis, Jr., Sept. 14, 1950, Box 5, DeVoto Papers.

46. *Saturday Evening Post,* July 22, 1950, 42; the article was reprinted later that year in *Reader's Digest* 57 (Nov. 1950): 18–24.

47. DeVoto to a Mr. Frank, July 29, 1950, Box 5, DeVoto Papers; news item, *Sierra Club Bulletin* 23 (Sept. 1950): 8.

48. Albright is quoted in Glenn Sandiford, "Bernard DeVoto and His Forgotten Contribution to Echo Park," *Utah Historical Quarterly* 59 (Winter 1991): 76.

49. "There's Plenty of Scenery for Us Western Natives," *Denver Post,* July 22, 1950; "Shall We Keep 'Unmarred Natural Spectacle' of Echo Park Hidden Away From Public?," *Salt Lake Tribune,* July 25, 1950.

50. DeVoto to *Denver Post* editor, Aug. 1, 1950, Box 5; Joe Penfold to DeVoto, Aug. 9, 1950, Box 17, DeVoto Papers.

51. Merle Crowell to G. E. Untermann, Nov. 10, 1950; Untermann to Frederick Champ, Feb. 16, 1954, Box 10, Untermann Papers.

52. Untermann's article appeared in the *Congressional Record Appendix,* 81st Cong., 2d sess., 1950, 96, pt. 18: A7559–A7561.

53. "Fight Goes On to Block Echo Park Project," *Salt Lake Tribune,* May 19, 1950; Carhart to DeVoto, Aug. 2, 1950, Box 9, DeVoto Papers.

54. Straus revealed his attitude toward preservationists in June 1950, when he remarked that the Interior Department's famous seal, decorated with a bison on the plains, should be changed to reveal a dinosaur in a canyon; B. Frank Ward to

Reva Beck Bosone, Dec. 15, 1950, NPS Records, 1945–1950, DNM; Stratton and Sirotkin, *Echo Park Controversy,* 49; Stegner, *Uneasy Chair,* 314.

55. Ronald A. Foresta, *America's National Parks and Their Keepers* (Washington, D.C.: Resources for the Future, 1984), 71–72; Drury to Straus, Sept. 6, 1950, Box 6, Drury Records, CCF/NPS, RG 79.

56. Drury to Straus, Sept. 6, 1950, Box 6, Drury Records, CCF/NPS, RG 79.

57. The Park Service developed a "Master Plan" for the monument, to take into account the new arrangements that must be made following construction of the dam. It called for Echo Park to become "the primary developed area proposed for visitor use in the Canyon section of the Monument," and included a lodge, store, service station, and other facilities to accommodate up to five hundred visitors, to be constructed far above Echo Park and with a spectacular view of the canyons. A copy of this Master Plan is in Box 14, Michael Straus Papers, Manuscript Division, Library of Congress.

58. Drury to Chapman, July 19, 1950, Box 6, Drury Records, CCF/NPS, RG 79.

59. Richardson, *Dams, Parks and Politics,* 62–63.

60. Drury remembered later that he was "a cat in a strange garret" as a Republican in a Democratic administration. See Amelia Fry, interview of Evison and Drury, 119, Bancroft Library; Drury to Straus, Aug. 7, 1950, June–December 1950, CCF/BR.

61. Murie to Chapman, Jan. 20, 1951, NPS Records, 1951–1953, DNM.

62. Fry, interview of Evison and Drury, 119, Bancroft Library; "Annual Report of the National Park Service," in *Annual Report of the Secretary of the Interior* (Washington, D.C.: U. S. Department of the Interior, 1950), 303–4; "Drury Warns of Peril to Park Program," *Washington Post,* Feb. 13, 1951; Murie to Morris Cooke, Apr. 8, 1950, "Wilderness Society" folder, Box 37, PWRPC Records.

63. Chapman to Drury, Dec. 13, 1950, Box 9, Doty Papers.

64. Ibid.; Drury to Chapman, Dec. 15, 1950, Box 14, DeVoto Papers; Drury to Chapman, Jan. 10, 1951, "National Park Service" folder, Box 9, Doty Papers; see also Waldo Leland's letter to the editor, *Washington Post,* Feb. 11, 1951.

65. Drury to Chapman, Jan. 19, 1951, Box 9, Doty Papers; Richardson, *Dams, Parks and Politics,* 63–64.

66. Leland quote in "Chapman Forced Parks Director Out, Friends Say," *Washington Daily News,* Feb. 5, 1951; see also the editorial "Parks for the People," *Washington Post,* Feb. 17, 1951, and Charles Sauers to Chapman, Feb. 5, 1951, Folder no. 9, Box 64, WS Papers.

67. Fred Packard to Rosalie Edge, Jan. 25, 1951; Irving Brant to Rosalie Edge, Feb. 12, 1951, Box 21, Brant Papers.

68. *Deseret News,* Feb. 17, 1951.

69. Clinton Anderson to DeVoto, Feb. 14, 1951, Box 61, Clinton Anderson Papers, Manuscript Division, Library of Congress (hereafter cited as Anderson Papers). See also Leland to DeVoto, Apr. 4, 1951; DeVoto to Leland, Apr. 5, 1951; Packard to Richard Leonard, Apr. 11, 1951, Box 102, Leland Papers; Leland to Chapman, Feb. 5, 1951, Folder no. 9, Box 64, WS Papers; DeVoto to Lee, May 17, 1951, Box 5, DeVoto Papers; Chapman to Charles Sauers, Mar. 27, 1951, "National Park Service" folder, Box 30, Joel Wolfsohn Papers, Harry S. Truman Library (hereafter cited as Wolfsohn Papers); "Utah Basin Plans Stir New

Dispute," *New York Times,* Feb. 11, 1951; Waldo Leland, "Newton Bishop Drury," *National Parks Magazine* 25 (Apr.–June 1951): 42–44, 62–66.

Another interesting twist to the Drury affair was that former Interior Secretary Ickes spoke out publicly during the uproar and said that he believed Drury should have been removed as Park Service director long before. Ickes believed Drury had never been an able defender of the national parks, citing as an example Drury's willingness to allow timber companies to cut sitka spruce from Olympic National Park during the war. Leland and other members of the Advisory Board strongly resented Ickes's attack, and it bolstered their determination to defend Drury. See Ickes's letter to the editor, *Washington Post,* Feb. 18, 1951; Leland to Charles Sauers, Feb. 22, 1951, Box 102, Leland Papers; Richardson, *Dams, Parks and Politics,* 64–65.

70. Chapman to Charles Sauers, Mar. 27, 1951, "National Park Service" file, Box 30, Wolfsohn Papers.

71. Drury later recalled that he had "asked for it," suggesting that he had irritated Chapman with various statements claiming that the primary threat to the park system now came from government agencies like the Bureau of Reclamation and state highway commissions. See Fry, interview of Evison and Drury, 107, Bancroft Library.

72. Journalist John Oakes of the *New York Times* published numerous articles about Echo Park Dam. One of the first appeared under the name "John B. Bertram," in a piece entitled "Debate over Dinosaur National Monument Raises Questions about Long-Range Policy," *New York Times,* Mar. 4, 1951; Tom Wallace to Waldo Leland, Feb. 15, 1951, Box 102, Leland Papers; DeVoto to Lee, May 17, 1951, Box 5, DeVoto Papers.

73. Copies of these letters are in a folder marked "National Park Service, 1948–1953," Box 54, Official File of Harry S. Truman, Truman Library. See also Arthur H. Carhart, "The Menaced Dinosaur Monument," *National Parks Magazine* 26 (Jan.–Mar. 1952): 3; "Nationwide Fight Opened on Echo Dam," *Denver Post,* Feb. 28, 1951.

74. Truman to Brant, Feb. 20, 1951, Box 21, Brant Papers.

75. Leland to Sauers, Feb. 22, 1951, Box 102, Leland Papers; Department of the Interior Press Release, Order No. 2618, Feb. 23, 1951, Box 50, Ickes Papers; Chapman to Zahniser, Aug. 30, 1951, Box 122, Chapman Papers.

Chapter 5

1. Luna Leopold to David Brower, Aug. 26, 1953, Box 114, HB Papers.

2. Voigt to Chapman, Nov. 30, 1950, CCF/BR.

3. Edwin Johnson to Michael Straus, Nov. 20, 1950, CCF/BR.

4. See Mark Harvey, "North Dakota, the Northern Plains, and the Missouri Valley Authority," *North Dakota History* 59 (Summer 1992): 28–39.

5. The best account of the rivalry between the Bureau and the Army Corps is Reisner, *Cadillac Desert,* 176–221.

6. U. S. Grant, III, "The Dinosaur Dam Sites Are Not Needed," *Living Wilderness* 15 (Autumn 1950): 17–24.

7. Ibid., 21.

8. See the Bureau's report, Mar. 24, 1950, 2, CCF/BR.

9. Grant, "Dinosaur Dam Sites Are Not Needed," 22.

10. Ibid., 19–22. Another valuable statement of Grant's position is found in Senate Committee on Interior and Insular Affairs, *Hearings on S. 1555 on CRSP* (1954), 471–72 [repr. in *Planning and Civic Comment* 20 (Mar. 1954)].

11. Grant, "Dinosaur Dam Sites Are Not Needed," 24.

12. Richard Allan Baker, *Conservation Politics: The Senate Career of Clinton P. Anderson* (Albuquerque: University of New Mexico Press, 1985), 65–68.

13. Fred Packard to President of Upper Colorado River Commission, Nov. 2, 1951, Box 3847, CCF/Interior, RG 48; Howard Zahniser to *Washington Post*, Feb. 20, 1951, Box 102, Leland Papers.

14. Grant testimony, Senate Committee on Interior and Insular Affairs, *Hearings on S. 1555 on CRSP* (1954), 472.

15. Ibid., 464–96; "Watkins Debunks Campaign against Dams," *Salt Lake Tribune*, Mar. 29, 1951. See also John G. Will, "Brief, Filed for the Upper Colorado River Commission," in "Reclamation—Colorado River" folder, Box 27, Hunt Papers.

16. "Supplementary Report by General Grant on the Dinosaur Controversy," *Planning and Civic Comment* 17 (Sept. 1951): 1–3.

17. Ibid, 3–4; Grant, "Dinosaur Dam Sites Are Not Needed," 24; Chapman quoted in John Will to Fred Packard, Dec. 3, 1951, Box 3847, CCF/Interior, RG 48.

18. Grant, "Supplementary Report," 4; see also Grant's "They Need Water—But They Don't Need Dinosaur Dams," *Sierra Club Bulletin* 37 (May 1952): 15–24.

19. Michael Straus to Oscar Chapman, Dec. 29, 1951, BRO, SLC.

20. Ibid.

21. "They Haven't Proved It," *Denver Post*, May 14, 1951.

22. Straus to Chapman, Dec. 29, 1951, BRO, SLC.

23. Grant, "Supplementary Report," 3.

24. Ibid., 7–8; U. S. Grant to Richard Bradley, Apr. 2, 1954, Richard Bradley personal files (copy in author's possession), hereafter cited as RB Files.

25. Dale Doty to Oscar Chapman, Jan. 22, 1952, Box 7, Doty Papers.

26. John Will spoke in Denver before the Colorado State Association of REA Cooperatives, Feb. 27, 1951. A copy is in Box 11, Untermann Papers.

27. "Congress Committee Hears Upper Basin Project Plea," *Salt Lake Tribune*, Dec. 10, 1951.

28. Les Miller, "The Battle that Squanders Billions," *Saturday Evening Post* 221 (May 14, 1949): 30–31, 160–62; Harold Ickes to Les Miller, July 16, 1949, and Miller to Ickes, July 25, 1949, "Reclamation 1946–1951" folder, Box 82, Ickes Papers.

29. Grant, "Supplementary Report," 3; Fred Packard to President of Upper Colorado River Commission, Nov. 2, 1951, Box 3847, CCF/Interior, RG 48.

30. Stratton and Sirotkin, *Echo Park Controversy,* 54–57.

31. Ibid., 54–56.

32. Quoted in ibid., 56.

33. Joseph C. O'Mahoney to Marshall Smith, July 7, 1951, Box 355, O'Mahoney Papers.

34. See comments of John G. Will in "Congress Committee Hears Upper Basin

Project Plea," *Salt Lake Tribune*, Dec. 10, 1951, and Joseph O'Mahoney's comment in the *Green River Star* (Wyo.) Dec. 7, 1951; Arthur Watkins to Bryant Stringham, Mar. 26, 1953, Box 3, Stringham Papers.

35. Arthur Carhart to Chapman, Nov. 29, 1951, Box 3847, CCF/Interior, RG 48; a copy of the Upper Colorado River Commission's resolution is in Box 27, Hunt Papers.

36. Quoted in "Interior Chief Wavering on Echo Park Plan," *Salt Lake Tribune*, Nov. 30, 1951.

37. Quoted in "Chapman Hopes to Avoid Dam in Dinosaur Area," *Rocky Mountain News*, Nov. 14, 1951; Arthur Carhart to Oscar Chapman, Nov. 29, 1951, Box 3847, CCF/Interior, RG 48; Chapman to George Kelly, Nov. 21, 1951, "Reading File 1951–52", Box 122, Chapman Papers; Richardson, *Dams, Parks and Politics*, 69.

38. Richardson, *Dams, Parks and Politics*, 69; Stratton and Sirotkin, *Echo Park Controversy*, 60; Bosone to Chapman, Oct. 23, 1950, Correspondence file, 1949–1953, Box 41, Chapman Papers.

39. Chapman held a press conference on Feb. 13, 1952, where he was questioned about the delays in approving the CRSP. A copy of the transcript is in Box 73, Chapman Papers.

40. Quoted in Stratton and Sirotkin, *Echo Park Controversy*, 63.

41. Ibid., 61–63; Watkins's bill was S. 3013, which he introduced on Apr. 16, 1952, 82d Cong., 2d sess.; a copy of the bill is in Box 355, O'Mahoney Papers.

42. Baker, *Conservation Politics*, 66.

43. Ibid., 63–66; Hunt to Watkins, Apr. 19, 1952, "Upper Colorado River Basin" folder, Box 19, Hunt Papers.

44. See Fred Packard's remarks at House Committee on Interior and Insular Affairs, *Hearings on H.R. 4449 et al. on CRSP* (1954), 803; Izaak Walton League Press Release, Jan. 1, 1953, Folder no. 10, Box 34, WS Papers; Stratton and Sirotkin, *Echo Park Controversy*, 63.

45. Ira Gabrielson made this point at a news conference in Washington in early 1954; "Opponents Open Fire On Echo Park Dam," *Salt Lake Tribune*, Jan. 5, 1954.

46. Arthur Watkins to Bryant Stringham, Mar. 26, 1953, Box 3, Stringham Papers.

47. Diary of Ralph Tudor, Apr. 19, 1953, Box 1, Ralph Tudor Papers, Dwight D. Eisenhower Library, Abilene, Kansas; C. J. Rogers to Douglas McKay, June 22, 1953, Box 15, Barrett Papers.

48. Robert F. Burk, *Dwight D. Eisenhower, Hero and Politician* (Boston: Twayne Publishers, 1986), 153.

49. Richardson, *Dams, Parks and Politics*, 71–74.

50. Quote from the *Idaho Statesman* in ibid., 85.

51. Richardson, *Dams, Parks and Politics*, 84.

52. *Annual Report of the Secretary of the Interior for 1953* (Washington, D.C.: Department of the Interior, 1953), xvi.

53. DeVoto to a Mr. McCarthy, Feb. 15, 1953, Box 6, DeVoto Papers; Morse quoted in Richardson, *Dams, Parks and Politics*, 86; see also Elmo Richardson, "The Interior Secretary as Conservation Villain: The Notorious Case of Douglas 'Giveaway' McKay," *Pacific Historical Review* 41 (Aug. 1972): 333–45.

54. Quoted in Wallace Stegner, "Battle for the Wilderness," *New Republic* 130

(Feb. 15, 1954): 15; John B. Oakes, "Conservation: Debate on National Parklands," *New York Times,* Feb. 7, 1954.

55. Richardson, *Dams, Parks and Politics,* 107–10.

56. The Republican party platform of 1952 opposed additional valley authorities. See ibid., 71–74; Gary W. Reichard, *The Reaffirmation of Republicanism: Eisenhower and the Eighty-Third Congress* (Knoxville: University of Tennessee Press, 1975), 175; Harvey, "North Dakota," passim.

57. Quoted in Richardson, *Dams, Parks and Politics,* 74; Burk, *Dwight D. Eisenhower,* 153–54.

58. Richardson, *Dams, Parks and Politics,* 115.

59. Ibid., 116–18. The Interior Department also indicated that it sought to encourage private industry to build power lines from existing dams like Grand Coulee and Bonneville.

60. Ibid., 157; see Neuberger's article, "Westerner against the West," *New Republic* 129 (Dec. 7, 1953): 11–12.

61. Stratton and Sirotkin, *Echo Park Controversy,* 70.

62. Wirth to McKay, Apr. 22, 1953, Box 3847, CCF/Interior, RG 48.

63. Richard Bradley, interview with author, June 17, 1988, Colorado Springs, Colo.; Fred Packard to J. H. Ratliff, Mar. 11, 1953, Box 10, Untermann Papers.

Chapter 6

1. Michael McCloskey, "Wilderness Movement at the Crossroads, 1945–1970," *Pacific Historical Review* 41 (Aug. 1972): 347–48.

2. Sierra Club Special Bulletin no. 4, Dec. 24, 1953, Box 208, Sierra Club Records, Bancroft Library (hereafter cited as Sierra Club Records).

3. Howard Zahniser to Olaus Murie, Nov. 21, 1946, Folder no. 25, Box 29; Murie to Wilderness Society Council, Oct. 1, 1948, Folder no. 21, Box 41, WS Papers.

4. Speaking at the Sierra Club's second annual wilderness conference in Oakland in 1951, Zahniser admitted that the postwar years had presented so many "test cases" in wilderness preservation that the movement had been placed on the defensive and lacked "the time or energy to pursue the all important positive program that alone can prevent the constant recurrence of these controversies." He made his remarks in a speech entitled "How Much Wilderness Can We Afford to Lose?" in Oakland, California, on Mar. 30, 1951. A copy of the speech is in Folder no. 21, Box 41, WS Papers; David R. Brower, interview with author, San Francisco, July 24, 1985; news item, *Living Wilderness* 16 (Autumn 1951): 30.

5. Benton MacKaye, "Dam Site v. Norm Site," *Scientific Monthly* 61 (Oct. 1950): 241–47.

6. Ibid., 241–42; see also Benton MacKaye to Morris L. Cooke, May 31, 1950, Folder no. 8, Box 171, MacKaye Family Papers, Dartmouth College Library. I am indebted to Larry Anderson, author of a forthcoming biography of MacKaye, for providing copies of MacKaye's correspondence from his own research and for helping explain the background of this article (Larry Anderson to author, Mar. 13, 1991).

7. MacKaye, "Dam Site v. Norm Site," 247.

8. Ibid., 243–44.

9. Susan Schrepfer, *The Fight to Save the Redwoods: A History of Environmental Reform, 1917–1978* (Madison: University of Wisconsin Press, 1983), 99.

10. Information about yearly visitation came from Dinosaur National Monument Visitor Center Records, Dinosaur National Monument, Jensen, Utah.

11. Margaret E. Murie, "A Matter of Choice," *Living Wilderness* 15 (Autumn 1950): 11–15; Margaret Murie, interview with author, Moose, Wyo., Aug. 5, 1988.

12. Murie, "A Matter of Choice," 12.

13. DeVoto, "Shall We Let Them Ruin Our National Parks?," *Saturday Evening Post,* July 22, 1950, 18; Butcher, "Stop the Dinosaur Power Grab," *National Parks Magazine* 24 (Apr.–June 1950): 61.

14. "The Menaced Dinosaur Monument," *National Parks Magazine* 26 (Jan.–Mar. 1952): 8; see also Philip Hyde, "Nature's Climax at Dinosaur," *Living Wilderness* 17 (Autumn 1952): 7–14.

15. Wallace Stegner, ed., *This Is Dinosaur: Echo Park Country and Its Magic Rivers* (New York: Alfred A. Knopf, 1955). The book was republished with a new introduction by Stegner by Roberts Rinehart, 1985.

16. George James to National Park Service Director, Jan. 5, 1953, Box 3847, CCF/Interior, RG 48; Webb, *If We Had a Boat,* 143–49.

17. Webb, *If We Had A Boat,* 143–48. In May 1951, Bus Hatch led a river trip for local chambers of commerce, and two boats flipped over and gear was lost; see Roy Webb, *Riverman: The Story of Bus Hatch* (Rock Springs, Wyo.: Labyrinth Publishers, 1991), 93.

18. Wirth to Secretary of the Interior, Feb. 28, 1952, Box 3847, CCF/Interior, RG 48.

19. Utah Congressman William Dawson, for example, raised the subject during the opening debate at the first congressional hearings on the CRSP in January 1954. He introduced several letters from residents of Vernal, testifying to the hazards of running the Green and Yampa rivers. See House Committee on Interior and Insular Affairs, *Hearings on H.R. 4449 et al. on CRSP* (1954), 853–55.

20. Cohen, *History of the Sierra Club,* 155.

21. Philip Hyde to G. E. Untermann, May 12, 1952, Box 10, Untermann Papers.

22. Untermann to Hyde, June 2, 1952, Box 10, Untermann Papers.

23. "Film in Color Supports Echo Park Dam Site," *Salt Lake Tribune,* Dec. 10, 1951.

24. Charles Sauers made this point before Oscar Chapman in the hearing at the Interior Department in April 1950; "Shall Dams Be Built in Dinosaur National Monument?," 37, Box 114, HB Papers.

25. John G. Will to Editor, *Christian Science Monitor,* July 24, 1953; G. E. Untermann to William Dawson, Mar. 17, 1953, Box 10, Untermann Papers. See also Conrad Wirth to Secretary Chapman, Feb. 28, 1952, Box 3847, CCF/Interior, RG 48.

26. Roscoe Fleming, "Dam Foes Tweet Alarm," *Denver Post,* Mar. 11, 1953; "Dams Will Open Canyon Wonderland," *Vernal Express,* July 16, 1953.

27. Stephen J. Bradley, "Folboats through Dinosaur," *Sierra Club Bulletin* 37 (Dec. 1952): 1.

28. Ibid., 4.

29. Ibid., 4, 8.

30. Richard Bradley, interview with author, June 17, 1988; Richard Bradley to author, Dec. 9, 1989 (in author's possession).

31. Harold Bradley, "The Dinosaur Case," *Garden Club Bulletin* (Mar. 1953): 85.

32. Ibid.

33. Cohen, *History of the Sierra Club,* 145–46.

34. David Brower, *For Earth's Sake: The Life and Times of David Brower* (Salt Lake City: Gibbs Smith, 1990), 175; Cohen, *History of the Sierra Club,* 146–47.

35. A list of groups to whom Bradley showed his film is in "Dinosaur National Monument" folder, Box 114, HB Papers.

36. "Hum O' the Hive," Bradley newsletter, Christmas 1953, quoted in Richard Bradley to author, Dec. 9, 1989; news item, *Sierra Club Bulletin* 39 (Feb. 1954): 2.

37. Brower, *For Earth's Sake,* 223.

38. Harold Bradley to family, May 30, 1953, Box 114, HB Papers.

39. News item, *Sierra Club Bulletin* 38 (Feb. 1953): 11–12.

40. Cohen, *History of the Sierra Club,* 149.

41. Joseph C. Bradley, "This, Your Land, Is in Danger," *Wisconsin State Journal,* Sept. 20, 1953; David Bradley, "Temple in Danger," *Valley News* (Hanover, N.H.), Apr. 15, 1954; Harold Bradley, "Danger to Dinosaur," *Pacific Discovery* 7 (Jan.–Feb. 1954): 5; Nathan Clark to author, Nov. 29, 1985.

42. Harold Bradley, "Danger to Dinosaur," 8; author's interview of Richard Bradley, June 17, 1988.

43. Actually, one boat did turn over in Split Mountain canyon, but no one drowned. Bradley must not have been aware of this incident, which was noted in Harry Robinson to Otis Marston, Aug. 11, 1955, Box 41, Otis Marston Papers, Manuscript Division, Huntington Library (hereafter cited as Marston Papers); Bradley, "Danger to Dinosaur," 4, and "Opponents Open Fire On Echo Park Dam," *Salt Lake Tribune,* Jan. 5, 1954.

44. Ruth Aiken to Harold Bradley, June 23, 1953, Box 114, HB Papers; Richard Bradley to author, Dec. 9, 1989; author's interview of Richard Bradley, June 17, 1988.

45. Brower, *For Earth's Sake,* 306; Nathan Clark also shot his own film on the Yampa in 1953, which he showed to Sierra Club groups and friends at his home, though not to "any large public audience." Nathan Clark to author, Nov. 29, 1985.

46. David Brower to "groups concerned with preserving the National Park System," Jan. 15, 1955, Folder no. 4, Box 34, WS Papers.

47. Author's interview of Richard Bradley, June 17, 1988; news item, *Sierra Club Bulletin* 39 (Feb. 1954): 2; David R. Brower, "Environmental Activist, Publicist, and Prophet," an oral history conducted 1974–1978 by Susan R. Schrepfer, Regional Oral History Office, The Bancroft Library, University of California, Berkeley), 113; Cohen, *History of the Sierra Club,* 158.

48. Cohen, *History of the Sierra Club,* 156.

49. Shrepfer interview of Brower, "Environmental Activist," 111.

50. John B. Oakes, interview with author by telephone, Feb. 27, 1986.

51. John Muir, "The Hetch-Hetchy Valley," *Sierra Club Bulletin* 6 (Jan. 1908): 216, 220.

52. David R. Brower, interview with author, Berkeley, Calif., Mar. 9, 1989.

53. Bradley, in "Hum O' the Hive," June 26, 1953, quoted in Richard Bradley to author, Dec. 9, 1989.

54. "Trouble in Dinosaur," *Sierra Club Bulletin* 39 (Feb. 1954); on Adams's genius, see Jonathan Spaulding, "The Natural Scene and the Social Good: The Artistic Education of Ansel Adams," *Pacific Historical Review* 60 (Feb. 1991): 15–42.

55. Joe Penfold, "Reclamation's Plan for Invasion," and Grant, "They Need Water—But They Don't Need Dinosaur Dams," *Sierra Club Bulletin* 37 (May 1952): 9–24.

56. Michael McCloskey, "Sierra Club Executive Director: The Evolving Club and the Environmental Movement, 1961–1981," an oral history conducted in 1981 by Susan R. Schrepfer, 77, 89–90, Regional Oral History Office, Bancroft Library, University of California, Berkeley, 1983.

57. Author's interview of Brower, Mar. 9, 1989.

58. DeVoto to Eugene McCarthy, Mar. 19, 1953, Box 6, DeVoto Papers.

59. Philip Hyde to G. E. Untermann, May 12, 1952, Box 10, Untermann Papers.

60. "Scenic, Recreational Values at Echo Park," *Salt Lake Tribune,* Jan. 7, 1954; "Echo Park Dam Project Is Necessary to Upper Colorado Basin Development," *Denver Post,* in *Congressional Record,* Mar. 9, 1984, 83d Cong., 2d sess., 1954, A2006–A2007; "Proposed Project Will Open New Fishing Region for Utahns," *Deseret News,* Dec. 14, 1946.

61. Joe Penfold testimony, House Committee on Interior and Insular Affairs, *Hearings on H.R. 4449 et al. on CRSP,* (1954), 767–87, (quotes on 768).

62. "News Items of Interest," *Living Wilderness* 20 (Winter 1955–56): 26–27.

63. Joe Penfold to Bernard DeVoto, Aug. 9, 1950, Box 17, DeVoto Papers.

64. Penfold to "Dinosaur Cooperators", Jan. 5, 1954, Folder no. 54, Box 292, Rosalie Edge Papers, Western History Department, Denver Public Library, (hereafter cited as Edge Papers).

65. Wirth to Harold Bradley, Aug. 20, 1952, Box 114, HB Papers.

66. N. B. Bennett, Jr., to Michael Straus, May 23, 1952; Wirth to Secretary Chapman, June 10, 1952, Box 3847, CCF/Interior, RG 48.

67. Wirth to Secretary of the Interior, Apr. 22, 1953, Box 3847, CCF/Interior, RG 48.

68. "Intramural Giveaway," *Harper's Magazine* 208 (Mar. 1954): 10.

69. Zahniser's remark appears in the minutes of the 1956 annual meeting of the Wilderness Society Council, Box 18, WS Papers.

70. David Canfield to National Park Service Region 2 Director, Apr. 23, 1952, NPS Records, 1951–1953, DNM; *Congressional Record,* 83d Cong., 1st sess., 1953, 99, 1: 140.

71. Sierra Club "special bulletin," Dec. 24, 1953, Box 208, Sierra Club Records. The Sierra Club Board of Directors first endorsed a Dinosaur national park in 1952; *Sierra Club Bulletin* 37 (Sept. 1952): 12–13; Bernard DeVoto to Congressman Eugene McCarthy, Feb. 15, 1953, Box 6, DeVoto Papers.

72. John Will to Douglas McKay, July 23, 1953, "Upper Colorado River Basin" folder, Box 3, Hunt Papers.

73. Quoted in Cohen, *History of the Sierra Club,* 156–57.

74. Fred Packard to J. H. Ratliff, Mar. 11, 1953, Box 10, Untermann Papers.

75. Richard Leonard to Jack Barnard et al., July 29, 1953, Folder no. 4, Box 34, WS Papers.

76. Leonard to Joe Penfold, Aug. 29, 1953, Box 114, HB Papers.

77. Howard Zahniser to *Washington Post,* Feb. 20, 1951, Box 102, Leland Papers; Fred Packard to J. H. Ratliff, Mar. 11, 1953, Box 10, Untermann Papers.

78. *Sierra Club Bulletin* 38 (September 1953): 4; author's interview of Brower, Mar. 9, 1989.

79. "Dam Foes Tweet Alarm," *Denver Post,* Mar. 11, 1953.

80. See editorial in the *Sheridan Press* (Wyo.), Mar. 12, 1954; Don Hatch to David Brower, Mar. 13, 1953, Box 114, HB Papers; "There Are No 'Alternatives' to Echo Park," *Denver Post,* Feb. 17, 1954.

81. "The Politics of River Basin Development," *Law and Contemporary Problems* 22 (Spring 1957): 268.

Chapter 7

1. Richardson, *Dams, Parks and Politics,* 130.

2. For Tudor's account of the trip, see his diary, "Notes Recorded While Under Secretary," issued under copyright of M. Lucile Tudor, Box 1, Ralph Tudor Papers, Dwight D. Eisenhower Library (hereafter cited as Tudor Papers).

3. Tudor to McKay, Nov. 27, 1953, Box 1, Tudor Papers.

4. Ibid.

5. See Tudor's diary for Jan. 17 and 24, 1954, Box 1, Tudor Papers.

6. Press Release, U.S. Department of the Interior, Dec. 12, 1953, Box 3, Hunt Papers; *New York Times,* Dec. 13, 1953.

7. For further coverage of this, see Richardson, "The Interior Secretary as Conservation Villain," 333–45; Martin Litton, "Upper Colorado Project Hits Eisenhower Snag," *Los Angeles Times,* Jan. 12, 1954.

8. Richard Bradley, interview with author, June 17, 1988.

9. Bishop to McKay, Dec. 21, 1953, "Colorado River" folder, Box 3, Frank Barrett Papers, Wyoming State Archives, Cheyenne.

10. Dawson to House members, Jan. 6, 1954, Box 114, HB Papers; "All Members of House Sent Dawson Letter on Echo," *Salt Lake Tribune,* Jan. 8, 1954; "Echo Park's Fossils 'Safe,' Critics Told," *Salt Lake Tribune,* Jan. 8, 1954.

11. Olaus Murie to Frank Barrett, Jan. 12, 1954, Box 15, Barrett Papers; Sierra Club Special Bulletin, Jan. 8, 1954, Box 114, HB Papers; Hugo G. Rodeck, "'Half-Truths' of Echo Dam Debate," *Denver Post,* Jan. 21, 1954.

12. Adams to Eisenhower, Dec. 28, 1953, Box 114, HB Papers.

13. Dec. 22, 1953.

14. The Whitewater dam site was ten miles southeast of Grand Junction, Colorado; Cross Mountain site was fifty miles west of Craig, on the Yampa; and Flaming Gorge site was in Utah, on the Green River north of Vernal. The combined storage of these dams would be 36 million acre-feet as opposed to 33 million acre-feet; "Opponents Open Fire on Echo Park Dam," *Salt Lake Tribune,* Jan. 5, 1954.

15. Thomas Maddock, Jr., to Richard Bradley, Nov. 30, 1954, RB files.

16. Grant quoted in *Salt Lake Tribune,* Jan. 5, 1954.

17. "More on Alternative Sites to Echo Park," *Salt Lake Tribune*, Jan. 12, 1954; "Water, Not Fossils, Seen Echo Foe Goal," *Salt Lake Tribune*, Jan. 15, 1954; "Facts Suffer in Drive Against Echo Park," *Salt Lake Tribune*, Jan. 3, 1954.

18. Barrett testimony, House Committee on Interior and Insular Affairs, *Hearings on H.R. 4449 et al. on CRSP* (1954), 234–37.

19. "Utah Delegation Presents Strong Case for Echo," *Salt Lake Tribune*, Jan. 23, 1954.

20. Stringfellow testimony, House Committee on Interior and Insular Affairs, *Hearings on H.R. 4449 et al. on CRSP* (1954), 265–71.

21. Watkins testimony, ibid., 239.

22. George D. Clyde, "The Story of the Upper Colorado and Preserving Utah's Last Water Hole," *Intermountain Industry* (Jan. 1954): 10–11, 26–27.

23. Baker, *Conservation Politics*, 68.

24. Stringfellow testimony, House Committee on Interior and Insular Affairs, *Hearings on H.R. 4449 et al. on CRSP* (1954), 265–71; Bennett testimony, ibid., 226–34.

25. Senate Committee on Interior and Insular Affairs, *Hearings on S. 1555 on CRSP* (1954); Fred Packard to David Brower, June 17, 1954, Folder no. 17, Box 33, WS Papers; Ralph Tudor diary, June 26, 1954, Tudor Papers.

26. Packard testimony, House Committee on Interior and Insular Affairs, *Hearings on H.R. 4449 et al. on CRSP* (1954), 799–801. Joe Penfold said similarly that "we endorse the fundamental purposes and objectives of the upper Colorado storage project." Penfold testimony, ibid., 768.

27. David Bradley testimony, ibid., 851.

28. Joe Penfold testimony, ibid., 769.

29. Ibid., 771.

30. Tudor testimony, ibid., 22.

31. Brower testimony, ibid., 792–93; Sierra Club Press Release, Jan. 27, 1954, Box 175, Sierra Club Records.

32. Douglas McKay to Charles Bennett, Apr. 13, 1954, Box 838, Official File of Dwight Eisenhower, Dwight D. Eisenhower Library, Abilene, Kansas.

33. Telegram of Charles Woodbury to Brower, Jan. 7, 1954, and Brower to Woodbury, Jan. 13, 1954, Folder no. 4, Box 34, WS Papers.

34. Tudor testimony, House Committee on Interior and Insular Affairs, *Hearings on H.R. 4449 et al. on CRSP* (1954), 15.

35. Ibid., 20–22.

36. Ibid., 22.

37. Some historians have erroneously stated that conservationists first suggested a "high" Glen Canyon Dam, but the Bureau of Reclamation had in fact done so earlier. See Mark Harvey, "Echo Park, Glen Canyon, and the Postwar Wilderness Movement," *Pacific Historical Review* 60 (Feb. 1991): 43–46.

38. By pointing out surface area, Tudor may have confused David Brower, who came to equate evaporation with the surface area of a reservoir.

39. Tudor testimony, House Committee on Interior and Insular Affairs, *Hearings on H.R. 4449 et al. on CRSP* (1954), 20–21.

40. Brower first shared his findings with General Grant, who decided to mention the multiplication error in his own testimony just a few hours before Brower. A quiet, courtly sort, Grant calmly mentioned the error but no one on the House

subcommittee seemed to notice, and in the questions that followed no one asked him to elaborate; Grant testimony, ibid., 708–16; Brower to Sirotkin, Sept. 15, 1955, Box 34, WS Papers.

41. Brower testimony, House Committee on Interior and Insular Affairs, *Hearings on H.R. 4449 et al. on CRSP* (1954), 795; Sierra Club Special Bulletin no. 6, Feb. 18, 1954, Box 208, Sierra Club Records.

42. Brower testimony, House Committee on Interior and Insular Affairs, *Hearings on H.R. 4449 et al. on CRSP* (1954), 824–25. Brower used simple multiplication with straight line ratios to argue that evaporation should be a function of the area of a reservoir:

a. Low Glen Reservoir Size ——— 153,000 acres
b. Low Glen Reservoir evaporation ——— 526,000 acre/ feet/ yr
c. High Glen Reservoir size ——— 186,000 acres

Using his "ninth grade arithmetic" he computed an evaporative rate for "high" Glen of 640,000 acre-feet:

$$\frac{153,000}{526,000} = \frac{186,000}{X}$$

(526,000 × 186,000 = 97,836,000,000; 97,836,000,000 divided by 153,000 = 639.45000 (rounded off to 640,000)

43. Ibid., 824.

44. Ibid., 826. The recent statement that Brower "acted with perfect discretion, not even seeming to state that the bureau was wrong," is inaccurate. Cohen, *History of the Sierra Club*, 183.

45. House Committee on Interior and Insular Affairs, *Hearings on H.R. 4449 et al. on CRSP* (1954), 785–86, 827.

46. Jacobsen testimony, ibid., 829.

47. Ibid.

48. Author's interview of Bradley, June 17, 1988; Bradley to George Clyde, Apr. 4, 1955, "Hum O' the Hive" newsletter, RB files.

49. Brower to Philip Sirotkin, Sept. 15, 1955, Box 34, WS Papers.

50. Author's interview of Bradley, June 17, 1988.

51. Quoted in news item, "Firing Begins in Dinosaur Fight," *Sierra Club Bulletin* 39 (Mar. 1954): 30.

52. Brower to Richard Bradley, Feb. 3, 1954, Box 114, HB Papers.

53. Author's interview of Bradley, June 17, 1988.

54. Richard Bradley to David Brower and Howard Zahniser, Mar. 26, 1954, Box 14, WS Papers.

55. Copies of the questionnaire and responses are in RB files; see also Bradley to Brower and Zahniser, Mar. 26, 1954, Box 14, WS Papers.

56. Bradley wrote two long essays on evaporation: "Is the Evaporation Argument Valid for Echo Park Dam?," Box 14, WS Papers, and "Evaporation and the Upper Colorado River Basin Storage Project," (the quote is from the latter), RB files; Brower to Richard Bradley, Feb. 7, 1955, Box 114, HB Papers; see also Richard Bradley's testimony, Senate Committee on Interior and Insular Affairs, *Hearings on S. 500 on CRSP*, 84th Cong., 1st sess., Feb. 28–Mar. 5, 1955, 393–403.

57. Bradley to Zahniser, Apr. 29, 1954; U. S. Grant, III to Zahniser, Apr. 12, 1954, Box 14, WS Papers; Luna Leopold to Bradley, May 13, 1954, RB files.

58. "Is the Evaporation Argument Valid?," Box 14, WS Papers.

59. Luna Leopold to Richard Bradley, May 13, 1954, RB files.

60. Lynn Thatcher [Utah State Department of Health] to Bradley, June 2, 1954, RB files.

61. House Committee on Interior and Insular Affairs, *Hearings on H.R. 4449 et al. on CRSP* (1954), 772–74.

62. "Is the Evaporation Argument Valid?," Box 14, WS Papers.

63. Ibid; See also Bradley's testimony in the Senate Committee on Interior and Insular Affairs, *Hearings on S. 500 on CRSP* (1955), 394–97.

64. "Evaporation and the Upper Colorado River Basin Storage Project," and U. S. Grant, III to Bradley, Apr. 2, 1954, RB files; Bradley to Howard Zahniser, Apr. 29, 1954, Box 14, WS Papers.

65. Richard Bradley to Harold Bradley, Mar. 2, 1954, in "Hum O' the Hive" family newsletter, RB files.

66. Romig to Richard Bradley, Apr. 27, 1954, RB files; Richard Bradley to Howard Zahniser, May 7, 1954, Box 14, WS Papers.

67. A story about the Aqualantes is found in the Logan, Utah, *Herald Journal,* Nov. 29, 1954; see also a fourteen-page essay by David Evans summarizing the case for the CRSP, Jan. 18, 1955, Box 15, Barrett Papers.

68. Richard Bradley to Douglas McKay, Feb. 22, 1954, RB files.

69. Dominy to Richard Bradley, Apr. 16, 1954, Box 114, HB Papers.

70. "U.S. Error Seen on Dam Location," *New York Times,* May 9, 1954; "Upper Colorado Storage Project: Evaporation-Loss Comparison," Apr. 28, 1954, Box 114, HB Papers; David Brower, "Evaporation—A Comedy of Errors," Apr. 26, 1954, RB files; David Brower, interview with author, San Francisco, July 24, 1985.

71. "Is the Storage Project Dead?," *Denver Post,* May 23, 1954.

72. Tudor to Fred Aandahl, May 14, 1954, Box 1; Tudor diary, June 6, 1954, Tudor Papers.

73. Tudor to Harrison, May 13, 1954, Box 1, Tudor Papers; Department of the Interior Press Release, May 14, 1954, Folder no. 2, Box 34, WS Papers; "Error Discovery Seen Blow for Echo Park," *Salt Lake Tribune,* May 15, 1954.

74. Wilderness Society Press Release (n.d), Box 17, WS Papers. In this statement, Zahniser is quoted as saying, "It is no wonder that conservationists have found the bureau's statements contradictory and have surmised that this evaporation argument is being used as a plausible pretext rather than a sound reason." See also Sierra Club Press Release, Oct. 18, 1954, Folder no. 4, Box 34, WS Papers; *New York Times,* June 6, 1954.

75. See Brower's letter to the editor in the *Washington Post* [n.d.], June 1954, clipping in Folder no. 1, Box 34, WS Papers; Brower to John Saylor, Apr. 23, 1954, Box 14, WS Papers.

76. Brower to Sherman Adams, May 26, 1954, Box 838, Official Files of Dwight Eisenhower, Eisenhower Library; Wilderness Society Press Release (n.d.), Box 17, WS Papers; "Figures that Evaporate," *Washington Post,* June 6, 1954.

77. Author's interviews of David R. Brower, July 24, 1985, San Francisco and Mar. 9, 1989, Berkeley, California.

78. Brower to Fred Packard, Jan. 21, 1955; Brower to Richard Bradley, Feb. 7, 1955, Box 114, HB Papers.

79. Ben Thompson to Newton Drury, Sept. 13, 1949, Box 6, Drury Records, CCF/NPS, RG 79.

80. Brower to Richard Bradley, Apr. 2, 1954, Box 175, Sierra Club Records; Sierra Club Memo, Sept. 8, 1954, Folder no. 4, Box 34, WS Papers.

Chapter 8

1. Liddle to John Oakes in *Congressional Record Appendix*, 83d Cong. 2d sess., July 29, 1954, A5700–A5701.

2. David Brower, "Dinosaurs, Parks, and Dams," *Pacific Spectator* 8 (Spring 1954): 158; Brower to Richard Bradley, Apr. 2, 1954, Box 175, Sierra Club Records; Brower to Douglas McKay, May 20, 1955, Folder no. 4, Box 34, WS Papers; Schrepfer, Interview of Brower, "Environmental Activist," 118–21.

3. Bernard DeVoto to Arthur Carhart, Jan. 4, 1955, Box 120, Carhart Papers; DeVoto to Raymond Moley, Feb. 8, 1955, Box 6, DeVoto Papers.

4. For DeVoto's views see his foreword to Wallace Stegner, *Beyond the Hundredth Meridian*, xv–xxiii, and "Intramural Giveaway," 10–11, 14–16.

5. DeVoto to Raymond Moley, Feb. 8, 1955, Box 6, DeVoto Papers.

6. Worster, *Rivers of Empire*, 266; Reisner, *Cadillac Desert*, 495–505.

7. Schrepfer, Interview of Brower, "Environmental Activist," 129.

8. DeVoto to Palmer Hoyt, Mar. 1, 1954, Box 120, Carhart Papers; DeVoto "Intramural Giveaway," 10–11, 14–16; "And Fractions Drive Me Mad," *Harper's Magazine* 209 (Sept. 1954): 11–19.

9. "Ike Throws Full Support to Echo, Central Utah Irrigation Projects," *Salt Lake Tribune*, Mar. 21, 1954. Eisenhower's statement was printed in the Senate Committee on Interior and Insular Affairs, *Hearings on S. 1555 on CRSP* (1954), 5–6.

10. Ben Moreell to Herbert Hoover, Apr. 19, 1954, Box 6, Colorado River Commission Records, Herbert Hoover Library, West Branch, Iowa (hereafter cited as Colorado River Commission Records.)

11. "Presidential Blessing," Casper *Tribune Herald*, Mar. 23, 1954; "Barlow Flays Partisanship on Water Storage Project," Wyoming *Star Tribune*, Mar. 28, 1954; Richardson, *Dams, Parks and Politics*, 129–30.

12. DeVoto, "Intramural Giveaway," 14.

13. On McCarthy, see David M. Oshinsky, *A Conspiracy So Immense: The World of Joe McCarthy* (New York: Free Press, 1983).

14. Author's interview of Richard Bradley, June 17, 1988; Rosalie Edge to John Brademas, Mar. 7, 1956, Box 21, Brant Papers; Richardson, *Dams, Parks and Politics,* 135–36.

15. Letters to Eisenhower protesting his approval of the project are found in Box 838, Official File of Dwight D. Eisenhower, Eisenhower Library, Abilene, Kansas; Harold Bradley to Richard Nixon, Mar. 29, 1954, Box 114, HB Papers.

16. Edge to Fairfield Osborn, Dec. 30, 1953; Brant to Edge, Dec. 4, 1953, Folder no. 64, Box 292, Edge Papers.

17. Quoted in "Ike Throws Full Support to Echo, Central Utah Irrigation Projects," *Salt Lake Tribune*, Mar. 21, 1954.

18. Richardson, *Dams, Parks and Politics,* 116–18; Neuberger's complaints about the administration's stand can be found in an article "Westerner against the West," *New Republic* 129 (Dec. 7, 1953): 11–12, and in his remarks in the *Congressional Record,* 84th Cong., 1st sess., Apr. 19, 1955, 101, pt. 4: 4652–53.

19. Gilbert Fite, *American Farmers: The New Minority* (Bloomington: Indiana University Press, 1981), 102–8; Raymond Moley, "Upper Colorado Defeat," *Newsweek* (Aug. 22, 1955), 96.

20. Stratton and Sirotkin, *Echo Park Controversy,* 83–85.

21. This quote comes from a reprint of Douglas's remarks, "Why the Upper Colorado River Project Is against the Public Interest," U.S. Congress, Senate, Apr. 18–19, 1955 (Washington DC: U.S. Government Printing Office, 1955), 7–8; his comments originally appeared in the *Congressional Record,* 84th Cong., 1st sess., Apr. 19, 1955, 101, pt. 4: 4634–41.

22. *Congressional Record,* 84th Cong., 1st sess., Apr. 19, 1955, 101, pt. 4: 4641.

23. Moley, "Gifts Along the Upper Colorado," *Los Angeles Times,* June 18, 1954; "McCarthy Row Cloaks Vital Issues, Moley Says," *Los Angeles Times,* May 7, 1954; "Irrigation—Hydropower's Expensive Partner," *Newsweek* (May 17, 1954); other Moley columns appeared in *Newsweek* on Apr. 12, Sept. 13, 1954, and May 9, 1955. See also John Taber to Richard Bradley, May 17, 1954, RB files; DeVoto to a Mr. Davoren, Mar. 9, 1955, Box 6, DeVoto Papers.

24. "Why the Upper Colorado River Project Is against the Public Interest," *Congressional Record,* 84th Cong., 1st sess., Apr. 19, 1955, 101, pt. 4: 4635–36; Bernard DeVoto to Paul Douglas, Apr. 26, 1955, Box 6, DeVoto Papers.

25. Bernard DeVoto to Arthur Schlesinger, Jr., Jan. 18, 1955, Box 6, DeVoto Papers; David Brower to Richard Bradley, Feb. 7, 1955, Box 114, HB Papers.

26. "Memorandum for Cooperators," June 14, 1954, Folder no. 2, Box 34, WS Papers; "Conservation and Development of National Water Resources," Box 6, Colorado River Commission Records; "Colorado River Storage Project," letter from Secretary of the Interior to House Committee on Interior and Insular Affairs, House Document No. 364, Apr. 6, 1954, 5.

27. "Engineers Question Upper River Policy," Casper *Tribune Herald,* July 2, 1954.

28. See the statement of the Engineers Joint Council, Senate Committee on Interior and Insular Affairs, *Hearings on S. 1555 on CRSP,* (1954), 681–85.

29. House Committee on Interior and Insular Affairs, *Hearings on H.R. 270, 2836, 3383, 3384, and 4488 on Colorado River Storage Project,* 84th Cong., 1st sess., Mar. 9–19, 28 and Apr. 18–22, 1955, 1070–77.

30. Cohen, *History of the Sierra Club,* 161–62; Brower's speech before thirtieth Annual Convention of the Associated Sportsmen of California, (Stockton, Calif.), Sept 17, 1954, Box 175, Sierra Club Records; Brower to Joe Penfold, Aug. 26, 1954, Folder no. 4, Box 34, WS Papers; Miller to Joseph C. O'Mahoney, Jan. 7, 1955, Box 356, O'Mahoney Papers; "Miller Supports Johnson's Views," Casper *Tribune Herald,* Jan. 10, 1955.

Critics also suspected that all of the hydropower from the CRSP dams could not be absorbed in the upper-basin market, despite all the rhetoric about growing population and industry. They charged that the Bureau overestimated the market for power, and argued that sufficient revenues to make the project pay out could only be generated if power were sold outside the upper basin. But that suggestion went nowhere with upper-basin lawmakers. Selling power on long-term contracts had been done in the case of Boulder Dam on the lower Colorado. The Boulder Canyon Project Act mandated that power be sold on long-term contracts, and these had been established with the city of Los Angeles. Now, as the power debate over the CRSP peaked, Herbert Hoover suggested that the current legislation contain a similar provision requiring that its hydropower be sold on long-term contracts in order to ensure full repayment of the CRSP. Yet if such a proposal made fiscal sense, it did not have enough political support in the upper Colorado basin, because industries like Geneva Steel in Utah and rural electrical cooperatives stridently resisted selling power outside the region; power companies and industries wanted a guarantee of adequate power to justify their own expansion, and they feared that they would not have it if CRSP mandated that power be sold elsewhere. Upper-basin lawmakers came under heavy pressure from both sides of this issue, with power companies wanting a plentiful supply of hydropower within the upper basin, and the Hoover Commission demanding long term contracts with outside markets. But the Hoover Commission could only recommend that this be done, and upper-basin lawmakers refused to accept long-term power contracts. For this aspect of the power issue, see Les Miller to Joseph C. O'Mahoney, Dec. 27, 1954, Box 356, O'Mahoney Papers; memorandum of Bureau of the Budget to Herbert Hoover, Apr. 15, 1954, Box 6, Colorado River Commission Records.

31. See Packard's testimony, Senate Committee on Interior and Insular Affairs, *Hearings on S. 500 on CRSP* (1955), 691–92.

32. Bureau of Reclamation, "Colorado River Storage Project and Participating Projects," Upper Colorado River Basin, Appendix "C" Power, Project Planning Report No. 4–8a. 81–0, (Salt Lake City: Bureau of Reclamation, Sept. 1950), 26–49, BRO, SLC; Ben H. Thompson to NPS Director, Sept. 13, 1949, NPS Records, 1945–1950, DNM.

33. Bureau of Reclamation, "Colorado River Storage Project," Appendix C, 14–25, BRO, SLC.

34. "Statement by Private Utilities Re Colorado River Storage Project," Feb. 1955, "Upper Colorado" folder, Box 355, O'Mahoney Papers; Senate Committee on Interior and Insular Affairs, *Hearings on S. 1555 on CRSP* (1954), 575–83.

35. Richard Bradley to David Brower, May 11, 1954, Box 14, WS Papers; John Saylor to House members, May 3, 1954, Folder no. 3, Box 18, WS Papers; Harold Bradley to John Saylor, Mar. 29, 1954, Box 114, HB Papers.

36. Casper *Star Tribune*, Apr. 12 and May 3, 1954.

37. Editorial, "Defeat for the Parks," *New York Times*, May 15, 1954.

38. National Wildlife Federation Press Release, May 14, 1954, Box 18, WS Papers. Aspinall, Dawson, Harrison, and congressmen from Nevada, Arizona, and Texas, as well as from other states voted against Saylor's amendment; John B. Oakes, "Conservation: The Echo Park Issue," *New York Times*, June 6, 1954; Casper *Star Tribune*, May 14 and May 18, 1954.

39. Baker, *Conservation Politics*, 74–76; "Notes on Meeting of Colorado Water Conservation Board," Oct. 14, 1954, distributed by the Sierra Club, RB files.

40. David Brower to Horace Albright, Aug. 27, 1954, Folder no. 4, Box 34, WS Papers.

41. Penfold to "Dinosaur Cooperators," Jan. 5, 1954, Folder no. 54, Box 292, Edge Papers.

42. Brower's testimony, Senate Committee on Interior and Insular Affairs, *Hearings on S. 1555 on CRSP* (1954), 506–7; Grant testimony, ibid., 464–96; Brower to Phillip Sirotkin, Sept. 15, 1955, Box 34, WS Papers.

43. See Harvey, "Echo Park, Glen Canyon," 43–67.

44. Assistant Secretary of the Interior to Joseph W. Martin, Jr., Apr. 6, 1954, "Colorado River Storage Project," House Document No. 364, 83d Cong., 2d sess., 113–18.

45. A copy of the transcript of this hearing is available in Box 114, HB Papers.

46. Brower to Saylor, Apr. 23, 1954, Box 14, WS Papers; Brower's letter to the editor, *Washington Post* [n.d.], June 1954, clipping in Box 34, WS Papers.

47. Brower to Penfold, Aug. 26, 1954, and June 3, 1955, Folder no. 4, Box 34, WS Papers.

48. Thomas Maddock, Jr., to Richard Bradley, Nov. 30, 1954, RB files.

49. Author's interview of Richard Bradley, June 17, 1988.

50. For more about the sacred meanings of Rainbow Bridge to the Navajo see Robert S. McPherson, *Sacred Land Sacred View* (Provo, Utah: Charles Redd Center for Western Studies, Brigham Young University, 1992), 31–32.

51. Brower statement, Senate Committee on Interior and Insular Affairs, *Hearings on S. 500 on CRSP* (1955), 634–43; Brower to Sirotkin, Sept. 15, 1955, Box 34, WS Papers.

52. Brower statement, Senate Committee on Interior and Insular Affairs, *Hearings on S. 500 on CRSP,* (1955), 635–40; Dexheimer statement, House Committee on Interior and Insular Affairs, *Hearings on H.R. 270 et al. on CRSP* (1955), 205; *Salt Lake Tribune* June 16, 1954; author's interview of David R. Brower, Mar. 9, 1989.

53. Brower statement, Senate Committee on Interior and Insular Affairs, *Hearings on S. 500 on CRSP* (1955), 635–40; Dexheimer statement, House Committee on Interior and Insular Affairs, *Hearings on H.R. 270 et al. on CRSP* (1955), 179–184; author's interview of Richard Bradley, June 17, 1988.

54. J. Neil Murdock statement, House Committee on Interior and Insular Affairs, *Hearings on H.R. 270 et al. on CRSP* (1955), 359–373.

55. Brower statement, Senate Committee on Interior and Insular Affairs, *Hearings on S. 500 on CRSP* (1955), 634–44, 645–52; Brower to Douglas McKay, May 20, 1955, Folder no. 4, Box 34, WS Papers.

56. Senate Committee on Interior and Insular Affairs, *Hearings on S. 500 on CRSP* (1954), 662–69.

57. Cohen, *History of the Sierra Club,* 161.

58. See the pamphlet published by the Colorado River Association of Los Angeles, "California and the Colorado River," Box 122, Anderson Papers; Hundley, *Great Thirst,* 224–32.

59. *New York Times,* June 28, 1954; Ely statement, Senate Committee on Interior and Insular Affairs, *Hearings on S. 1555 on CRSP* (1954), 584–646.

60. Remarks of Congressman John Shelley, *Congressional Record,* 83d Cong., 2d sess., July 28, 1954, 100, pt. 9: 12482–84.

61. Under the Compact, the upper basin was required to send 75 million acre-feet to the lower basin every ten years. According to Ely, the Bureau operated on the premise that the upper basin could use 9 million acre-feet one year, 6 the next, and thereby meet the terms of the Compact by sending an *average* amount downstream. Ely and Senator Thomas Kuchel preferred a maximum of 7.5 mil-

lion acre-feet of use per year in the upper basin so that the lower basin's annual requirements would be met. See House Committee on Interior and Insular Affairs, *Hearings on H.R. 4449 et al. on CRSP* (1954), 703; Senate Committee on Interior and Insular Affairs, *Hearings on S. 1555 on CRSP* (1954), 589. See also the essay by John Will, "Comments on Senator Kuchel's Minority Views," in "Upper Colorado River" folder, Box 2, Crippa Papers.

62. Senate Report no. 1983, "Authorizing the Secretary of the Interior to Construct, Operate, and Maintain the Colorado River Storage Project and Participating Projects," July 26, 1954, 25–30.

63. Norris Hundley Jr., "Clio Nods: Arizona v. California and the Boulder Canyon Act—A Reassessment," *Western Historical Quarterly* 3 (Jan. 1972): 17–51; Hundley, *Great Thirst*, 300–2.

64. Hundley, "Clio Nods," 30–31; John Geoffrey Will, "Report of the Secretary, Upper Colorado River Commission, Sept. 20, 1954, in "Upper Colorado River" folder, Box 2, Crippa Papers; George Clyde to Frederick Champ, July 21, 1954, Box C24F4, Water and Power Board Records, USA.

65. Earl C. Behrens, "Dinosaur Dam Plan Would Cost Californians $93,000,000 in Taxes," *San Francisco Chronicle,* July 27, 1954.

66. Remarks of California Congressman Harry Sheppard, *Congressional Record,* 83d Cong., 2d sess., Aug. 20, 1954, 100, pt. 12: 15749–51.

67. Remarks of Wayne Aspinall, *Congressional Record,* 83d Cong., 2d sess., Aug. 16, 1954, 100, pt. 11: 14700.

68. Author's interview of Brower, Mar. 9, 1989; Schrepfer interview of Brower, "Environmental Activist," 121–22.

69. Coury made his remarks before the Upper Colorado River Commission meeting on Sept. 20, 1954, Official Record of the Upper Colorado River Commission, William Robertson Coe Library, University of Wyoming; see also the editorial in the *Salt Lake Tribune,* June 11, 1954.

70. See Mark W. T. Harvey, "Echo Park Dam: An Old Problem of Federalism," *Annals of Wyoming* 55 (Fall 1983): 9–18.

71. Casper *Tribune Herald,* Aug. 29, 1954; *Deseret News,* Jan. 5, 1954; *Sheridan Press,* Mar. 12, 1954; "Please Take a Look at Echo Park," *Deseret News,* Apr. 17, 1954; "California Opposition to Upper Basin Plan," *Salt Lake Tribune,* June 11, 1954.

72. Editorial, "Echo Park Opponents Grow More Strident," *Salt Lake Tribune,* Feb. 21, 1954; Don Hatch to Brower, Mar. 13, 1953, Box 114, HB Papers.

73. Anderson's remark appears in Howard Zahniser's statement, Senate Committee on Interior and Insular Affairs, *Hearings on S. 500 on CRSP* (1955), 706.

74. Members of the Upper Colorado River Commission discussed this point at their Sept. meeting. See "Minutes of Sept. 20, 1954," Official Record of the Upper Colorado River Commission, William Robertson Coe Library, University of Wyoming.

75. Packard to Brower, June 17, 1954; Olson to Packard, Nov. 5, 1954, Box 33, WS Papers.

Chapter 9

1. Brower, "Dinosaurs, Parks, and Dams," 155–56.
2. John B. Oakes, "Visit to Vanishing America," *New York Times,* July 17, 1955.
3. Zahniser testimony, Senate Committee on Interior and Insular Affairs, *Hearings on S. 1555 on CRSP* (1954), 546.
4. Ibid., 544–45.
5. Jack Breed, "Shooting Rapids in Dinosaur Country," *National Geographic Magazine* 105 (Mar. 1954): 363–90; Harold Bradley, "Danger to Dinosaur," 3–8; "River Boat Run through Dinosaur National Monument," *Sunset* 112 (March 1954): 22–23; Stegner, "Battle for the Wilderness," 13–15; "Trouble in Dinosaur," *Sierra Club Bulletin* 39 (Feb. 1954): 2–12; "Firing Begins in Dinosaur Fight," Ibid., 39 (March 1954): 25–31.
6. Mrs. Roosevelt's column appeared in the *Washington Daily News,* July 13, 1954; DeVoto, "Parks and Pictures," *Harper's Magazine* 208 (Feb. 1954): 12–17; "Intramural Giveaway," 10–11, 14–16; "And Fractions Drive Me Mad," 10–14; DeVoto to Palmer Hoyt, Mar. 1, 1954, Box 120, Carhart Papers.
7. Perlman's articles ran from June 28 to July 5 in the *San Francisco Chronicle,* with the last one entitled "Figures Fail to Prove Dinosaur Dam Essential." The *Chronicle* editorialized against the dam on July 7, 1954, and the *Christian Science Monitor* did so in "Dinosaur: Battle for the Future," Mar. 17, 1954.
8. "Should Water Developments Invade National Parks?," *Washington Star,* May 23, 1954. The Council of Conservationists published a pamphlet filled with many newspaper editorials, Box 114, HB Papers.
9. *New York Times,* July 17, 1955; John Oakes, interview with author, Feb. 27, 1986; Oakes to author, Jan. 6, 1992.
10. This quote is that of Brower, recalling Peterson's remark in later years; Schrepfer interview of Brower, "Environmental Activist," 125; Brower, interview with author, July 24, 1985.
11. "Conservation: Debate on National Parklands," *New York Times,* Feb. 7, 1954; "Partisan Feeling Running High on the Colorado River Project," ibid., June 14, 1955; "Conservation: Congress' Record," ibid., Sept. 4, 1955; "Conservation: The Year in Review," ibid., Jan. 1, 1956.
12. Sierra Club memorandum on "Dinosaur Film" Jan. 6, 1955, Folder no. 4, Box 34, WS Papers.
13. David Brower to "groups concerned with preserving the National Park System," Jan. 15, 1955, Folder no. 4, Box 34, WS Papers.
14. Brower, *For Earth's Sake,* 225.
15. Schrepfer, interview of Brower, "Environmental Activist," 128.
16. *Sierra Club Bulletin* 39 (June 1954): 11–14; Brower to Phillip Sirotkin, Sept. 15, 1955, Box 34, WS Papers.
17. Brower testimony, Senate Committee on Interior and Insular Affairs, *Hearings on S. 1555 on CRSP* (1954), 503.
18. Nash, *Wilderness,* 141–60.
19. See, for instance, Devereux Butcher to Howard Zahniser, July 26, 1958, Folder no. 26, Box 58, WS Papers.
20. A recent examination of this central thread of Park Service history is in

Michael Frome, *Regreening the National Parks* (Tucson: University of Arizona Press, 1992).

21. Levi's letter to the editor appeared on May 13, 1954.

22. Douglas quoted in J. Ronald Engel, *Sacred Sands: The Struggle for Community in the Indiana Dunes* (Middletown, Conn.: Wesleyan University Press, 1983), 237.

23. David Perlman, "Dinosaur—Old Rocks and New Needs," *San Francisco Chronicle*, June 30, 1954.

24. The phrase "eons of time" is found in Olaus Murie and Joseph W. Penfold, "The Natural World of Dinosaur," in Stegner, *This Is Dinosaur*, 2d ed., 31.

25. "Dinosaur, Hot Battle, White Water," *San Francisco Chronicle*, June 29, 1954.

26. "Trouble in Dinosaur," 1.

27. Stegner, *This Is Dinosaur*, 2d ed., 15; Fred Packard testimony, Senate Committee on Interior and Insular Affairs, *Hearings on S. 1555 on CRSP* (1954), 552.

28. Runte, *National Parks*, 11–47.

29. "Danger to Dinosaur," 8.

30. See Roderick Nash, *The Rights of Nature: A History of Environmental Ethics* (Madison: University of Wisconsin Press, 1989); Runte, *National Parks*, 138.

31. See, for instance, Howard Zahniser's speech, "The Need for Wilderness Areas," June 1, 1955, *Congressional Record*, 84th Cong., 1st sess, A3809.

32. Olson statement, Senate Committee on Interior and Insular Affairs, *Hearings on S. 500 on CRSP* (1955), 680.

33. Kenneth T. Jackson, *Crabgrass Frontier: The Suburbanization of the United States* (New York: Oxford University Press, 1985); Charles Callison statement, Senate Committee on Interior and Insular Affairs, *Hearings on S. 500 on CRSP* (1955), 569–71.

34. "Why Not Write a Letter?," *Salt Lake Tribune*, Mar. 14, 1954.

35. "Reclamation Foes Grow More Strident," *Salt Lake Tribune*, June 20, 1955.

36. Liddle's letter to Oakes found its way via Congressman Dawson into the *Congressional Record Appendix*, Aug. 3, 1954, 83d Cong., 2d sess., A5700–A5701.

37. These remarks of Untermann are found in the Official Record of the Upper Colorado River Commission, "Minutes of March 20, 1950," 129–133, Box 44, Bashore Papers.

38. Untermann presented this statement at the House subcommittee in January 1954. See House Committee on Interior and Insular Affairs, *Hearings on H.R. 4449 et al. on CRSP* (1954), 197–227.

38. Ibid.; "Let's Tell Our Congressmen What We Want," *Rock Springs Daily Rocket*, Mar. 11, 1954.

40. Barbara and Richard Schall to the Sierra Club, May 7, 1954, Box 114, HB Papers. The phrase "narrow escape" comes from the diary of Harry B. Robinson, former ranger at Dinosaur Monument, May 13, 1954. Mr. Robinson made available his diary to the author during an interview, July 4, 1985, Columbia, Mo.

41. Fred Packard statement, Senate Committee on Interior and Insular Affairs, *Hearings on S. 1555 on CRSP* (1954), 552–67, and Senate Committee on Interior and Insular Affairs, *Hearings on S. 500 on CRSP* (1955), 694–95.

42. "Who Says Echo Park 'Unprecedented?'" *Denver Post,* Mar. 9, 1954; editorial, "Precedent," *Deseret News,* Apr. 17, 1954; John Geoffrey Will to Friend of the Upper Colorado River Commission, Mar. 5, 1954, "Echo Park-Split Mountain Dam" folder, Box 3847, CCF/Interior, RG 48; Neel, "Irreconcilable Differences," 165–77.

43. Packard to Philip Hyde, Apr. 30, 1952, Box 114, HB Papers; Packard to Arthur Carhart, Apr. 5, 1954, Box 120, Carhart Papers.

44. Packard to Arthur Carhart, Apr. 5, 1954, Box 120, Carhart Papers; Susan Neel investigated Packard's charge that the Park Service fired Madsen but found no evidence to prove it. Instead numerous letters in Park Service records indicate praise for Madsen. See Neel, "Irreconcilable Differences," 169, f.n.

45. Wilbur Dexheimer statement, Senate Committee on Interior and Insular Affairs, *Hearings on S. 500 on CRSP* (1955), 18–19; Michael Straus to Oscar Chapman, Mar. 28, 1951, Box 3847, CCF/Interior, RG 48.

46. Grant to Watkins, Apr. 14, 1955, "Colorado River Project," folder, Paul Douglas Papers, Chicago Historical Society (hereafter cited as Paul Douglas Papers).

47. Senate Committee on Interior and Insular Affairs, *Hearings on S. 500 on CRSP* (1955), 694; Packard to President, Upper Colorado River Commission, Nov. 2, 1951, Box 3847, CCF/Interior, RG 48.

48. Packard to Carhart, Apr. 5, 1954, Box 120, Carhart Papers.

49. Packard to Edge, Apr. 19, 1954, Box 120, Carhart Papers; Richardson, *Dams, Parks and Politics,* 141.

50. Watkins's remarks appeared in the *Congressional Record,* 84th Cong., 1st sess., Mar. 28, 1955, 101, 3: 3806–19.

51. See Daniel Hale et al., to Estes Kefauver, Apr. 24, 1951, CCF/BR.

52. *Congressional Record,* Mar. 28, 1955, 101, 3: 3806–19; George Clyde to George M. Laing, May 10, 1955, Box 14, Stringham Papers.

53. Howard Zahniser's letter to the editor in the *Washington Post,* May 25, 1955, and Smith to Watkins, Apr. 16, 1955, in Folder no. 540, Box 15, Barrett Papers.

54. Smith to Watkins, Apr. 16, 1955, Box 15, Barrett Papers.

55. Grant to Watkins, Apr. 14, 1955, "Colorado River Project" folder, Paul Douglas Papers.

56. Saylor, "National Park System Threatened," *Congressional Record,* 84th Cong., 1st sess., June 8, 1955, 101, 6: 7919; William Norris to Howard Zahniser, June 6, 1955 and Norris's legal brief found in Folder no. 2, Box 34, WS Papers; Richard Leonard to Watkins, Apr. 15, 1955, Box 4, Theodore F. Stevens Records, 1956–1960, Central Classified Files of the Department of the Interior, Record Group 48, National Archives.

57. Casper *Tribune Herald,* Apr. 20, and June 16, 1955.

58. Watkins's remarks appeared in the *Congressional Record,* 83d Cong., 2d sess., Mar. 31, 1954, 100, 3: 4206–7.

59. Mar. 21, 1954.

60. "Dams v. Dinosaurs," *Time,* (Jan. 31, 1955): 14–15.

61. "In summary," Brower wrote to Luce, "the best statement in this article is where *Time* says: 'The people of the area want water; how they get it less important.' Substitute for the word 'water' the term 'multi-billion dollar boom at everyone else's expense' and you come a little bit closer." Brower to Luce, Feb. 9,

1955, Sierra Club Records; Fred Smith to Roy Larsen, Mar. 31, 1955, Box 66, DeVoto Papers; see also a guest editorial by Hugo G. Rodeck, "'Half-Truths' of Echo Dam Debate," *Denver Post,* Jan. 21, 1954.

62. Joseph C. O'Mahoney, "Echo Park Is Not a National Park," U. S. Senate, Apr. 18 and 20, 1955, copy of speech in Box 356, O'Mahoney Papers. See also Sierra Club "Special Bulletin," Jan. 8, 1954, Box 114, HB Papers.

63. Wirth to Bradley, Mar. 15, 1954, Box 114, HB Papers.

64. Harry Robinson, interview with author, July 4, 1985; Robinson to Robert Athearn, Mar. 26, 1956, Harry Robinson personal files (hereafter cited as Robinson Files). See also Linda West and Dan Chure, *Dinosaur: The Dinosaur National Monument Quarry* (Jensen, Utah: Dinosaur Nature Association, 1984), 13.

65. Robinson to Walter Keller, Apr. 26, 1954, and Robinson to Adrian Reynolds, Mar. 7, 1955, Robinson files.

66. "Job Figures Estimated for Echo Park Dam," *Salt Lake Tribune,* Apr. 4, 1954.

67. Harry Robinson, interview with author, July 4, 1985; Robinson diary, June 4, 1954, Robinson files.

68. Mar. 27, 1954, Robinson files.

69. Packard testimony, Senate Committee on Interior and Insular Affairs, *Hearings on S. 500 on CRSP* (1955), 692.

70. Righter, *Crucible for Conservation,* 43–65, 103–25.

71. O'Mahoney's remarks in *Congressional Record,* 84th Cong., 1st sess., Apr. 19, 1955, 101, 4: 4654.

72. Ibid.; O'Mahoney to Murie, May 14, 1955, Box 356, O'Mahoney Papers.

73. Murie to O'Mahoney, May 10, 1955, Box 356, O'Mahoney Papers.

74. "National Monuments to National Parks: The Use of the Antiquities Act of 1906," *Western Historical Quarterly* 20 (Aug. 1989): 281–301.

75. Sierra Club Press Release, Mar. 31, 1955, Folder no. 4, Box 34, WS Papers.

76. Stegner, *Beyond the Hundredth Meridian,* 361; author's interview of David Brower, July 24, 1985.

77. Stegner to Robinson, Feb. 15, 1955, Robinson files.

78. Stegner, ed., *This Is Dinosaur.* First published by Alfred A. Knopf in 1955, the book was recently reissued with a new introduction by Stegner from Roberts Rinehart Inc., (Boulder, Colorado, 1985.)

79. Ibid., 2d ed., 85.

80. Ibid., 1st ed., v.

81. Knopf to Marston, Apr. 21, 1955, Folder no. 16, Box 113, Marston Papers.

82. *This Is Dinosaur,* 1st ed., brochure insert.

83. Ibid., 17.

84. Cohen, *History of the Sierra Club,* 163–66; David Brower to Phillip Sirotkin, Sept. 15, 1955, Box 34, WS Papers; Richard Leonard, "We Defend the Parks," *Sierra Club Bulletin* 40 (Jan. 1955): 3–5.

85. Fond of mountaineering, Mallinckrodt had been a friend of Francis Farquhar of the Sierra Club; author's interview of David Brower, Mar. 9, 1989; Nash, *Wilderness,* 213.

86. Schrepfer, interview of Brower, "Environmental Activist," 130; "Third Battle of Dinosaur," *National Parks Magazine* 29 (Jan.–Mar. 1955): 21–22.

87. Quoted in Howard Zahniser's conversation with Carl Gustafson, Dec. 20,

1955, on reel to reel tape, Folder no. 2, Box 37, WS Papers; "News Items of Interest," *Living Wilderness* 20 (Winter–Spring 1955–56): 26–27.

88. "Foes Seen Dooming Echo Park Project," *Denver Post,* Sept. 14, 1954; Olson to Packard, Nov. 5, 1954, Box 33, WS Papers; Frank Gregg to Howard Zahniser, Sept. 10, 1954, Folder no. 2, Box 34, WS Papers. Gregg, editor of *Colorado Conservation,* admitted to Zahniser in this letter that Echo Park Dam was dead and asked that conservationists support the remainder of the project.

89. Olson's statement, Nov. 17, 1954, is in Box 175, "Echo Park Dam" files, Sierra Club Records.

90. Fred Packard's statement, Senate Committee on Interior and Insular Affairs, *Hearings on S. 500 on CRSP* (1955), 689–90.

Chapter 10

1. Besser to Frank Barrett, Mar. 7, 1955, Box 16, Barrett Papers.

2. *Congressional Record,* 84th Cong., 1st sess., Apr. 19, 1955, 101, 4: 4640.

3. John Geoffrey Will analyzed these problems at a meeting of the Upper Colorado River Commission, Sept. 20, 1954; see the Report of the Secretary of the Upper Colorado Commission, Box 13, Stringham Papers.

4. Watkins's speech appeared in the *Congressional Record,* 84th Cong., 1st sess., Mar. 28, 1955, 101, 3: 3806–19.

5. "Look for President's Big Guns to Boom Colorado River Plan," *Salt Lake Tribune,* Jan. 19, 1955.

6. *Salt Lake Tribune,* "Empire Edition," Jan. 19, 1955; see also the *Rock Springs Miner,* Feb. 27, 1955 and Wyoming Senator Joseph O'Mahoney's remarks in "Are You For or against the Echo Park Dam?" *Collier's* 135 (Feb. 18, 1955): 78.

7. See Bolack's testimony, House Committee on Interior and Insular Affairs, *Hearings on H.R. 270 et al. on CRSP* (1955), 491.

8. Several of the Aqualante pamphlets are in Boxes 355 and 356, O'Mahoney Papers; see also "Minutes of Feb. 25, 1955," Official Record of the Upper Colorado River Commission, 46–52, William Robertson Coe Library, University of Wyoming; Richardson, *Dams, Parks and Politics,* 137.

9. Simpson's statement appears in the House Committee on Interior and Insular Affairs, *Hearings on H.R. 270 et al. on CRSP* (1955), 619; Ralph Conwell to Joseph C. O'Mahoney, "Upper Colorado" folder, Box 356, O'Mahoney Papers.

10. "10 Senators Introduce Basin Bill," *Salt Lake Tribune,* Jan. 19, 1955; Anderson Press Release, Feb. 5, 1955, Box 16, Barrett Papers; Baker, *Conservation Politics,* 77.

11. Baker, *Conservation Politics,* 79.

12. Mary Grace Fitzpatrick to David Brower, Feb. 18, 1955, Box 175, Sierra Club Records; "Dollars into Dust," *Readers Digest* 66 (May 1955): 109; "The Reclamation Follies," *Chicago Daily Tribune,* Feb. 19, 1955.

13. Anderson's speech appears in the *Congressional Record,* 84th Cong., 1st sess., Apr. 19, 1955, 101, 4: 4642–50; Baker, *Conservation Politics,* 79. The Aqualantes also published a pamphlet, "Food for the Future," Box 355, O'Mahoney Papers.

14. *Salt Lake Tribune,* Jan. 19, 1955.

15. Hansen is quoted in Casper *Tribune Herald,* June 7, 1955; see also Emil Gradert's statement in House Committee on Interior and Insular Affairs, *Hearings on H.R. 270 et al. on CRSP* (1955), 628.

16. "Senate Approves Big River Project," *New York Times,* Apr. 21, 1955.

17. This quote comes from a form letter distributed by the Council of Conservationists, Folder no. 19, Box 17; Fred Smith to Joe Penfold et al., May 27, 1955, Folder no. 2, Box 34; Zahniser to Harvey Broome, May 6, 1955, Folder no. 12, Box 79, WS Papers.

18. The Council reprinted editorials such as "Bananas on Pike's Peak?" *New York Times,* Apr. 24, 1955; "New Plan Urged on the Colorado," *New York Times,* May 15, 1955.

19. Outdoor News Bulletin (published by Wildlife Management Institute), Feb. 11, 1955, Box 175, Sierra Club Records; Sierra Club Press Release, Apr. 15, 1955, RB files.

20. Author's interview of David Brower, Mar. 9, 1989.

21. Brower to Richard Bradley, Feb. 7, 1955; Brower to Fred Packard, Jan. 21, 1955, Box 114, HB Papers.

22. Wallace Stegner to author, July 14, 1984.

23. Howard Zahniser to Wilderness Society members, May 4, 1955, Box 17, WS Papers.

24. Council of Conservationists, "Why Echo Park Dam Must Be Stopped," (New York, N. Y., 1955), "Upper Colorado River Storage Project" folder, Box 16, Barrett Papers; copies of the Council's flyer of editorials appears in Box 114, HB Papers.

25. The pamphlet was published by the Council of Conservationists (New York, 1955), and is found in "Upper Colorado River Storage Project" folder, Box 16, Barrett Papers.

26. Mrs. A. Paul Hartz testimony, House Committee on Interior and Insular Affairs, *Hearings on H.R. 270 et al. on CRSP* (1955), 1106; Mrs. Edward R. McPherson, Jr., to Harry Robinson, Mar. 10, 1955, Robinson files.

27. Testimony of Charles Callison, House Committee on Interior and Insular Affairs, *Hearings on H.R. 270 et al. on CRSP* (1955), 1069–70.

28. Nash, *Wilderness,* 216.

29. O'Mahoney to Adrian Reynolds, May 18, 1955, Folder no. 2, Box 202, O'Mahoney Papers.

30. Thomson to Milward Simpson, et al., May 2, 1955; Frank Barrett to Edward Crippa, Apr. 21, 1955, "Upper Colorado River" folder, Box 16, Barrett Papers; Richardson, *Dams, Parks and Politics,* 147–48; "Senate Approves Big River Project," *New York Times,* Apr. 21, 1955.

31. "House Unit Takes Vote in Secret," *Denver Post,* June 9, 1955.

32. "Echo Park Is Dropped from River Project," Casper *Tribune Herald,* June 10, 1955; News Release, Trustees for Conservation, July 13, 1955, Folder no. 2, Box 34, WS Papers.

33. "House Unit Takes Vote in Secret," *Denver Post,* June 9, 1955.

34. "Work on Basin Bill Nears End," *Denver Post,* June 13, 1955; "River Bill Okayed," *Denver Post,* June 14, 1955.

35. Press Release of Trustees for Conservation, "No Matter How Thin They Slice It . . . Echo Park Dam Is Still in the Upper Colorado Project," July 13, 1955, Folder no. 2, Box 34, WS Papers.

36. Citizens Committee on Natural Resources to Joseph C. O'Mahoney, July 15, 1955, "Upper Colorado" folder, Box 355, O'Mahoney Papers; *Deseret News,* June 9, 1955; see also Congressman Donald L. Jackson's remarks, *Congressional Record,* 84th Cong., 1st sess., July 18, 1955, 101, 8: 10868–69, and those of Congressman Lee Metcalf, *Congressional Record,* 84th Cong., 1st sess., 101, pt. 7: 9304–5; Fred Packard, "Echo Park Dam? Not By A Damsite!," *National Parks Magazine* 29 (July–Sept. 1955): 99–100; Bill Losh to Howard Zahniser and Fred Packard, July 13, 1955, and Trustees for Conservation press release, Box 34, WS Papers.

37. See the editorial "And Now for Fair Consideration in House," *Salt Lake Tribune,* July 22, 1955.

38. "Echo Dam Proposal Hit by Two," *Denver Post,* July 20, 1955.

39. "Friends Tell Concern for River Dams," *Salt Lake Tribune,* July 4, 1955.

40. *Salt Lake Tribune,* July 22, 1955.

41. Hosmer's remarks in *Congressional Record,* 84th Cong., 1st sess., July 13, 1955, 101, 8: 10481–83.

42. "House Sidetracks Basin Bill for Another Try in January," *Salt Lake Tribune,* July 27, 1955; Fred Smith to Frank Barrett, Oct. 29, 1955, Box 15, Barrett Papers.

43. "'Greatest Water Steal' in Western History," *Salt Lake Tribune,* Aug. 2, 1955, clipping in Folder no. 1, Box 34, WS Papers.

44. "West Needs New Faith, Unified Approach," *Salt Lake Tribune,* July 27, 1955.

45. "Johnson Has Plan to Share Benefits," *Denver Post,* Sept. 1, 1955; see also Johnson's testimony, Senate Committee on Interior and Insular Affairs, *Hearings on S. 500 on CRSP* (1955), 152–78. See also Milward Simpson to Frank Barrett, Jan. 6, 1955, Box 16, Barrett Papers; "Governors Chart 'Unified Front' for Upper River," Casper *Tribune Herald,* Jan. 4, 1955; see also editorial, "Agreed On Basin Program," Casper *Tribune Herald,* Jan. 6, 1955.

46. News Release of Barrett's remarks, Oct. 4, 1955, Box 16, Barrett Papers; "Senator Asks States to Reach Agreement," *Wyoming State Tribune,* Oct. 4, 1955; Richardson, *Dams, Parks and Politics,* 150–51.

47. Baker, *Conservation Politics,* 81–82.

48. "Echo Park Dam 'Dead' on Eve of Strategy Parley," *Rocky Mountain News,* Nov. 1, 1955.

49. Fred Smith to Frank Barrett, Oct. 29, 1955, Box 15, Barrett Papers; Brower to Raymond Moley, Oct. 31, 1955, Folder no. 4, Box 34, WS Papers.

50. Schrepfer, interview of Brower, "Environmental Activist," 129; "An Open Letter to the 'Strategy Committee' of the Upper Colorado Project," *Denver Post,* Oct. 31, 1955.

51. Baker, *Conservation Politics,* 82.

52. Fred Smith to Frank Barrett, Oct. 29, 1955, Box 16, Barrett Papers.

53. Howard Zahniser to Olaus Murie, Nov. 4, 1955, Folder no. 11, Box 79, WS Papers; "Actions of Conference on Upper Colorado River Legislation," copy in Box 356, O'Mahoney Papers; Anderson to Smith, Nov. 3, 1955, Box 114, HB Papers; "McKay Says Officially: Echo Park Dam Out," *Rocky Mountain News,* Nov. 30, 1955; Joseph Penfold to Joseph O'Mahoney, Jan. 9, 1956, Box 356, O'Mahoney Papers. An account of the final settlement is found under "News Items of Interest," *Living Wilderness* 20 (Winter–Spring 1955–56): 23–43.

54. "News Items of Interest," *Living Wilderness,* 30.

55. Harold Bradley to Anderson, Nov. 3, 1955; Howard Zahniser to Harold Bradley, Nov. 14, 1955, Box 114, HB Papers; Council of Conservationists to Joseph O'Mahoney, Dec. 15, 1955, Box 356, O'Mahoney Papers; Brower to Ben Thompson, Dec. 28, 1955, Box 320, Marston Papers.

56. "Feeding Fires of Preservationists' Doubts," *Salt Lake Tribune,* Dec. 3, 1955.

57. Zahniser recalled his visit with Congressman Dawson two days afterwards, in a telephone conversation with Carl Gustafson. A copy of that conversation is on a reel to reel tape, Folder no. 2, Box 37, WS Papers.

58. Council of Conservationists to Joseph C. O'Mahoney, Dec. 15, 1955, "Upper Colorado" folder, Box 356, O'Mahoney Papers.

59. Quoted from taped conversation of Zahniser and Gustafson, Dec. 22, 1955, in Folder no. 2, Box 37, WS Papers.

60. Brower to Penfold, Aug. 26, 1954, Folder no. 4, Box 34, WS Papers.

61. Author's interview of David Brower, Mar. 9, 1989.

62. Schrepfer, interview of Brower, "Environmental Activist," 130.

63. For an alternate interpretation see Russell Martin, *A Story that Stands Like a Dam: Glen Canyon and the Struggle for the Soul of the West* (New York: Henry Holt and Co., 1989); see also Harvey, "Echo Park, Glen Canyon," 43–67.

64. Anderson quoted in "News Items of Interest," *Living Wilderness,* 31; Zahniser to Anderson, Dec. 28, 1955, Folder no. 5, Box 20, WS Papers.

65. Horace Albright, Ira Gabrielson, and Howard Zahniser to Usher Burdick, Jan. 23, 1956, Box 8, Usher Burdick Papers, Special Collections, University of North Dakota, Grand Forks (hereafter cited as Burdick Papers); Wayne Aspinall and William Dawson to Edgar Wayburn, Jan. 26, 1956, Box 114, HB Papers; News Items of Interest, *Living Wilderness,* 23.

66. Wilderness Society News Release, Jan. 23, 1956, Box 34, WS Papers; "Colorado Fight Ends," *New York Times,* Jan. 25, 1956; Albright, Gabrielson, and Zahniser to Usher Burdick, Jan. 23, 1956, Box 8, Burdick Papers; Wayne Aspinall and William Dawson to Edgar Wayburn, Jan. 26, 1956, Box 114, HB Papers; "News Items of Interest," *Living Wilderness,* 33–34.

67. "Is Echo Park Reservoir Abandoned,?" *Los Angeles Times,* Nov. 3, 1955; "California Water Lobby Loses Rallying Cry," *Salt Lake Tribune,* Jan. 25, 1956; Wilderness Society News Release, Jan. 23, 1956, Box 34, WS Papers.

68. Quoted in "Lawmaker Hurls Rock into Fight against Dam," *Los Angeles Times,* Jan. 30, 1956.

69. Editorial, "Desperation Marks New California Attack," *Salt Lake Tribune,* Feb. 1, 1956.

70. "Senator Hurls 'Big Lie' at California Group," *Los Angeles Times,* Jan. 31, 1956.

71. "Shale Debunked as Peril to Dam," *Salt Lake Tribune,* Feb. 3, 1956.

72. Anderson quoted in "Conferees Set to Drop Echo Park Project," *New York Times,* Mar. 9, 1956; *Washington Post,* Mar. 4, 1956.

73. "Plan to Develop Colorado's Basin Passed by House," *New York Times,* Mar. 2, 1956; see also "Congress Approves Colorado Basin Bill, *New York Times,* Mar. 29, 1956.

Chapter 11

1. Hundley, *Great Thirst* 352.

2. Runte, *National Parks*, 189–96; Francois Leydet, *Time and the River Flowing: Grand Canyon* (San Francisco: Sierra Club Books, 1964).

3. Devereux Butcher to Howard Zahniser, July 26, 1958, Folder no. 26, Box 58, WS Papers.

4. *Wilderness*, 212.

5. Brant to Truman, Feb. 16, 1951; Brant to Stevenson, Sept. 7, 1957, Box 21, Brant Papers.

6. Harold Bradley to Fred Smith, Nov. 12, 1957, Box 114, HB Papers; "Conservation: The Year in Review," *New York Times*, Jan. 1, 1956.

7. John Oakes, interview with author by telephone, Feb. 27, 1986.

8. Fred Packard to Hubert H. Humphrey, Apr. 3, 1957; Richard Goodwin to Humphrey, Apr. 22, 1957, Folder no. 18, Box 64, WS Papers.

9. Packard to Brower, June 17, 1954, Box 33, WS Papers.

10. Runte, *National Parks*, 11–47.

11. Webb, *If We Had A Boat*, 136.

12. Stegner to author, July 14, 1984 (in author's possession).

13. "Short-Changed?," *Denver Post*, Apr. 30, 1956; "Developing Utah's Sealed-Up Wonderlands," *Salt Lake Tribune*, Apr. 10, 1956; see also the Council of Conservationists full page advertisement in *Denver Post*, Apr. 27, 1956.

14. Telegram from Citizens Committee on Natural Resources, Council of Conservationists and Trustees for Conservation to editor of *Salt Lake Tribune*, Apr. 24, 1956, Folder no. 3; Carl Gustafson to John Saylor, Apr. 24, 1956, Folder no. 4, Box 34, WS Papers; "Why the Unseemly Haste to 'Wrap It Up'?," *Salt Lake Tribune*, Apr. 29, 1956; L. C. Bishop to Watkins, May 8, 1956, "Upper Colorado" folder, Box 355, O'Mahoney Papers.

15. Gordon Allott to Ethel Larsen, Aug. 23, 1957, Box 34, WS Papers.

16. Brower to Fred Smith, Sept. 19, 1957; John Saylor to Charles Porter, Aug. 11, 1959, Folder no. 3, Box 34, WS Papers.

17. "Short-Changed?," *Denver Post*, Apr. 30, 1956; John B. Oakes, "Conservation: A New Difficulty," *New York Times*, Oct. 6, 1957.

18. Smith to Brower, Sept. 23, 1957, Folder no. 9, Box 34, WS Papers.

19. Packard to Fred Seaton, Aug. 15, 1957, Folder no. 3, Box 34, WS Papers.

20. Murie to Board of Trustees of National Parks Association, Aug. 12, 1957, Box 265, Murie Papers; Smith to Zahniser (telegram), Aug. 5, 1957, Folder no. 9; Smith to Packard, Aug. 5, 1957, Folder no. 3, Box 34, WS Papers; *Sierra Club Bulletin* 42 (Sept. 1957): 4.

21. "Right Now It's an Unwelcome Echo," *Denver Post*, Jan. 5, 1959.

22. Stewart Udall to Wayne Aspinall, Aug. 27, 1960, Box 72, Stewart Udall Papers, Department of Special Collections, University of Arizona (hereafter cited as S. Udall Papers).

23. Thomas Kuchel to Otis Marston, Feb. 6, 1962, Box 320, Marston Papers; Floyd Dominy to Wallace Bennett, July 19, 1961, Folder no. 12, Box 253, Carl Hayden Papers, Department of Archives and Manuscripts, Arizona State University.

24. Howard Zahniser to Stewart Udall, May 23, 1961, Box 20, WS Papers.

25. *Sierra Club Bulletin* 46 (Oct. 1961): 19–20; Brower to Stewart Udall, Dec.

7, 1961, Box 190, S. Udall Papers; editorial, "Rainbow Bridge: Record and Requiem," *National Parks Magazine* (May 1963): 18–19.

26. Randall Henderson to Otis Marston, Apr. 10, 1957, Box 38, Marston Papers.

27. Harold Bradley to Conrad Wirth, Sept. 6, 1955, Folder no. 12, Box 79, WS Papers.

28. Eggert to David Brower, Jan. 31, 1956, Box 320, Marston Papers.

29. Eliot Porter, *The Place No One Knew: Glen Canyon on the Colorado* (San Francisco: Sierra Club Books, 1963).

30. "Requiem for a Canyon," *Pacific Discovery* 17 (May–June 1964): 9.

31. David Brower, interview with author, Mar. 9, 1989.

32. David Brower to Stewart Udall, June 22, 1963, Box 20, WS Papers; Nash, *Wilderness,* 227–37; Cohen, *History of the Sierra Club,* 357–65.

33. Luna Leopold, "Probability Analysis Applied to a Water-Supply Problem," (Washington, D.C.: USGS, Geological Survey Circular 410, 1959); Walter Langbein, "Water Yield and Reservoir Storage in the United States," (Washington, D.C.: USGS, Geological Survey Circular 409, 1959).

Bibliography

I. Archival Materials

A. National Archives, Washington D.C.
    Record Group 48, Records of the Department of the Interior
        Central Classified Files of the Secretary of the Interior, 1937–1953
    Official Files of Secretary Douglas McKay, 1953–1956
    Theodore F. Stevens Records, 1956–1960
    Record Group 79, Records of the National Park Service (includes Central Classified Files, 1933–1949, and Records of Newton B. Drury, 1940–1951)
    Record Group 115, Records of the Bureau of Reclamation, 1930–1945
    Records of the Bureau of Reclamation, 1945–1951
B. Dinosaur National Monument Library, Jensen, Utah
    Official Records of the National Park Service, 1945–1950, and 1951–1953
C. Dwight D. Eisenhower Library, Abilene, Kansas
    Ralph Tudor Papers
    Official File of Dwight D. Eisenhower
D. Harry S. Truman Library, Independence, Missouri
    Official File of Harry S. Truman
    Oscar Chapman Papers
    Dale E. Doty Papers
    Joel Wolfsohn Papers
    Records of the President's Water Resources Policy Commission
E. Herbert Hoover Library, West Branch, Iowa
    Colorado River Commission Records
F. Utah State Archives, Salt Lake City, Utah
    Utah Water and Power Board Records
G. Bureau of Reclamation Regional Office, Salt Lake City
    Records of the Bureau of Reclamation
H. Wyoming State Archives, Cheyenne, Wyoming
    Frank Barrett Papers

II. *Manuscript Collections*

Clinton P. Anderson Papers, Manuscript Division, Library of Congress, Washington, D.C.

Frank Barrett Papers, American Heritage Center, University of Wyoming, Laramie, Wyoming

Harry Bashore Papers, American Heritage Center, University of Wyoming, Laramie, Wyoming

Harold Bradley Papers, Bancroft Library, Berkeley, California

Irving Brant Papers, Manuscript Division, Library of Congress, Washington, D.C.

Usher Burdick Papers, Department of Special Collections, University of North Dakota, Grand Forks, North Dakota

Arthur Carhart Papers, Western History Department, Denver Public Library, Denver, Colorado

Edward C. Crippa Papers, American Heritage Center, University of Wyoming, Laramie, Wyoming

Bernard DeVoto Papers, Department of Special Collections, Stanford University, Palo Alto, California

Paul Douglas Papers, Chicago Historical Society, Chicago, Illinois

Rosalie Edge Papers, Western History Department, Denver Public Library, Denver, Colorado

Carl V. Hayden Papers, Department of Archives and Manuscripts, Arizona State University, Tempe, Arizona

Lester Hunt Papers, American Heritage Center, University of Wyoming, Laramie, Wyoming

Harold Ickes Papers, Manuscript Division, Library of Congress, Washington, D.C.

Waldo Leland Papers, Manuscript Division, Library of Congress, Washington, D.C.

MacKaye Family Papers, Special Collections, Dartmouth College Library, Hanover, New Hampshire

Joseph C. O'Mahoney Papers, American Heritage Center, University of Wyoming, Laramie, Wyoming

Otis Marston Papers, Huntington Library, San Marino, California

Olaus Murie Papers, Western History Department, Denver Public Library, Denver, Colorado

Sierra Club Records, Bancroft Library, Berkeley, California

Michael Straus Papers, Manuscript Division, Library of Congress, Washington, D.C.

Bryant Stringham Papers, Department of Special Collections, Utah State University, Logan, Utah

Stewart Udall Papers, Department of Special Collections, University of Arizona, Tucson, Arizona

G. E. Untermann Papers, Department of Special Collections, University of Utah, Salt Lake City, Utah

Wilderness Society Papers, Western History Department, Denver Public Library, Denver, Colorado

III. Dissertations

Flores, Dan L. "Islands in the Desert: An Environmental Interpretation of the Rocky Mountain Frontier." Texas A & M University, 1978.

Kathka, David. "The Bureau of Reclamation in the Truman Administration: Personnel, Politics, and Policy." University of Missouri, 1976.

Koppes, Clayton R. "Oscar L. Chapman: A Liberal at the Interior Department, 1933–1953." University of Kansas, 1974.

Mehls, Steven F. "David H. Moffat, Jr.: Early Colorado Business Leader." University of Colorado, 1982.

Neel, Susan Rhoades. "Irreconcilable Differences: Reclamation, Preservation and the Origins of the Echo Park Dam Controversy." University of California-Los Angeles, 1990.

IV. Published Government Documents

Congressional Record, 1950–1956

Congressional Record Appendix, 1950–1956

U.S. Congress, House. Committee on Public Lands. *Hearings Before a Subcommittee on Irrigation and Reclamation.* 81st Cong., 1st sess., 1949.

U.S. Congress, House. Committee on Interior and Insular Affairs. *Hearings on H.R. 270, 2836, 3383, 3384, and 4488 on Colorado River Storage Project.* 84th Cong., 1st sess., 1955.

U.S. Congress, House. Committee on Interior and Insular Affairs. *Hearings on H.R. 4449, 4443, and 4463 on Colorado River Storage Project.* 83d Cong., 2d sess., 1954.

U.S. Congress, Senate. Committee on Interior and Insular Affairs. *Hearings on S. 1555 on Colorado River Storage Project.* 83d Cong., 2d sess., 1954.

U.S. Congress, Senate. *Hearings Before the Special Committee on Postwar Economic Policy and Planning.* 78th Cong., 2d sess., 1944.

U.S. Congress, Senate. Committee on Interior and Insular Affairs. *Hearings on S. 500 on Colorado River Storage Project.* 84th Cong., 1st sess., 1955.

U. S. Congress. Senate. Committee on Interior and Insular Affairs. *Authorizing the Secretary of the Interior to Construct, Operate and Maintain the Colorado River Storage Project and Participating Projects.* 83d Cong., 2d sess., 1954, S. Rept. 1983.

U.S. Department of the Interior. National Park Service. *Annual Reports* of the Director of the National Park Service to the Secretary of the Department of the Interior, 1942–1953.

U.S. Department of the Interior. Bureau of Reclamation. *The Colorado River: "A Natural Menace Becomes A Natural Resource," A Comprehensive Report on the Development of the Water Resources of the Colorado River Basin for Irrigation, Power Production, and Other Beneficial Uses in Arizona, California, Colorado, Nevada, New Mexico, Utah, and Wyoming,* 1946.

U.S. Department of the Interior. *A Survey of the Recreational Resources of the Colorado River Basin,* 1950.

U.S. Geological Survey. "Probability Analysis Applied to a Water-Supply Problem." USGS Circular 410. Washington, D.C., 1959.

U.S. Geological Survey. "Water Yield and Reservoir Storage in the United States."
    USGS Circular 409. Washington, D.C., 1959.

V. *Newspapers*

Casper *Tribune Herald*
Chicago *Daily Tribune*
Christian Science Monitor
Craig, Colorado *Empire Courier*
Denver Post
Deseret News
Green River Star (Wyoming)
Los Angeles Times
Moffat County Courier
New York Times
Rock Springs, Wyoming *Daily Rocket*
Rocky Mountain News
Salt Lake Tribune
San Francisco Chronicle
Sheridan Press (Wyoming)
Vernal Express
Washington Post
Washington Star
Wyoming *State Tribune*

VI. *Interviews*

Bishop, Floyd. Interview with author. Laramie, Wyoming, February 9, 1982.
Bradley, Richard. Interview with author. Colorado Springs, Colorado, June 17,
    1988.
Brower, David R. Interviews with author, San Francisco and Berkeley, California,
    July 24, 1985, and March 9, 1989.
Brower, David R. Interview by Susan R. Schrepfer, "Environmental Activist, Pub-
    licist, and Prophet." Regional Oral History Office, Bancroft Library, Univer-
    sity of California, Berkeley, 1974–1978.
Drury, Newton B. Interview by Amelia R. Fry and Susan R. Schrepfer. "Parks and
    Redwoods, 1919–1971." Regional Oral History Office, Bancroft Library,
    University of California, Berkeley, 1972.
Evison, Herbert, and Newton B. Drury. Interview by Amelia Fry, "The National
    Park Service and the Civilian Conservation Corps." Regional Oral History
    Office, Bancroft Library, University of California, Berkeley, 1963.
McCloskey, Michael. Interview by Susan R. Schrepfer, "Sierra Club Executive
    Director: The Evolving Club and the Environmental Movement, 1961–
    1981." Regional Oral History Office, Bancroft Library, University of Califor-
    nia, Berkeley, 1983.
Murie, Margaret. Interview with author. Moose, Wyoming, August 5, 1988.
Oakes, John B. Interview with author by telephone. February 27, 1986.

Rechard, Paul. Interview with author. Laramie, Wyoming, October 26, 1981.

Robinson, Harry B. Interview with author. Columbia, Missouri, July 4, 1985.

VII. *Miscellaneous Records, Publications, Sources*

Richard Bradley Personal Files

Mann, Dean, Weatherford, Gary, and Nichols, Phillip. "Legal-Political History of Water Resource Development in the Upper Colorado River Basin." Lake Powell Research Project Bulletin No. 4. Los Angeles, National Science Foundation, 1974.

Official Record of the Upper Colorado River Commission, general collection, William Robertson Coe Library, University of Wyoming, Laramie.

Harry B. Robinson Personal Files

Stratton, Owen, and Sirotkin, Phillip. *The Echo Park Controversy.* University of Alabama, 1959.

VIII. *Articles*

"Are You For or Against the Echo Park Dam?" *Collier's* 135 (February 18, 1955): 76–83.

Bates, J. Leonard. "Fulfilling American Democracy: The Conservation Movement, 1907–1921." *Mississippi Valley Historical Review* 44 (June 1957): 29–57.

Bradley, Harold. "Danger to Dinosaur." *Pacific Discovery* 7 (January–February 1954): 3–8.

Bradley, Richard. "Requiem For A Canyon." *Pacific Discovery* 17 (May–June 1964): 2–9.

Bradley, Stephen. "Folboats Through Dinosaur." *Sierra Club Bulletin* 37 (December 1952): 1–8.

Breitenstein, Jean S. "The Upper Colorado River Basin Compact." *State Government* 22 (1949): 214–16, 225.

Breed, Jack. "Shooting Rapids in Dinosaur Country." *National Geographic Magazine* 105 (March 1954): 363–90.

Brower, David. "Dinosaurs, Parks, and Dams." *Pacific Spectator* 8 (Spring 1954): 151–60.

———. "To Dam Or Not To Dam." *Sierra Club Bulletin* 33 (September–October 1948): 3–4.

Butcher, Devereux. "Stop the Dinosaur Power Grab." *National Parks Magazine* 24 (April–June 1950): 61–65.

———. "This Is Dinosaur." *National Parks Magazine* 24 (October–December 1950): 122–36.

Carhart, Arthur. "The Menaced Dinosaur Monument." *National Parks Magazine* 26 (January–March 1952): 19–30.

———. "Our Public Lands in Jeopardy." *Journal of Forestry* 46 (June 1948): 409–16.

Clements, Kendrick. "Politics and the Park: San Francisco's Fight for Hetch Hetchy, 1908–1913." *Pacific Historical Review* 48 (May 1979): 185–215.

Clyde, George. "The Story of the Upper Colorado and Preserving Utah's Last Water Hole." *Intermountain Industry* (January 1954): 10–11, 26–27.

"Dams v. Dinosaurs." *Time* 65 (January 31, 1955): 14–15.

DeVoto, Bernard. "The Western Land Grab." *Harper's Magazine* 194 (June 1947): 543–46.

———. "Shall We Let Them Ruin Our National Parks?" *Saturday Evening Post* 223 (July 22, 1950): 17–19, 42–48.

———. "One Hundred Year Plan." *Harper's Magazine* 201 (August 1950): 60–64.

———. "Let's Close the National Parks." *Harper's Magazine* 207 (October 1953): 49–52.

———. "Parks and Pictures." *Harper's Magazine* 208 (February 1954): 12–17.

———. "Intramural Giveaway." *Harper's Magazine* 208 (March 1954): 10–11, 14–16.

———. "And Fractions Drive Me Mad." *Harper's Magazine* 209 (September 1954): 11–19.

"Echo Park Controversy Resolved." *Living Wilderness* 20 (Winter–Spring 1955–1956): 23–43.

Foote, James A. "Shall Public Or Private Interest Rule?" *National Parks Bulletin* 13 (June 1937): 3–5.

———. "Mr. Ickes—Your National Parks." *National Parks Bulletin* 13 (December 1937): 3–5, 26.

Gottlieb, Robert, and Wiley, Peter. "The CUP Runneth Over." *Utah Holiday Magazine* 15 (November 1985): 54–60, 108–10.

Grant III, U. S. "The Dinosaur Dam Sites Are Not Needed." *Living Wilderness* 15 (Autumn 1950): 17–24.

———. "Supplementary Report by General Grant on the Dinosaur Controversy." *Planning and Civic Comment* 17 (September 1951): 1–10, 21.

———. "They Need Water—But They Don't Need Dinosaur Dams." *Sierra Club Bulletin* 37 (May 1952): 15–24.

Griffith, Robert. "Dwight D. Eisenhower and the Corporate Commonwealth." *American Historical Review* 87 (February 1982): 87–122.

Hampton, H. Duane. "Opposition to National Parks." *Journal of Forest History* 25 (January 1981): 36–45.

Harvey, Mark W. T. "Echo Park Dam: An Old Problem of Federalism." *Annals of Wyoming* 55 (Fall 1983): 9–18.

———. "Echo Park, Glen Canyon, and the Postwar Wilderness Movement." *Pacific Historical Review* 60 (February 1991): 43–67.

———. "Utah, the National Park Service, and Dinosaur National Monument, 1909–56." *Utah Historical Quarterly* 59 (Summer 1991): 243–63.

———. "North Dakota, the Northern Plains, and the Missouri Valley Authority." *North Dakota History* 59 (Summer 1992): 28–39.

Hays, Samuel P. "From Conservation to Environment: Environmental Politics Since World War Two." *Environmental Review* 6 (Fall 1982): 14–41.

Heald, Weldon. "The Squeeze Is On the National Parks." *National Parks Magazine* 24 (January–March 1950): 3–4.

Hundley, Norris Jr. "Clio-Nods: *Arizona v. California* and the Boulder Canyon Act—A Reassessment." *Western Historical Quarterly* 3 (January 1972): 17–51.

Hyde, Philip. "Nature's Climax At Dinosaur." *Living Wilderness* 17 (Autumn 1952): 7–14.

Jeancon, Jean Allard. "Antiquities of Moffat County, Colorado." *Colorado Magazine* 4 (January 1927): 18–27.

"Jackson Hole National Monument." *National Parks Bulletin* 74 (July–September 1943): 25–26.

Koppes, Clayton R. "Environmental Policy and American Liberalism: The Department of the Interior, 1933–1953." *Environmental Review* 7 (Spring 1983): 17–41.

"Krug Stimulates Cooperation." *National Parks Magazine* 22 (January–March 1948): 3.

Leland, Waldo. "Newton Bishop Drury." *National Parks Magazine* 25 (April–June 1951): 42–44, 62–66.

Leonard, Richard. "We Defend the Parks." *Sierra Club Bulletin* 40 (January 1955): 3–5.

MacKaye, Benton. "Dam Site v. Norm Site." *Scientific Monthly* 61 (October 1950): 241–47.

Mackintosh, Barry. "Harold L. Ickes and the National Park Service." *Journal of Forest History* 29 (April 1985): 78–84.

Maw, Herbert. "More Power to Utah From the Multi-Million Dollar Central Utah Project." *Reclamation Era* 33 (February 1947): 30–32.

McCloskey, Michael. "Wilderness Movement at the Crossroads, 1945–1970." *Pacific Historical Review* 41 (August 1972): 346–61.

McConnel, Grant. "The Conservation Movement—Past and Present." *Western Political Quarterly* 7 (September 1954): 463–78.

McMechen, Edgar C. "The Hermit of Pat's Hole." *Colorado Magazine* 19 (May 1942): 91–98.

Miller, Les. "The Battle That Squanders Billions." *Saturday Evening Post* 221 (May 14, 1949): 30–31, 160–62.

———. "Dollars Into Dust." *Reader's Digest* 66 (May 1955): 109–14.

Moley, Raymond. "Irrigation—Hydropower's Expensive Partner." *Newsweek* 43 (May 17, 1954): 84–88.

Muir, John. "The Hetch-Hetchy Valley." *Sierra Club Bulletin* 6 (January 1908): 216–22.

Murie, Margaret. "A Matter of Choice." *Living Wilderness* 15 (Autumn 1950): 11–15.

Murie, Olaus. "Primitive Area on Trial." *Living Wilderness* 13 (Spring 1948): 16–25.

"National Primeval Park Standards." *National Parks Magazine* 19 (October–December 1945): 6–11.

Neuberger, Richard L. "Westerner Against the West." *New Republic* 129 (December 7, 1953): 11–12.

Olson, Sigurd. "The Preservation of Wilderness." *Living Wilderness* 13 (Autumn 1948): 1–8.

Packard, Fred. "Grand Canyon Park and Dinosaur Monument in Danger." *National Parks Magazine* 23 (October–December 1949): 11–13.

———. "Echo Park Dam? Not By A Damsite!" *National Parks Magazine* 29 (July–September 1955): 99–100, 122.

Penfold, Joe. "Reclamation's Plan for Invasion." *Sierra Club Bulletin* 37 (May 1952): 10–14.

Pisani, Donald J. "State vs. Nation: Federal Reclamation and Water Rights in the Progressive Era." *Pacific Historical Review* 51 (August 1982): 265–82.

———. "Reclamation and Social Engineering in the Progressive Era." *Agricultural History* 57 (January 1983): 46–83.

———. "Deep and Troubled Waters: A New Field of Western History?" *New Mexico Historical Review* 63 (October 1988): 311–31.

Powell, John Wesley. "Institutions for the Arid Lands." *Century Magazine* 40 (May 1890): 111–16.

Reid, Kenneth. "Your Heritage: A Statement Clarifying the Threatened Federal Landgrab." *National Parks Magazine* 21 (July–September 1947): 8–13.

Richardson, Elmo R. "Federal Park Policy in Utah: The Escalante National Monument Controversy of 1935–1940." *Utah Historical Quarterly* 33 (Spring 1965): 109–33.

———. "The Interior Secretary as Conservation Villain: The Notorious Case of Douglas 'Giveaway' McKay." *Pacific Historical Review* 41 (August 1972): 333–45.

Righter, Robert. "National Monuments to National Parks: The Use of the Antiquities Act of 1906." *Western Historical Quarterly* 20 (August 1989): 281–301.

"River Boat Run Through Dinosaur National Monument." *Sunset* 112 (March 1954): 22–23.

Roth, Dennis. "The National Forests and the Campaign for Wilderness Legislation." *Journal of Forest History* 28 (July 1984): 112–25.

Sandiford, Glenn. "Bernard DeVoto and His Forgotten Contribution to Echo Park." *Utah Historical Quarterly* 59 (Winter 1991): 73–86.

Sellars, Richard W. "National Parks: Worthless Lands or Competing Land Values?" *Journal of Forest History* 27 (July 1983): 130–34.

Spaulding, Jonathan. "The Natural Scene and the Social Good: the Artistic Education of Ansel Adams." *Pacific Historical Review* 60 (February 1991): 15–42.

Stegner, Wallace. "Battle for the Wilderness." *New Republic* 130 (February 15, 1954): 13–15.

Story, Isabelle F. "Water Recreation in the Desert." *National Parks Magazine* 22 (January–March 1948): 23–26.

Swain, Donald C. "The Passage of the National Park Service Act of 1916." *Wisconsin Magazine of History* 50 (Autumn 1966): 4–17.

———. "The National Park Service and the New Deal, 1933–1940." *Pacific Historical Review* 41 (August 1972): 312–32.

———. "The Bureau of Reclamation and the New Deal, 1933–1940." *Pacific Northwest Quarterly* 61 (July 1970): 137–46.

"Third Battle of Dinosaur." *National Parks Magazine* 29 (January–March 1955): 21–22.

"Trouble in Dinosaur." *Sierra Club Bulletin* 39 (February 1954): 3–12.

Turner, Tom. "The Dinosaur Story." *Sierra* 76 (November/December 1991): 28–29.

Wharton, William P. "The National Primeval Parks." *National Parks Bulletin* 13 (February 1937): 3–5.

"What Is A Sound Business Decision on Dinosaur?" *Planning and Civic Comment* 20 (December 1954): 2–4.

White, Richard. "American Environmental History: The Development of a New Historical Field." *Pacific Historical Review* 54 (1985): 297–335.

Yard, Robert Sterling. "Historical Basis of National Park Standards." *National Parks Bulletin* 10 (November 1929): 3–6.

———. "Losing Our Primeval System in Vast Expansion." *National Parks Bulletin* 13 (February 1936): 1–4.

Zahniser, Howard. "Why We Cherish the San Gorgonio Primitive." *Living Wilderness* 12 (March 1947): 1–7.

Zahniser, Howard, and Baker, F. S. "We Certainly Need A Sound Philosophy." *Living Wilderness* 12 (Winter 1947–1948): 1–5.

IX. Books

Athearn, Frederic J. *An Isolated Empire: A History of Northwest Colorado.* Denver: Bureau of Land Management, 1977.

Athearn, Robert. *Rebel of the Rockies: A History of the Denver and Rio Grande Western Railroad.* New Haven: Yale University Press, 1962.

———. *The Coloradoans.* Albuquerque: University of New Mexico Press, 1976.

———. *The Mythic West in Twentieth Century America.* Lawrence: University Press of Kansas, 1986.

Baker, Richard Allan. *Conservation Politics: The Senate Career of Clinton P. Anderson.* Albuquerque: University of New Mexico Press, 1985.

Brower, David. *For Earth's Sake: The Life and Times of David Brower.* Salt Lake City: Gibbs Smith, 1990.

Buchholtz, C. W. *Rocky Mountain National Park: A History.* Boulder: Colorado Associated University Press, 1983.

Burk, Robert F. *Dwight D. Eisenhower: Hero and Politician.* Boston: Twayne Publishers, 1986.

Burroughs, John Rolfe. *Where the Old West Stayed Young.* New York: William Morrow & Co., 1962.

Chronic, Halka. *Pages of Stone: Geology of Western National Parks and Monuments.* Seattle: The Mountaineers, 1984.

Chure, Dan, and West, Linda. *Dinosaur: The Dinosaur National Monument Quarry.* Jensen, Utah: Dinosaur Nature Association, 1984.

Cohen, Michael P. *The History of the Sierra Club, 1892–1970.* San Francisco: Sierra Club Books, 1988.

Darrah, William Culp, *Powell of the Colorado.* Princeton: Princeton University Press, 1951.

DeVoto, Bernard. *The Easy Chair.* Boston: Houghton Mifflin Co., 1955.

Diggins, John. *The Proud Decades: America in War and Peace, 1941–1960.* New York: W. W. Norton & Co., 1988.

Donovan, Robert J. *Tumultuous Years: The Presidency of Harry S. Truman, 1949–1953.* New York: W. W. Norton & Co., 1982.

Dunbar, Robert. *Forging New Rights in Western Waters.* Lincoln: University of Nebraska Press, 1983.

Dunlap, Thomas R. *Saving America's Wildlife.* Princeton: Princeton University Press, 1988.

Engel, J. Donald. *Sacred Sands: The Struggle for Community in the Indiana Dunes.* Middletown, Connecticut: Wesleyan University Press, 1983.

Fite, Gilbert. *American Farmers: The New Minority.* Bloomington: Indiana University Press, 1981.

Foresta, Ronald A. *America's National Parks and Their Keepers.* Washington D. C.: Resources for the Future, 1984.

Fox, Stephen. *John Muir and His Legacy: The American Conservation Movement.* Boston: Little, Brown, and Co., 1981. Reprint. Madison: University of Wisconsin Press, 1985.

Fradkin, Philip. *A River No More: The Colorado River and the West.* Tucson: University of Arizona Press, 1981.

Frome, Michael. *Regreening the National Parks.* Tucson: University of Arizona Press, 1992.

Garnsey, Morris. *America's New Frontier: The Mountain West.* New York: Alfred A. Knopf, 1950.

Goetzmann, William H. *Exploration and Empire: The Explorer and the Scientist in the Winning of the American West.* New York: W. W. Norton & Co., 1966.

Hays, Samuel P. *Conservation and the Gospel of Efficiency: The Progressive Conservation Movement.* Cambridge: Harvard University Press, 1959. Reprint. New York: Atheneum, 1975.

———. *Beauty, Health and Permanence: Environmental Politics in the United States, 1955–1985.* New York: Cambridge University Press, 1987.

Hughes, J. Donald. *American Indians in Colorado.* Boulder, Colorado: Pruett Publishing Co., 1977.

Hundley, Jr., Norris. *The Great Thirst: Californians and Water, 1770s–1990s.* Berkeley and Los Angeles: University of California Press, 1992.

———. *Water and the West: The Colorado River Compact and the Politics of Water in the American West.* Berkeley and Los Angeles: University of California Press, 1975.

Ise, John. *Our National Park Policy: A Critical History.* Baltimore: Johns Hopkins Press, 1961.

Jackson, Kenneth T. *Crabgrass Frontier: The Suburbanization of the United States.* New York: Oxford University Press, 1985.

Jones, Holway. *John Muir and the Sierra Club: The Battle for Yosemite.* San Francisco: Sierra Club, 1965.

Lacey, Michael J., ed. *Government and Environmental Politics: Essays on Historical Developments Since World War Two.* Washington, D.C.: Woodrow Wilson Center Press; Baltimore: Johns Hopkins University Press, 1991.

Larson, T. A. *History of Wyoming.* 2d ed. Lincoln: University of Nebraska Press, 1978.

Leopold, Aldo. *A Sand County Almanac and Sketches Here and There.* New York: Oxford University Press, 1949.

Leydet, Francois. *Time and the River Flowing: Grand Canyon.* San Francisco, Sierra Club Books, 1964.

Lister, Robert H., and Lister, Florence C. *Those Who Came Before.* Tucson: University of Arizona Press, 1983.

Lowitt, Richard. *The New Deal and the West.* Bloomington: Indiana University Press, 1984.

Martin, Russel. *A Story That Stands Like A Dam: Glen Canyon and the Struggle for the Soul of the West*. New York: Henry Holt & Co., 1989.

McCarthy, G. Michael. *Hour of Trial: The Conservation Conflict in Colorado and the West, 1891–1907*. Norman: University of Oklahoma Press, 1977.

McCoy, Donald. *The Presidency of Harry S. Truman*. Lawrence: University Press of Kansas, 1984.

McDonald, Michael, and Wheeler, William Bruce. *TVA and the Tellico Dam, 1936–1979: A Bureaucratic Crisis in Post-Industrial America*. Knoxville: University of Tennessee Press, 1986.

McPhee, John. *Encounters With the Archdruid*. New York: Farrar, Straus, and Giroux, 1971.

Mackintosh, Barry. *The National Parks: Shaping the System*. 1st ed., Washington, D. C.: Department of the Interior, 1985; 2d ed., Washington, D. C.: Department of the Interior, 1991.

Malone, Michael P., Roeder, Richard B., and Lang, William L. *Montana: A History of Two Centuries*. 2d ed. Seattle: University of Washington Press, 1991.

Malone, Michael P., and Etulain, Richard. *The American West: A Twentieth Century History*. Lincoln: University of Nebraska Press, 1989.

Meine, Curt. *Aldo Leopold: His Life and Work*. Madison: University of Wisconsin Press, 1988.

Moehring, Eugene P. *Resort City in the Sunbelt: Las Vegas, 1930–1970*. Reno and Las Vegas: University of Nevada Press, 1989.

Nash, Gerald D. *The American West Transformed: The Impact of the Second World War*. Bloomington: Indiana University Press, 1985.

Nash, Roderick. *Wilderness and the American Mind*. 3rd ed. New Haven: Yale University Press, 1982.

———. *The Rights of Nature: A History of Environmental Ethics*. Madison: University of Wisconsin Press, 1989.

Nixon, Edgar B. ed. *Franklin D. Roosevelt and Conservation, 1911–1945*. 2 vols. Hyde Park: Franklin D. Roosevelt Library, 1957.

O'Neill, William. *American High: The Years of Confidence*. New York: The Free Press, 1986.

Osgood, Ernest Staples, *The Day of the Cattleman*. Chicago: University of Chicago Press, 1929.

Oshinsky, David M. *A Conspiracy So Immense: The World of Joe McCarthy*. New York: The Free Press, 1983.

Pisani, Donald J. *To Reclaim A Divided West: Water, Law, and Public Policy 1848–1902*. Albuquerque: University of New Mexico Press, 1992.

Poll, Richard, Alexander, Thomas G., Campbell, Eugene E., and Miller, David E. eds. *Utah's History*. Provo: Brigham Young University Press, 1978.

Porter, Eliot. *Down the Colorado*. New York: E. P. Dutton, 1969.

———. *The Place No One Knew: Glen Canyon on the Colorado*. San Francisco: Sierra Club Books, 1963.

Reisner, Marc. *Cadillac Desert: The American West and Its Disappearing Water*. New York: Viking Penguin Inc., 1986.

Richardson, Elmo. *Dams, Parks and Politics: Resource Development and Preservation in the Truman-Eisenhower Era*. Lexington: University Press of Kentucky, 1973.

————. *The Presidency of Dwight D. Eisenhower.* Lawrence: University Press of Kansas, 1979.

Righter, Robert. *Crucible for Conservation: The Creation of Grand Teton National Park.* Boulder: Colorado Associated University Press, 1982.

Robinson, Michael C. *Water for the West: The Bureau of Reclamation, 1902–1977.* Chicago: Public Works Historical Society, 1979.

Rothman, Hal. *Preserving Different Pasts: The American National Monuments.* Urbana: University of Illinois Press, 1989.

Runte, Alfred. *National Parks: The American Experience.* 2d. ed. Lincoln: University of Nebraska Press, 1987.

Schrepfer, Susan. *The Fight to Save the Redwoods: A History of Environmental Reform, 1917–1978.* Madison: University of Wisconsin Press, 1983.

Shankland, Robert. *Steve Mather of the National Parks.* 3rd ed. New York: Alfred A. Knopf, 1970.

Smythe, William E. *The Conquest of Arid America.* New York: n. p., 1905.

————, *The Conquest of Arid America,* reprint, ed. Lawrence B. Lee. Seattle: University of Washington Press, 1969.

Stegner, Wallace. *Beyond the Hundredth Meridian: John Wesley Powell and the Second Opening of the West.* Boston: Houghton Mifflin Co., 1954.

————. *The Uneasy Chair: A Biography of Bernard DeVoto.* Garden City, New York: Doubleday & Co., 1974.

————, ed. *This Is Dinosaur: Echo Park Country and Its Magic Rivers.* 1st ed. New York: Alfred A. Knopf, 1955; 2d ed. Boulder, Colorado: Roberts Rinehart, Inc., 1985.

Stevens, Joseph E. *Hoover Dam: An American Adventure.* Norman: University of Oklahoma Press, 1988.

Vandenbusche, Duane, and Smith, Duane. *A Land Alone: Colorado's Western Slope.* Boulder, Colorado: Pruett Publishing Co., 1981.

Watkins, T. H. *Righteous Pilgrim: The Life and Times of Harold Ickes.* New York: Henry Holt & Co., 1990.

Weatherford, Gary, and Brown, F. Lee. *New Courses for the Colorado River: Major Issues for the Next Century.* Albuquerque: University of New Mexico Press, 1986.

Webb, Roy. *If We Had A Boat: Green River Explorers, Adventurers, and Runners.* Salt Lake City: University of Utah Press, 1986.

Wiley, Peter, and Gottlieb, Robert. *Empires in the Sun: The Rise of the New American West.* Tucson: University of Arizona Press, 1982.

Worster, Donald. *Rivers of Empire: Water, Aridity, and the Growth of the American West.* New York: Pantheon Books, 1985. Reprint, New York: Oxford University Press, 1992.

————, ed. *The Ends of the Earth: Perspectives on Modern Environmental History.* New York: Cambridge University Press, 1988.

Zaslowsky, Dyan, and the Wilderness Society. *These American Lands.* New York: Henry Holt & Co., 1986.

Index